TRANSFORMERS FOR TUBE AMPLIFIERS

HOW TO DESIGN, CONSTRUCT & USE
POWER, OUTPUT & INTERSTAGE TRANSFORMERS AND CHOKES
IN AUDIOPHILE AND GUITAR TUBE AMPLIFIERS

IGOR S. POPOVICH, B. Sc. (El. Eng.)

DISCLAIMER & INDEMNITY NOTICE

Published by Career Professionals Australia

Revised & corrected, 2022

Bulk purchases

This book may be purchased in larger quantities for educational, business or promotional use. Please e-mail us at sales@careerprofessionals.com.au

National Library of Australia Cataloguing-in-Publication Data:

Popovich, Igor S.

Transformers for Tube Amplifiers

ISBN: 978-0-9806223-8-6

1. Electrical engineering 2. Electronics

I Igor S. Popovich II Title III Index

621.3

CONTENTS

CONTENTS, cont.

AUDIOPHILE TUBE AMPLIFIER BOOKS BY IGOR S. POPOVICH

Available from all major online bookstores

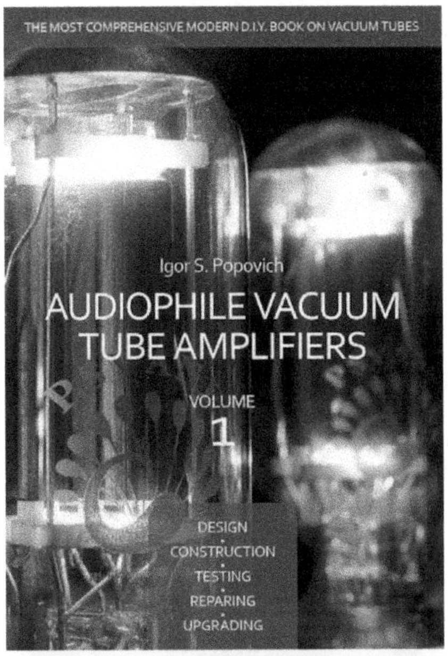

Audiophile Vacuum Tube Amplifiers, Vol. 1
ISBN: 978-0-9806223-2-4

- BASIC ELECTRONIC CIRCUIT THEORY
- ELECTRONIC COMPONENTS
- AUDIO FREQUENCY AMPLIFIERS
- PHYSICAL FUNDAMENTALS OF VACUUM TUBE OPERATION
- VOLTAGE AMPLIFICATION WITH TRIODES - THE COMMON CATHODE STAGE
- OTHER VOLTAGE AMPLIFICATION STAGES WITH TRIODES
- TETRODES AND PENTODES AS VOLTAGE AMPLIFIERS
- FREQUENCY RESPONSE OF VACUUM TUBE AMPLIFIERS
- IMPEDANCE-COUPLED STAGES AND INTERSTAGE TRANSFORMERS
- NEGATIVE FEEDBACK
- TONE CONTROLS, ACTIVE CROSSOVERS AND OTHER CIRCUITS
- PRACTICAL LINE-LEVEL PREAMPLIFIER DESIGNS
- PHONO PREAMPLIFIERS
- SINGLE-ENDED TRIODE OUTPUT STAGE
- PRACTICAL SINGLE-ENDED TRIODE AMPLIFIER DESIGNS
- PRACTICAL SINGLE-ENDED PSEUDO-TRIODE DESIGNS
- SINGLE-ENDED PENTODE AND ULTRALINEAR OUTPUT STAGES

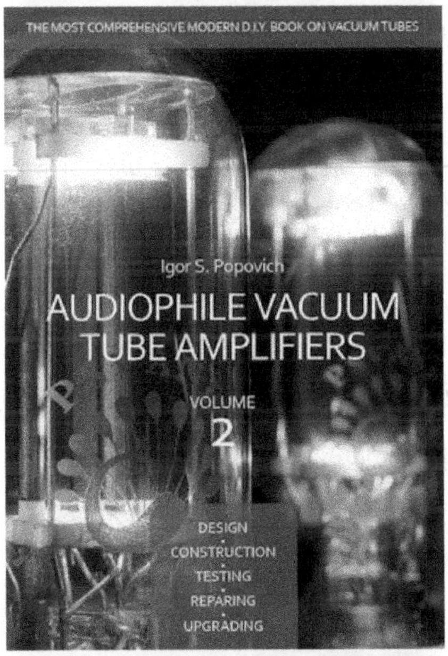

Audiophile Vacuum Tube Amplifiers, Vol. 2
ISBN: 978-0-9806223-3-1

- PRACTICAL SINGLE-ENDED PENTODE AND ULTRALINEAR DESIGNS
- PUSH-PULL OUTPUT STAGES
- PRACTICAL PUSH-PULL AMPLIFIER DESIGNS
- BALANCED, BRIDGE AND OTL (OUTPUT TRANSFORMERLESS) AMPLIFIERS
- THE DESIGN PROCESS
- FUNDAMENTALS OF MAGNETIC CIRCUITS AND TRANSFORMERS
- MAINS TRANSFORMERS AND FILTERING CHOKES
- POWER SUPPLIES FOR TUBE AMPLIFIERS
- AUDIO TRANSFORMERS
- TROUBLESHOOTING AND REPAIRING TUBE AMPLIFIERS
- UPGRADING & IMPROVING TUBE AMPLIFIERS
- SOUND CONSTRUCTION PRACTICES
- AUDIO TESTS & MEASUREMENTS
- TESTING & MATCHING VACUUM TUBES

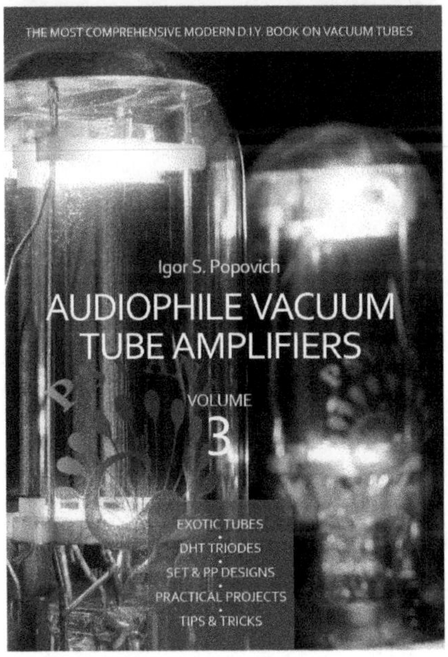

Audiophile Vacuum Tube Amplifiers, Vol. 3
ISBN: 978-0-9806223-4-8

- THE FRONT-END: SUPERIOR INPUT & DRIVER STAGES
- FROM SHOCKING TO SUBLIME: LESSONS FROM COMMERCIAL LINE STAGES
- DIY LINE-LEVEL PREAMPLIFIERS: $10,000 SOUND ON $500-$1,000 BUDGET
- THE STARS OF THE AUDION ERA: ANCIENT TUBES IN MODERN AMPS
- CHEAP & CHEERFUL: PREAMP & DRIVER TUBES FOR AUDIO EXPLORERS
- SLEEPING GIANTS: OUTPUT TUBES FOR THOSE WHO WANT TO BE DIFFERENT
- THE QUEEN OF HEARTS: SINGLE-ENDED AMPLIFIERS WITH 300B TRIODES
- TRIODES, PENTODES AND BEAM TUBES: MORE SINGLE-ENDED DESIGNS
- BIG BOTTLES: SET AMPLIFIERS WITH HIGH VOLTAGE TRANSMITTING TUBES
- THE WAY IT USED TO BE: VINTAGE PUSH-PULL AMPLIFIERS
- NEW? IMPROVED? MODERN PUSH-PULL AMPLIFIER DESIGNS
- CUTE, CLEVER OR CONTROVERSIAL? INTERESTING IDEAS FROM TUBE AUDIO'S PAST AND PRESENT
- THRIFTY TIPS & TRICKS: TIME & MONEY SAVING IDEAS
- OUTPUT AND INTERSTAGE TRANSFORMERS: FROM COMMERCIAL BENCHMARKS TO YOUR OWN DESIGNS
- MEASUREMENTS VERSUS LISTENING AND OTHER AUDIO DESIGN DILEMMAS

GUITAR TUBE AMPLIFIER AND TUBE TESTER BOOKS BY IGOR S. POPOVICH

Available from all major online bookstores

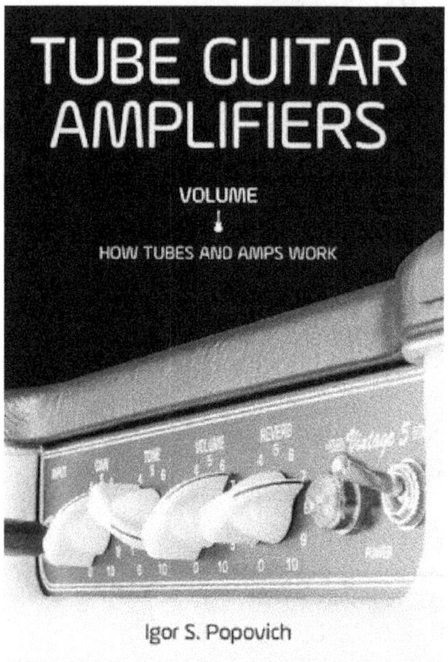

Tube Guitar Amplifiers Volume 1: How Tubes and Amps Work
ISBN: 978-0-9806223-5-5

- BASIC ELECTRONIC CIRCUIT THEORY
- AUDIO AMPLIFIERS
- ELECTRONIC COMPONENTS
- PHYSICAL FUNDAMENTALS OF VACUUM TUBE OPERATION
- TRIODES AS VOLTAGE AMPLIFIERS
- TETRODES, PENTODES AND BEAM-POWER TUBES
- INPUT CIRCUITS AND STAGES
- TONE CONTROLS
- ANALOG EFFECTS (TREMOLO, VIBRATO, REVERB) AND EFFECTS LOOPS
- POWER SUPPLIES FOR TUBE AMPLIFIERS
- SINGLE-ENDED TRIODE, PENTODE AND ULTRALINEAR OUTPUT STAGES
- PHASE SPLITTERS OR INVERTERS
- PUSH-PULL OUTPUT STAGES
- NEGATIVE FEEDBACK
- TRANSISTOR AND HYBRID GUITAR AMPLIFIERS

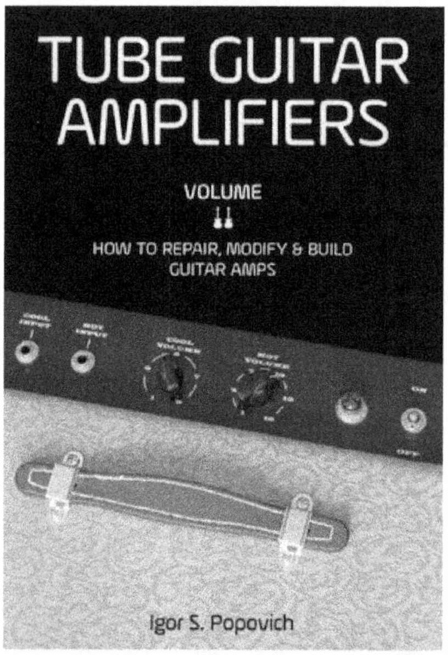

Tube Guitar Amplifiers Volume 2: How to Repair, Modify & Build Guitar Amps
ISBN: 978-0-9806223-6-2

- OUTPUT AND INTERSTAGE TRANSFORMERS FOR TUBE GUITAR AMPS
- LOUDSPEAKERS, OUTPUT ATTENUATORS & HEADPHONE CIRCUITS
- TROUBLESHOOTING AND REPAIRING TUBE GUITAR AMPLIFIERS
- WIRING, SOLDERING & MODIFICATION PRACTICES
- POWER SUPPLY MODIFICATIONS AND IMPROVEMENTS
- TONE TWEAKS
- MODERN PUSH-PULL AMPS
- DIY PROJECTS: CONVERTING SOLID STATE GUITAR AMPS TO TUBES
- DIY PROJECTS: ULTRA-SMALL AMPS
- REBUILDING COMMERCIAL AMPS IN A HANDWIRED (POINT-TO-POINT) FASHION
- DIY PROJECTS: QUIRKY & UNUSUAL DESIGNS
- CONVERTING VINTAGE TUBE GEAR INTO GUITAR AMPS

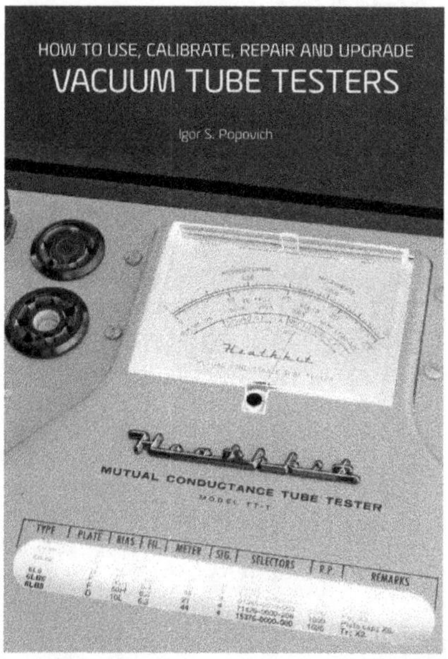

How to Use, Calibrate, Repair and Upgrade Vacuum Tube Testers
ISBN: 978-0-9806223-7-9

- HOW VACUUM TUBES WORK
- TESTING & MATCHING VACUUM TUBES
- EMISSION TESTERS
- GRID CIRCUIT TESTERS
- DYNAMIC CONDUCTANCE TESTERS
- PROPORTIONAL MUTUAL CONDUCTANCE TESTERS
- HICKOK-TYPE TESTERS
- TRUE MUTUAL CONDUCTANCE TESTERS
- REPAIRING & UPGRADING VINTAGE TUBE TESTERS
- TESTING & MATCHING TUBES WITHOUT A TUBE TESTER

INTRODUCTION

- Why this unique manual?
- How is this book structured
- Getting in touch with us
- How to read this book
- Symbols used
- Abbreviations used

1

Why this unique manual?

After producing dozen or so DVD workshops about tube amplifiers, testers, audio measurements, and related topics, we've received quite a few inquiries about the design side of transformer making. "How to Make Your Own Chokes, Power and Output Transformers For Tube Amplifiers," our first DVD program, started the whole series of workshops in 2005. A few years later, we released the "Transformer Tests & Measurements" DVD. The only thing missing was the first step, explaining the physical background of transformer operation and how to go about designing them.

Small transformer design and construction is considered a specialist subject, which could explain an almost total lack of practical and useful literature on the subject. A few vintage books, mostly in German, Italian and French, deal with the design of small power transformers but mention audio transformers in passing only.

Electronic transformer books in the English language are less practical, a few have useful snippets of information, but most are next to useless to the DIY amp and transformer builder. We hope that this book will fill this void.

Dr. Bob Popovich, a physics professor by day and an electronics guru by night, has been building receivers, guitar, and hi-fi amplifiers, all using tube technology, for more than six decades. Bob has designed and wound hundreds, if not thousands of power and audio transformers along his long and exciting audio journey. Bob is still a master transformer winder even in his ninth decade of life!

His son Igor started with the theory first, completing his university degree in electronics in 1986, and then learned the practicalities in his various engineering positions in Australia.

RIGHT: Dr. Slobodan (Bob) Popovich as a budding student of physics in 1958 in Yugoslavia, financed his university studies by building and repairing tube guitar and hi-fi amplifiers and radios.

FAR RIGHT: Your author, Igor S. Popovich, in 2018

How is this book structured

We will always start with the theoretical background, explaining the fundamental concepts of magnetism and transformer operational principles, followed by an overview of various types of magnetic laminations and cores.

Filtering chokes and power (mains) transformers are the next subjects. They operate on a single frequency and can be considered a special case of output transformers, a much simpler version. So, methodologically, we start with simpler issues and move towards the more complex ones. Chokes are the least critical and are the easiest magnetic components to design and construct, so if you are a novice in this field, they should be your first project.

Finally, we look at the (almost always conflicting) demands placed on audio transformers, two kinds, in particular, interstage transformers and output transformers, and how to go about meeting those contradictory demands.

Electronics in general and transformer design, in particular, is the art of compromise. There is no free lunch; there are always consequences whatever you do. Furthermore, transformers are very nonlinear devices, meaning that linear models and simple formulas are of limited use in their design, which is thus not purely a rational science but also an intuitive art.

Getting in touch with us

If you've liked the book and benefited from it, the best way to repay a favor is to recommend it to your friends and to write a glowing online review. Also, if you spot an error or an omission or should you have any constructive criticism of the book, I'd like to hear from you, so we can fix it together.

If you want to contribute ideas or projects for the next edition, or if you have ideas on how to make the next edition better, please let me know. My e-mail is igorpop@careerprofessionals.com.au

I hope this book has answered some of your questions about audio transformer design and building. I wish you every success on your audio journey!

Igor S. Popovich

How to read this book

One way to read a technical book like this one is to immediately go to a section or topic that interests you and then to keep jumping back-and-forth to the related issues and chapters. This will, I suspect, be the way the more experienced designers and constructors will approach this book.

A more systematic approach is sequential, starting from the beginning and reading in order. This is what I would recommend. Although it seems more time-consuming (since you will read about many issues you may already know a lot about), paradoxically, this approach is often faster. You will not miss anything, and you will not waste time flipping forward and backward trying to clarify an issue that you've overlooked and perhaps not fully understood.

Whatever you do, don't treat this book as Holy Scripture. Underline or highlight the important parts, write your thoughts and ideas on its margins, sketch diagrams and circuits in its blank spaces.

MEASURED RESULTS:

All designs in this book have either been tried in practice as DIY projects, or are commercial designs. This type of frame summarizes the measured parameters of a finished transformer or choke.

RULE-OF-THUMB

For those who don't want to bother with high-level maths, models and similar highbrow concepts, these Rules-of-Thumb are simple shortcuts, valid under certain assumed conditions.

IMPORTANT FORMULA

The calculator symbol indicates an important or often-used formula.

DIY PROJECT

Small projects framed and marked with this soldering-iron symbol deal mostly with DIY test jigs and relatively simple test instruments.

A CRITICAL QUESTION

While there are no silly questions (only silly answers!), some questions are of far more importance than others. Some of those are answered in frames with this symbol.

MANUFACTURER'S SPECIFICATION

The checklist symbolizes a list of technical parameters of commercial transformers or amplifiers, provided by their manufacturer.

A WARNING OR A VERY IMPORTANT POINT!

Some issues, myths and warnings are so important that they warrant being emphasized in a frame of this kind.

FURTHER INFORMATION

Catalogues, patent documents and magazine articles contain a wealth of information, most of which is available online. A box of this kind points you to such "further reading" sources.

QUICK CALCULATION

Some numerical examples, calculations and design problems are emphasized in a frame with the pencil symbol.

An example of test results "box":

MEASURED RESULTS:

- $R_P = 177\Omega$, $R_S = 176\ \Omega$
- $L_{P120} = 12.4H$, $L_{P1000} = 11.3$ H
- $L_L = 0.40$ mH, $C_{PS1} = 64$ nF, $C_{PS2} = 2.2$ nF

Primary inductance is of more importance at low frequencies so it should be measured at 120Hz or the lowest frequency an LCR meter is capable of. However, the difference between results at high (1,000Hz) and low test frequency (120Hz) points to the core's or laminations' quality and suitability for the audio applications. The less difference the better!

Usually there is a difference between the primary-secondary parasitic capacitance figures, depending on which transformer's terminals were used as test points for LCR meter connection.

Leakage inductance and parasitic capacitances should be measured at the highest frequency an LCR meter is capable of, in our case 1kHz. Some models also use a 10kHz test signal.

Symbols used

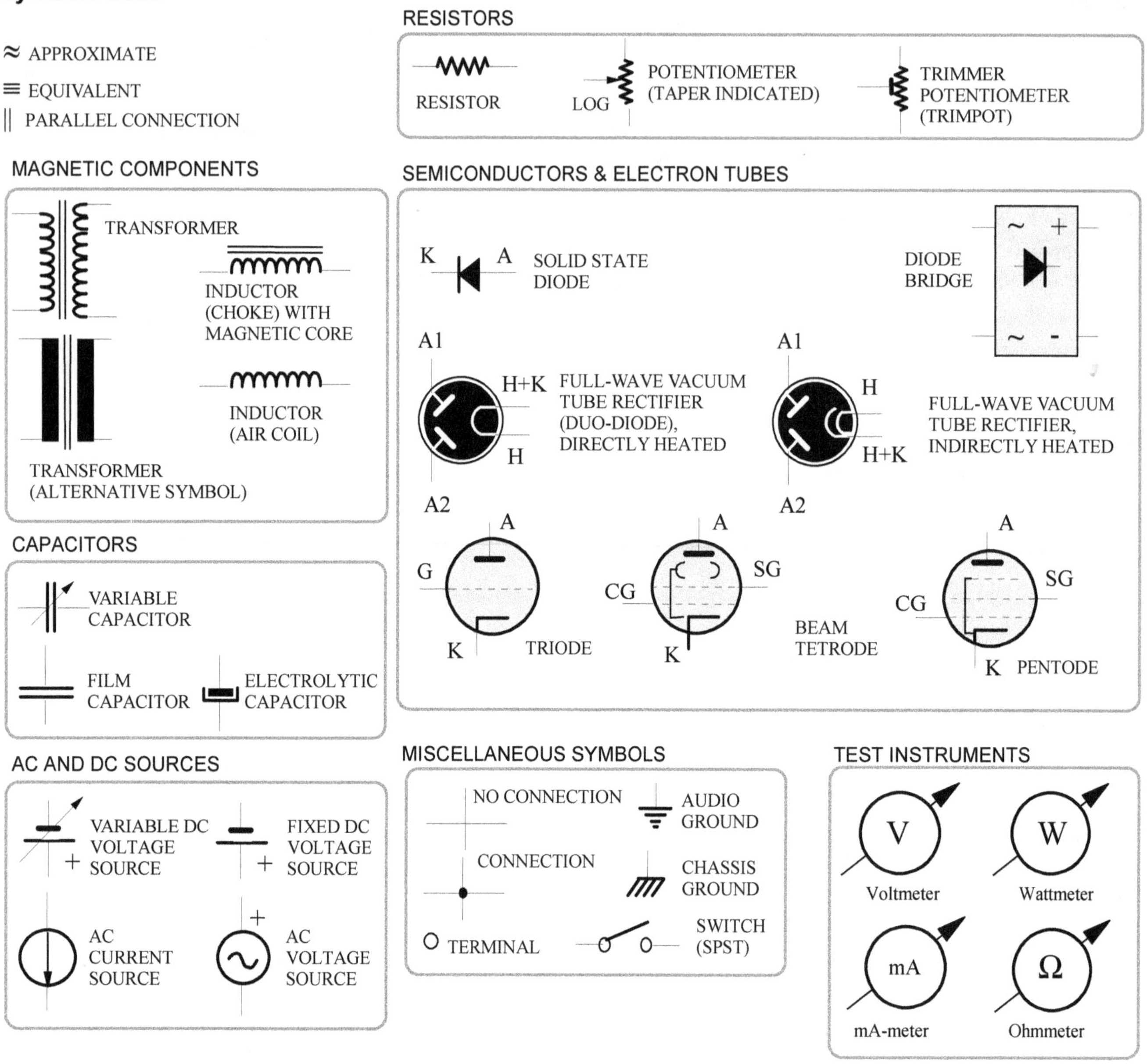

Abbreviations used

AC	Alternating current	MAX	Maximum	GND	Ground terminal
DC	Direct current	MIN	Minimum	COM	Common terminal
THD	Total harmonic distortion	RMS	Root-Mean-Square	AWG	American Wire Gauge
DCR	DC resistance	PP	Peak-to-peak (AC signal)	REG	Regulation
PP	Push-pull (amplifier)	HF	High frequency	NFB	Negative feedback
MC	Moving coil (phono cartridge)	LF	Low frequency	AF	Audio frequency
LOG	Logarithmic scale or taper (potentiometer)	VA	volt-ampere (power) rating of a transformer	LIN	Linear scale or taper (potentiometer)
TR, VR	Turns or voltage ratio (of a transformer)	E/S	Electrostatic (field, interference or shield)	SET, PSET	Single-ended triode, parallel SET
IR	Impedance ratio (of a transformer)	RC	Resistor-capacitor coupling between stages	LC	Inductive-capacitive coupling between stages
TPL	Turns per layer	HFF	Horizontal fill factor	MLT	Mean Length of Turn
TPV	Turns-per-volt	GOSS	Grain-oriented silicon steel	RFI	Radio frequency interference
WL	Window length	WH	Window height	CL	Coil length
CH	Coil height	EMF	Electromotive force	CT	Center tap (of a transformer)
RF	Reduction factor (of an autotransformer)	U/L	Ultralinear taps or circuit	DAC	Digital-to-Analog Converter

PHYSICAL FUNDAMENTALS OF MAGNETIC CIRCUITS AND TRANSFORMERS

- OPERATIONAL PRINCIPLES & KEY PARAMETERS OF TRANSFORMERS & INDUCTORS
- MAGNETIC PROPERTIES OF THE CORE
- EDDY CURRENTS
- LAMINATION AND CORE TYPES
- THE OVERVIEW OF UNITS, SYMBOLS AND CONVERSION FACTORS
- FROM THEORETICAL BACKGROUND TO PRACTICAL DESIGNS

2

OPERATIONAL PRINCIPLES & KEY PARAMETERS OF TRANSFORMERS & INDUCTORS

The ideal transformer

A source of AC voltage $v_1(t)$ is connected to the transformer's primary winding with N_1 turns of wire around its magnetic core, causing the primary AC current to flow (i_1). This current produces magnetic flux $\Phi_1(t)$ in the core, which couples the primary and secondary winding. As a result of electromagnetic induction, this varying (AC) flux $\Phi_1(t)$ induces an alternating voltage v_2 in the secondary winding.

Once load R_L is connected, the secondary current I_2 flows and produces its own magnetic flux $\Phi_2(t)$, which will be of the opposite direction to Φ_1, until eventually, under steady-state conditions, the two fluxes balance each other.

All these voltages and currents are alternating, usually depicted by lower case italic symbols such as v(t), meaning a certain voltage v is a function of time. A transformer can only pass through AC voltages and currents; DC voltages & currents can flow through the primary, the secondary, or both windings but are not transmitted between the primary and secondary!

The diagram depicts one point in time when the upper side of v1 is positive. The "dot convention" says that at any moment, voltages in points with the dot are of the same polarity or "in phase." Since *v*1 is a generator producing voltage, the current *i*1 comes out of its positive terminal and flows "into" the dot. *v*2 is the induced voltage, and *i*2 is the load current, so it flows out of the point with the dot (out of the plus side of *v*2).

For simplicity's sake and the ease of reading, we will use capital letters for both AC and DC voltages, currents, fluxes, and other physical entities.

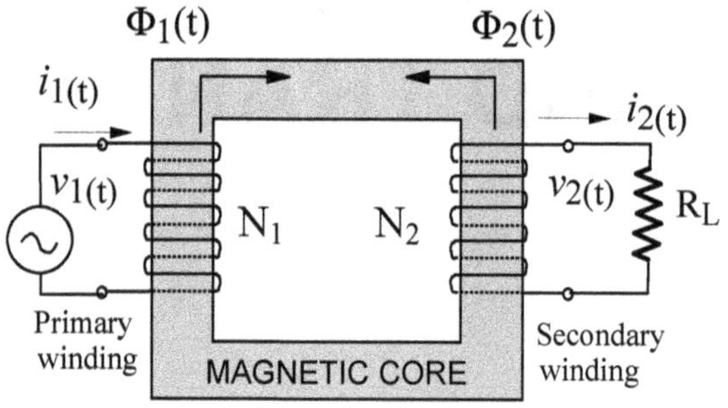

ABOVE: The magnetic flux Φ_1 through the ferromagnetic core provides magnetic coupling between transformer's primary and secondary windings

The right-hand rule for coils

This rule applies to both DC and AC current flow at any instant. If the current through a winding or a coil flows in the direction of the four fingers, then the thumb indicates the direction of the magnetic flux produced.

The right-hand rule is valid only for the international definition of current (SI-system), where current is considered the flow of positive charges (opposite of the flow of electrons). If the current is defined as the flow of electrons, as in most older and even some current American books, then use your left hand.

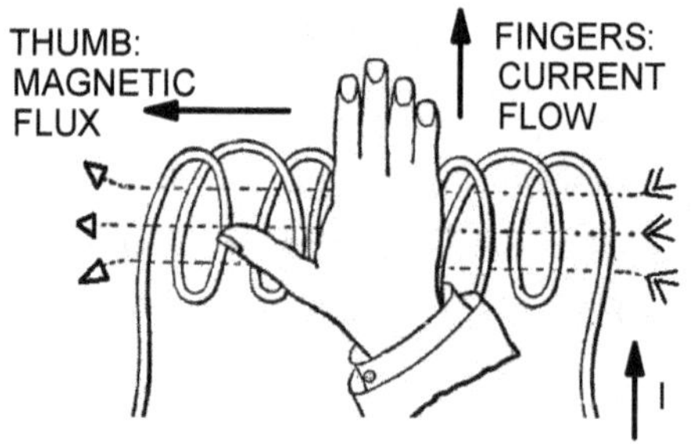

ABOVE: The right-hand rule for magnetic coils

Turns & impedance ratio

The primary and secondary voltages of a transformer are in direct proportion to the ratio of the transformer's primary and secondary turns, so we call it voltage ratio or turns ratio, TR for short: TR = $N_1/N_2 = v_1/v_2$.

Since energy must be conserved, the primary and secondary powers are equal for an ideal transformer: $P_1=P_2$ or $I_1v_1=I_2v_2$ Written another way, $I_1/I_2 = v_2/v_1 = 1/TR$. This means that currents are inversely proportional to the TR!

The resistance "seen" by the voltage source $v_1(t)$ driving the primary (called "secondary or *load* resistance reflected to the primary") is $R_{LR}=R_L(N_1/N_2)^2$ or $R_{LR}=R_LTR^2 = R_LIR$! This square of the TR is called *Impedance Ratio* or IR for short.

High primary and low secondary turns mean high primary and low secondary voltage, so the transformer is of a *step-down* kind. Output transformers for tube amps are step-down transformers. A few primary and many secondary turns mean the transformer is of a *step-up* kind, that it will transform a low input voltage to a higher output voltage.

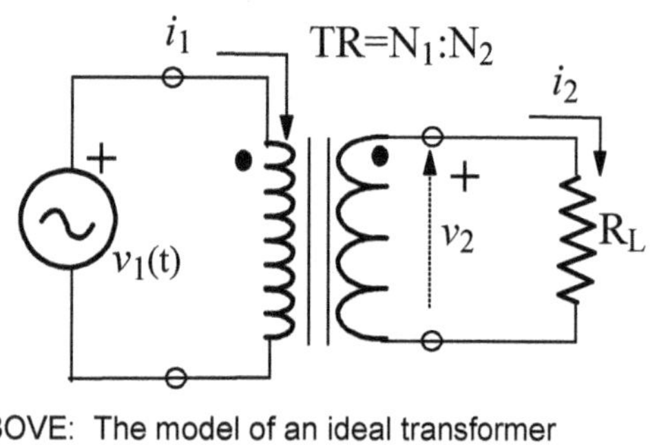

ABOVE: The model of an ideal transformer

IDEAL TRANSFORMER RATIOS

VOLTAGE OR TURNS RATIO:
$VR = TR = N_1/N_2 = v_1/v_2 = i_2/i_1$
IMPEDANCE RATIO:
$IR = TR^2 = (N_1/N_2)^2 = (v_1/v_2)^2 = (i_2/i_1)^2$
POWER: $P_1 = P_2$ so $i_1v_1 = i_2v_2$

SECONDARY (LOAD) IMPEDANCE REFLECTED ONTO THE PRIMARY SIDE

$R_{LR}=R_LIR = R_LTR^2 = R_L(N_1/N_2)^2$

MC (moving coil) transformers transform small signals from an MC turntable cartridge (0.1 mV) into 10 or 20 times higher output voltage 10-20 mV. The current ratios are the opposite; a voltage step-up transformer is a current step-down transformer and vice versa.

There are no power losses or signal attenuation of any kind in an ideal transformer, which is thus a convenient fantasy but still useful in understanding the operational principles of real, lossy transformers. The primary and secondary wires have resistance. There is a parasitic capacitance between turns, layers, and sections of each winding and between different windings. Most importantly, the magnetic core consumes some energy needed to magnetize and demagnetize it. More on all these phenomena soon.

THE ARROW CONVENTION

When looking at the cross-section of a wire or a coil (the wire direction is perpendicular to the paper), we can imagine the current flow as a flying arrow. The current flowing away "into the paper" is marked with a cross (the tail of the arrow), and the current flowing "from the paper" towards you is marked with a point (the tip of the arrow).

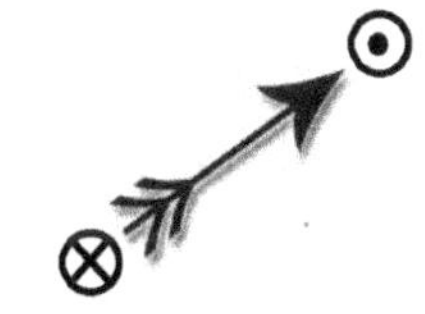

BASIC PARAMETERS OF AUDIO TRANSFORMERS

ANALYSIS

A SET amplifier with 300B output triodes is rated at 8 Watts of output power. If the nominal load (speaker) impedance at a certain frequency is 8Ω, calculate the voltage & impedance ratios of the output transformers and the primary and secondary voltages & currents. The transformers have 3,500Ω primary impedance.

The output transformers reflect 8Ω load as 3,500Ω to the primary side. Thus, their impedance ratio is IR=3,500/8 = 437.5 The voltage and turns ratio is VR=TR=√IR = √437.5 = 20.9.

Since $P_{OUT}=V_2^2/R_L$, the secondary AC (load) voltage is $V_2=\sqrt{(P_{OUT}R_L)}=\sqrt{(8*8)}$ = 8 V, while the primary signal voltage is $V_1=TR*V_2$= 20.92*8 = 167V. The secondary current is $I_2= P_{OUT}/V_2$= 8/8 = 1 A and the primary AC (signal) current is $I_1= I_2/TR$= 1/20.92 = 47.8 mA.

The inductance equation

What if a transformer had only a primary winding? There is still a magnetic flux through the core caused by the current flowing through the winding, only this time such a varying flux will induce an opposing voltage in the primary winding itself. This type of induction is called self-induction, and such a "transformer" with only one coil is called an *inductor* or a "choke" because the self-induced voltage tries to oppose or "choke" any changes in the current. Let's now look at the fundamental equations:

$H(t)=i(t)N/\ell_{MP}$ (H= magnetizing force or "magnetic field strength", N= number of turns of an inductor or choke, ℓ_{MP} = the length of the mean magnetic path)

$B(t)= \mu H(t)$ (flux density = permeability * magnetizing force) The SI unit for magnetic flux density B is Tesla, the older, now obsolete unit is G (Gauss). 1T = 10,000 G = 10^4 G!

$\mu= \mu_R\mu_0$ (permeability = relative permeability of the material * absolute permeability) Absolute permeability in SI system of units is $\mu_0 = 4\pi 10^{-7} = 1.257*10^{-8}$, while **$\Phi(t) = B(t)A$** (magnetic flux = flux density * cross-sectional area of the core). The Greek letter Φ is pronounced "fi". By combining all of the above we get **$\Phi(t) = i(t)N\mu_R\mu_0/\ell_{MP}$**

The induced voltage is proportional to the rate of change (derivative) of the magnetic flux but (-) signifies that it is of the opposite polarity - it tries to counteract the magnetic flux produced by the excitation voltage **$v(t) = -Nd\Phi(t)/dt$**. The relationship between the self-induced voltage and the current through the coil is **$v(t) = N^2\mu_R\mu_0A/\ell_{MP}\, di(t)/dt$** or **$v(t)= Ldi(t)/dt$**. In the first approximation, the factor L is constant, so the induced voltage is proportional to the derivative or the rate of change of the AC current flowing through the winding.

Factor L is called *inductance* (of a choke) or *primary inductance* (of a transformer): **$L= N^2\mu_R\mu_0A/\ell_{MP}$** The unit for inductance L is [H] or Henry.

Inductance increases with a square of the number of turns, so the easiest way to achieve higher inductance is to increase the number of turns. Obviously, a winding window is of fixed dimensions, and only so many turns can fit, so there is a limit to how far that strategy can be pursued.

INDUCTANCE OF A CHOKE OR TRANSFORMER'S PRIMARY

$L= N^2\mu_R\mu_0A/\ell_{MP}$

(A in cm^2 and ℓ_{MP} in cm)

The inductance also increases with the magnetic laminations stack's cross-sectional area "A." This would go well with the previous requirement (increasing the number of turns) since increasing the window size would allow for both; larger laminations have a wider center leg (resulting in a higher A) and larger windows. However, physically larger cores also have a longer magnetic path ℓ_{MP}, reducing the inductance, leading to a contradictory situation.

So, a designer has to determine what is more important, a larger cross-sectional area (which also achieves an increased power rating of the core) or a shorter length of the magnetic path, which, in audio transformers, leads to wider bandwidth and better performance at frequency extremes (bass and treble regions). The choice will depend on the type of transformer in question and its expected performance.

Inductance also increases if magnetic materials with high relative permeability μ_R are used. However, such materials are much more expensive than ordinary or even GOSS silicon steel. Furthermore, their maximum density of magnetic flux or "B" is much lower, meaning they saturate at much lower amplitudes of audio signals. Thus, another contradictory or compromising situation ensues.

Unfortunately, since μ_R depends on the signal's frequency, as we will soon see, L is not constant at all. It changes with both the signal frequency and the level of DC current through the choke or transformer's primary! It is important to keep in mind that many formulas in this book should, strictly speaking, have the approximate sign (l) instead of the equal sign (=).

The fundamental transformer equation

The self-induced AC voltage in the choke's winding or in a transformer's primary winding is $v(t) = -Nd\Phi(t)/dt$.

For sinusoidal flux $\Phi(t) = \Phi_{MAX}\sin(\omega t)$ we get $v(t) = -N\Phi_{MAX}\omega\cos\omega t = -V_{MAX}\omega\cos\omega t$

Since $\omega=2\pi f$, the effective or RMS value of the induced voltage is $V_{EF}=0.707V_{MAX}= \underline{0.707*2*\pi}*f*N*\Phi_{MAX} = \underline{4.44}fN\Phi_{MAX}$. Substituting $A_{EF}B_{MAX}$ for Φ_{MAX} we get $V_{EF} = 4.44fNA_{EF}B_{MAX}$.

If A_{EF} is in cm^2 instead of m^2 (unit too large), we need to include the 10^{-4} factor: $\mathbf{V_{EF} = 4.44fNA_{EF}B_{MAX}*10^{-4}}$.

B is in T (Tesla), A_{EF} in cm^2, f in Hz (cycles-per-second) and the number of turns (n) has no dimension. From this equation we can express turns-per-volt as $\mathbf{TPV = N_1/V_1=10^4/(4.44fB_{MAX}A_{EF})}$.

Note that the 4.44 factor ($0.707*2*\pi$) is only valid for sinusoidal waveforms, which is fine for mains and audio transformers, but is not valid for transformers used in DC-DC converters for instance, which work with square waveforms. Square waveform has the crest and form factors of 1, since the peak of a square wave is the same as its RMS and average value.

TRANSFORMER EQUATION

$V_{EF}=4.44fNA_{EF}B_{MAX}10^{-4}$

V in Volts, A_{EF} in cm^2, B in Tesla, f in Hz

$TPV = N_1/V_1=10^4/(4.44fB_{MAX}A_{EF})$.

The magnetic path

When placed in a uniform magnetic field, some materials become weakly magnetized in a direction opposite the applied field, which means their relative magnetic permeability is less than 1. Such materials, most notable of which are water, alcohol, air, hydrogen, and metals such as antimony, bismuth, copper, gold, and mercury, are called diamagnetic materials.

The relative magnetic permeability of paramagnetic materials is slightly higher than 1. In contrast to diamagnetic materials, the induced magnetic fields in paramagnets are in the direction of the applied magnetic field. Tungsten, aluminium, lithium and magnesium, are just some examples of paramagnetic materials.

Ferromagnetic materials have a relative permeability much higher than one. As the name suggests, iron (ferro), cobalt, nickel, and their alloys are ferromagnetic. The illustration below shows what happens when a ferromagnetic bar is placed in a uniform external magnetic field.

The induced magnetic field of the iron bar itself combines with the external magnetic field so that magnetic flux lines take a magnetic path of a lesser resistance through the ferromagnetic bar instead of the surrounding air.

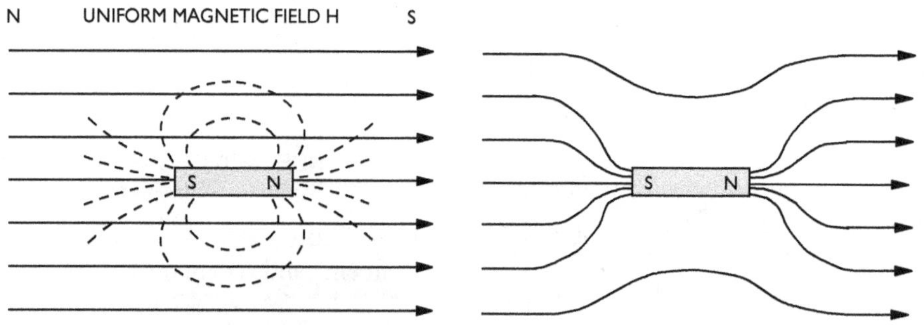

FAR LEFT: A steel- or iron-bar placed in a uniform magnetic field becomes magnetized by that field through magnetic induction. The inducing field lines are full lines and the induced lines are dotted.

LEFT: Combined fields show "crowding" of magnetic flux lines through the ferromagnetic bar, which, due to its high permeability, represents a path of lower resistivity for magnetic flux compared to high permeability surrounding air.

MAGNETIC PROPERTIES OF THE CORE

Hysteresis loops

For previously non-magnetized or demagnetized ferromagnetic core, the B-H curve follows the S-shaped path, called the curve of the first magnetization. The "curve" in its name foreshadows the fact that the relationship between B (the dependent variable or "effect") and H, the independent variable or the "cause," is nonlinear. This complicates things immensely.

If we increase H to reach $+H_S$, we will saturate the core. A further increase in H will not increase B. The return path will be different if we start reducing H (by reducing the DC current through the coil); the curve will follow the upper S-curve.

Even when the current through the coil and its magnetic field H_S drop to zero, there is still some flux through the core, and this B_R is called remnant flux density. To bring B to zero, we need to reverse the direction of the current and "force" B to zero.

B will eventually drop to zero at some value $-H_C$ (negative coercive force). A further increase in the magnetizing current will again saturate the core at the value $-H_S$. By now reducing & reversing the current again, we will complete the whole major hysteresis loop. This means that magnetic components are not just nonlinear but also directional!

The level of magnetic flux density B depends not only on the level of the magnetizing force or "magnetic field strength" H but also on the previous magnetization history of the magnetic material.

The hysteresis illustration also shows three minor hysteresis loops. For smaller variations in H around any point on the major curve, the B-H relationship follows a smaller or minor loop. The three loops all have the same B range (the height between their maximum and minimum, or ΔB), but notice how the H range (the width between their leftmost and rightmost points) or ΔH is the smallest for loop A, larger for loop B and the largest for loop C!

Three areas can be identified along the magnetizing curve. Initially, μ is low (point A), then it raises and reaches its maximum value, and then it drops, so in the saturation area around point C it is low again.

In addition to a varying permeability for DC currents, we can see that the minor loops have a different permeability from the major loop, meaning that the permeability of the core for AC currents is different again and that it varies with the amplitude of such current signal.

This is a depressing piece of news because it means that magnetic materials are extremely nonlinear, so, as a consequence, magnetic circuits, transformer, and choke designs cannot easily be calculated or analyzed.

There are ways to simplify complex magnetic circuit formulas and turn them into rules-of-thumb, but these are limited to only certain frequencies, specific magnetic materials, and small or modest signal amplitudes.

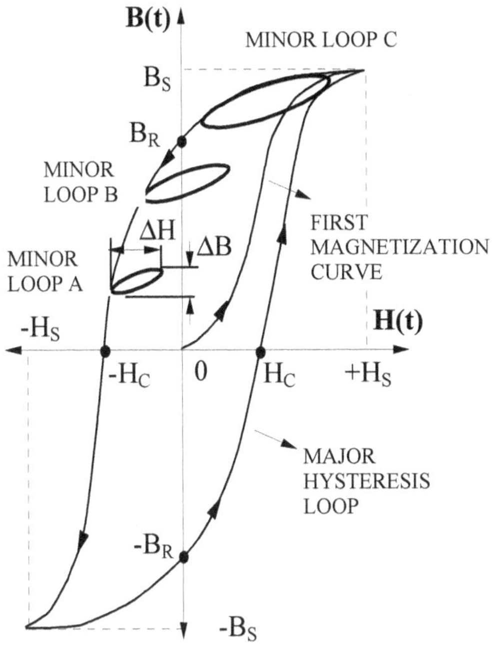

B - density of magnetic flux, B_S - saturation flux density, B_R - remnant flux density, H_C - coercive force

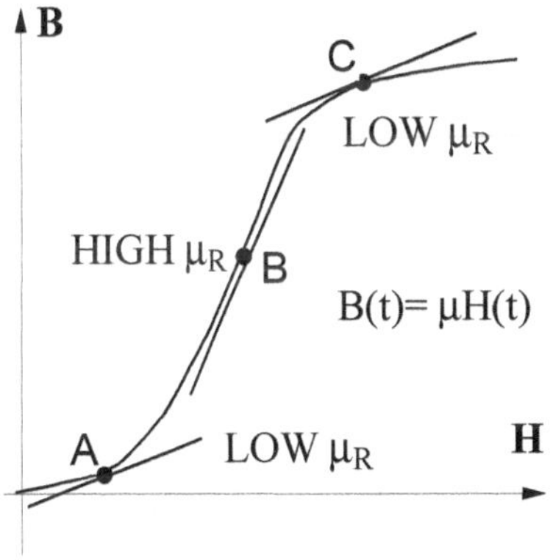

ABOVE: The relative permeability is the slope of the B-H curve in any particular point.

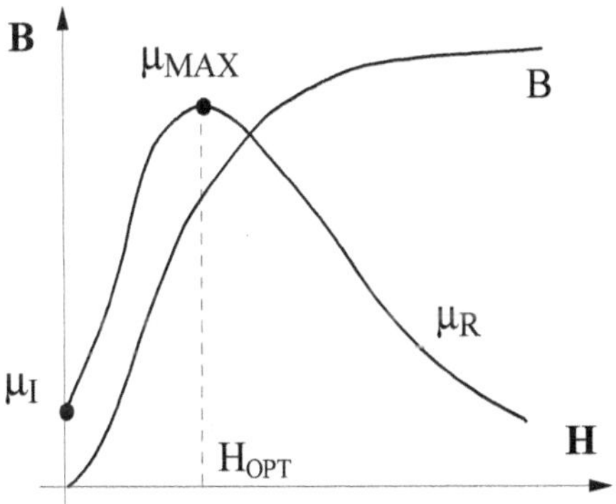

ABOVE: The variation of relative permeability μ_R with H and B. μ_I is initial permeability (in one point only, for H=0!), do not confuse it with μ_{INC}, the incremental permeability!

Relative and incremental permeability

As the DC component of magnetizing current increases (from B to C), a greater variation in H is required to produce the same variation in B. That means that the inductance L of such a coil decreases as the DC current through such coil is increased! This is caused by the reduction in the effective permeability of the magnetic core due to the presence of DC current. This permeability is called "incremental" or permeability for small AC signals (increments), and the related AC inductance is called incremental inductance. Do not confuse it with μ_I, the initial permeability (for H=0).

The last illustration shows a pronounced maximum in relative permeability μ_{MAX} for magnetic field H_{OPT}. Any further increase in H causes a sharp drop in permeability coupled with very little increase in B, a condition known as the saturation of the magnetic core.

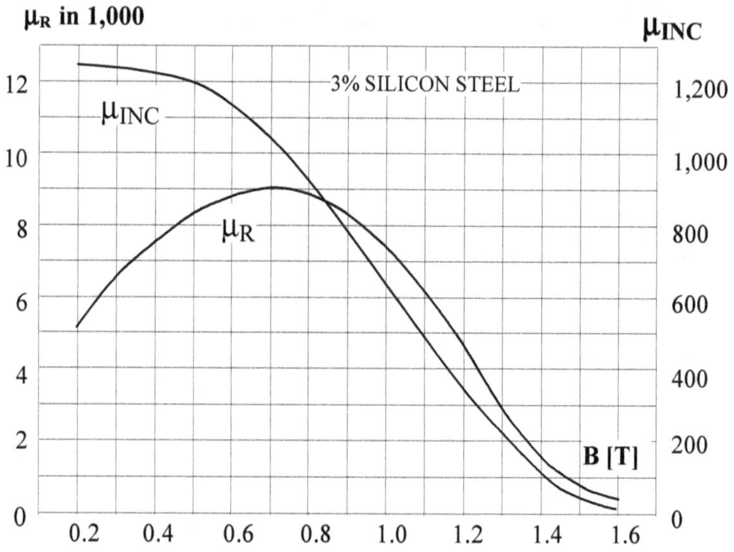

RIGHT: Relative permeability μ_R and incremental permeability μ_{INC} for 3% silicon steel (non-oriented) as a function of magnetic flux density B

Magnetic materials

Two main materials have been widely used in magnetic components for tube amps and preamps, ordinary 3% silicon steel and GOSS or grain-oriented silicon steel. Over the past 70 years, many superior magnetic materials have been developed. Manufacturers gave them proprietary names such as Hipersil, Microsil, Superflux, and Permendur.

"Permalloy" *(Perm*eability *Alloy*) is an alloy with about 20% iron and 80% nickel content. "Supermalloy" has 5% molybdenum, 79% Ni and 15% iron. Both belong to the same family of materials with around 80% nickel content. These materials saturate early, at 0.7 to 0.8 Tesla, so the usable induction (the flux level at which the incremental permeability has substantially decreased) is 0.65 to 0.75 Tesla, much lower than silicon steel at around 1.6 T.

The other common group has 49-50% nickel content, and higher useable induction levels, up to 1.2T. The higher the nickel content, the sooner the material reaches saturation.

These magnetic materials are much more expensive than GOSS, are not widely available to the general public and DIY community, and are therefore only used for special applications, such as step-up moving coil transformers and microphone input transformers. Interstage transformers would also benefit from these high nickel content laminations.

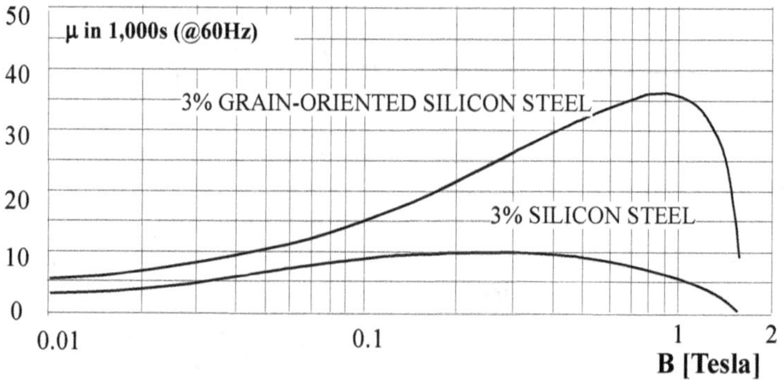

ABOVE: Grain-oriented materials have a much higher permeability compared to ordinary silicon steel

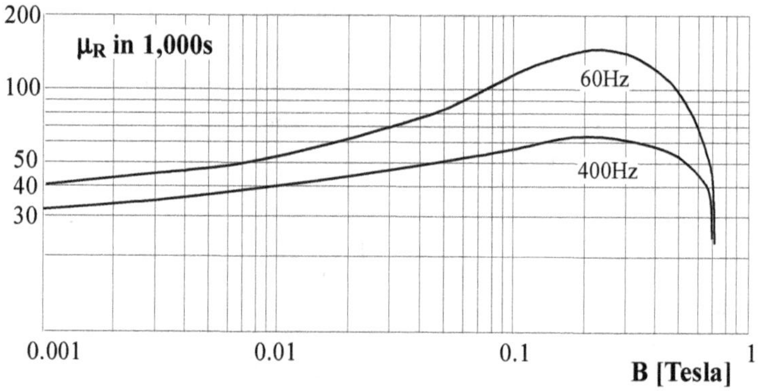

ABOVE: Relative permeability versus flux density for Permalloy 80

The analogy between magnetic and electric circuits

We have already mentioned that $H=NI/\ell_{MP}$ (H= magnetizing force or "magnetic field strength", N= number of turns of an inductor or choke, ℓ_{MP} = the length of the mean magnetic path). The product of excitation current "I" and the number of turns "N" that current flows through is known as "Ampere-Turns" or AT for short. The "proper" physical name for Ampere-Turns is Magnetomotive Force: MMF=NI [Ampereturns]. It is also called "total field".

This "force" that creates the magnetic flux Φ through the magnetic core is equivalent to an electromotive force (voltage) producing electric current through a load composed of one or more resistances. Magnetomotive force should not be confused with the magnetizing force H, but the two are related by $H = MMF/\ell_{MP}$ or $MMF = H*\ell_{MP}$!

Ohm's Law (V=I*R in electric circuits) also applies to magnetic circuits: $MMF=\Phi*\mathfrak{R}$, where $\mathfrak{R}$ is "reluctance", an equivalent of resistance in electric circuits. To distinguish the two, we will use $\mathfrak{R}$ for reluctance and R for resistance. Although the analogy is not complete or physically equivalent (only phenomenologically), the current flow through an electric circuit is akin to the flow of magnetic flux through a magnetic core and its air gaps.

There is only one reluctance for a magnetic core without a gap, that of the core itself. Reluctance depends on the physical core properties such as the cross-sectional area A and ℓ_{MP} (length of the magnetic path) but also on the permeability of the core material: $\mathfrak{R}=\ell_{MP}/(\mu A)$

Thus, high permeability materials have low reluctance; the core presents a very low magnetic resistance to the flux. Very little flux will escape the core into the surrounding air, which has a very high magnetic reluctance compared to the core. Thus, the leakage inductance and magnetic radiation from a choke or transformer made on such a core will be very low.

The inverse of reluctance is called permeance and is a property of a specific core: $\mathcal{P}=1/\mathfrak{R}=\mu A/\ell_{MP}$. If the number of turns N is known and inductance L is measured, permeance can easily be deduced since $L=N^2\mathcal{P}$.

Since MMF is constant, very low reluctance also means a very high flux through the core (just as constant voltage means the current through a low resistance will be high), which can take the core into saturation sooner than for a higher reluctance core. This is why cores with high μ materials such as Permalloy, with much lower reluctance than silicon steel, saturate at half the flux levels than silicon steel (0.8 versus 1.6 Tesla).

In chokes and transformers with DC current, combining the signal or AC flux and bias or DC flux would saturate the core much sooner. Thus, the reluctance of the magnetic circuit is deliberately increased by introducing an air gap. The larger the gap, the more the magnetizing curve approaches that of the air coil (straight line) and the lower the magnetic flux density B!

The hysteresis curve will be "flattened". But, there is a price to pay - the inductance L of the choke or primary inductance of a transformer with DC current (as in SE amplifiers) will also be reduced.

The gap reluctance $\mathfrak{R}_G$ is much larger than core reluctance $\mathfrak{R}_C$ because the reluctance of air (or paper used as a gap) is much higher than the reluctance of any magnetic material. With the same MMF, this will reduce the magnitude of the flux (current) through the core (circuit) and thus prevent saturation.

The air gap width is g. The actual gap between the E and I pieces is g/2 since there are two gaps - one breaks the center leg, and the other breaks the two outer legs. The magnetic flux in each of the two loops crosses the physical air gap (g/2) wide twice; thus, the total gap is "g".

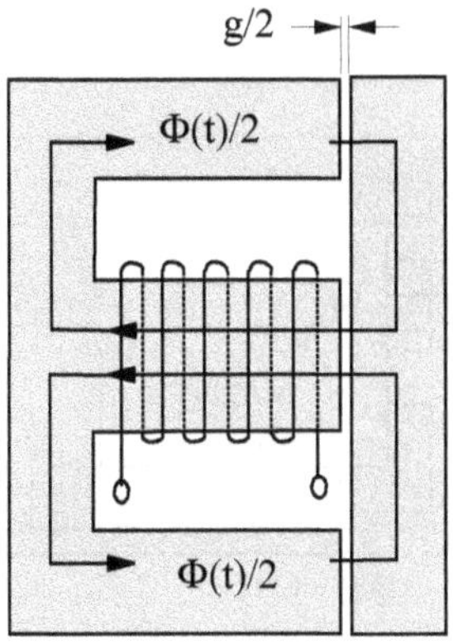

ABOVE: How the magnetic flux splits into two in an inductor or transformer with EI laminations

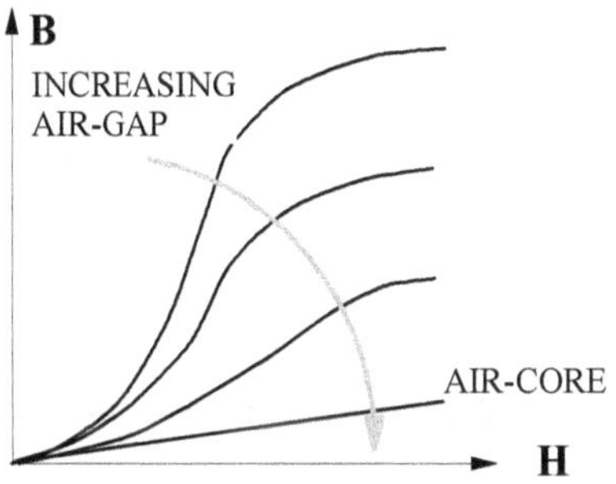

ABOVE: Air gap reduces μ (permeability) and B (density of magnetic flux)

BELOW: Hysteresis loops for a typical gapless magnetic core (solid line) and the same core with an air gap (dotted line).

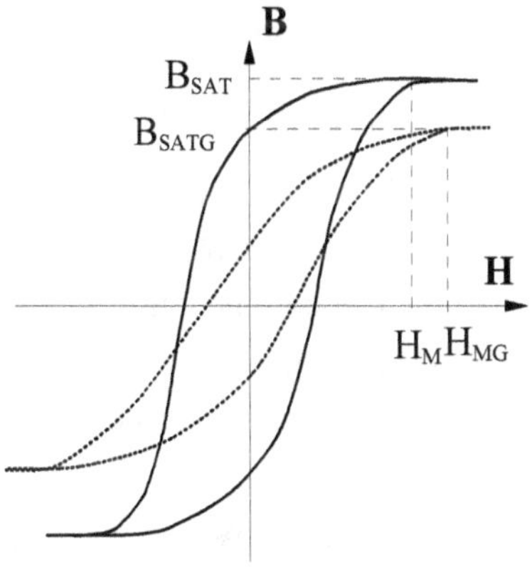

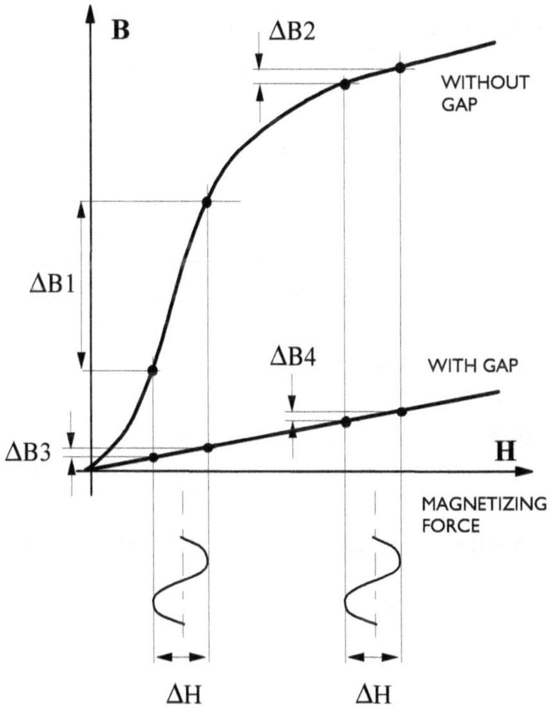

LEFT: Due to nonlinear magnetizing curve without air gap, the same change in magnetizing force ΔH produces a large ΔB1 at low H levels and a much smaller ΔB2 at higher H levels. With air gap the B-H curve is almost linear, so the same change in magnetizing force ΔH produces equal ΔB3 = ΔB4 at all H levels.

As with tube rectifiers and other nonlinear loads, the operating point Q can also be determined graphically (next page) or using basic algebra in simple cases.

For a simple EI core with one gap, the MMF equals the sum of the "voltage drops" (magnetic potential differences) in the series circuit: $MMF=NI=H_G*g + H_C*\ell_{MP}$.

Since $H=B/\mu = \Phi/(\mu A)$, we have:

$MMF=NI=\Phi*(\mathfrak{R}_G+\mathfrak{R}_C)= \Phi[g/(\mu_0 A_G) + \ell_{MP}/(\mu_0\mu_R A_C)]$.

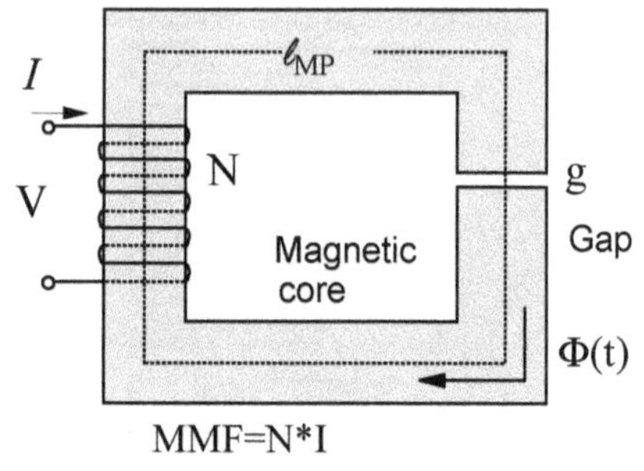

a) Magnetic circuit (the size of *g* is grossly exaggerated for illustration purposes)

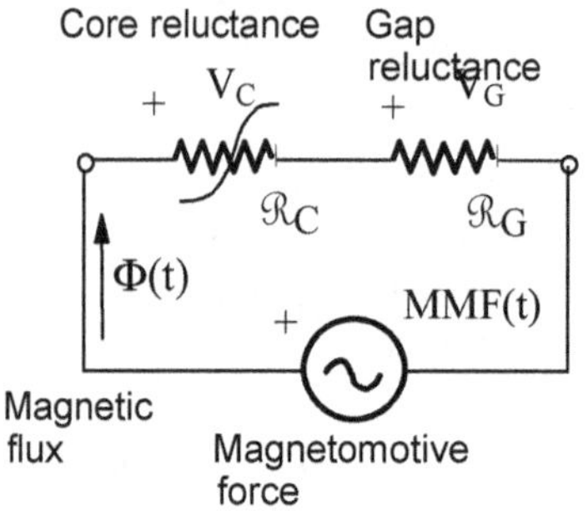

b) Equivalent electric circuit

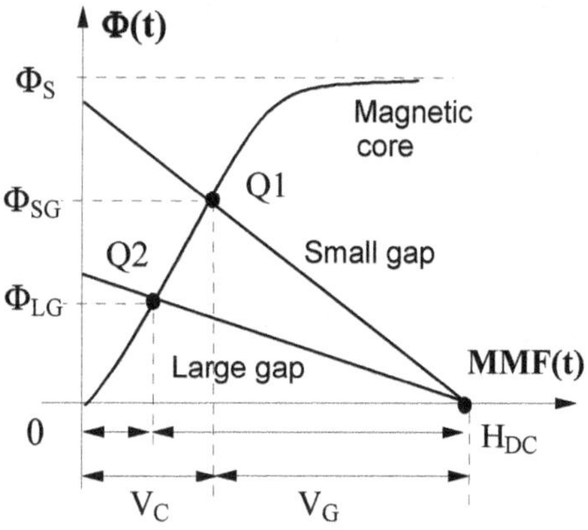

c) Graphical solution

The expression in square brackets is a sum of the gap reluctance $\mathcal{R}_G$=g/($\mu_0 A_G$) and core reluctance $\mathcal{R}_C$=ℓ_{MP}/($\mu_0\mu_R A_C$).

The larger the air gap "g," the more pronounced the fringing effect, illustrated on the right. The flux lines at the fringes of the gap take a longer path through the air, effectively increasing the air gap cross-sectional area A_G. Some pedantic designers use A_G=1.2A_C; however, in most cases, we can simply assume that A_G=A_C!

RIGHT: Flux lines across a large gap showing the fringing effect

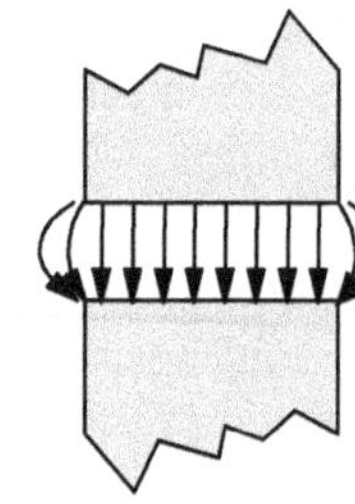

In that case NI = Φ[g/(μ_0A) + ℓ_{MP}/($\mu_0\mu_R$A)] = Φ/A*[g/μ_0 + ℓ_{MP}/$\mu_0\mu_R$] = B[g/μ_0 + ℓ_{MP}/$\mu_0\mu_R$].

Dividing the whole equation by ℓ_{MP} we get **H = NI/ℓ_{MP} = B[1/μ_0 * g/ℓ_{MP} + 1/$\mu_0\mu_R$]** and since H=B/μ_{EF} we can see that **1/μ_{EF} = 1/μ_0 * g/ℓ_{MP} + 1/$\mu_0\mu_R$**! μ_{EF} is the effective permeability of the core with air gap. Going back to the graphical solution, the equation of a loadline for air gap starting in point H_{DC} (MMF=NI) is MMF-Φ_G= NI-Φ_G.

The load line for a smaller gap intersects the magnetizing curve of the core in the operating point Q1, producing flux Φ_{SG}. For a larger gap (higher reluctance and lower flux Φ_{LG}), the reluctance line is more horizontal (less flux for the same MMF). The total reluctance of this series circuit is $\mathcal{R}$= $\mathcal{R}_C$ + $\mathcal{R}_G$ = ℓ_{MP}/(μA)+ g/A (μ for air is 1).

The effective permeability with air gap

Say you have EI66 laminations (22mm center leg) and assemble a core with 31 mm stack thickness S, to make a filtering choke. The cross-sectional area is A=aS=2.2*3.1= 6.8 cm^2 and the magnetic path length is ℓ_{MP}=11.4*a/2= 12.54 cm. The core is non-oriented 4% silicon steel with μ=10,000, and the gap is g=0.15mm.

The total reluctance of this series circuit is $\mathcal{R}$= $\mathcal{R}_C$ + $\mathcal{R}_G$= ℓ_{MP}/(μA)+g/A=12.54/(10,000*6.8)+ 0.015/6.8= 0.184*10^{-3} +2.2*10^{-3}. We used cm since we are not interested in the absolute but relative values, we can see that $\mathcal{R}_G$ is much larger (more than 10 times) than $\mathcal{R}_C$ ($\mathcal{R}_G$ = 11.9$\mathcal{R}_C$)!

The same case but using GOSS laminations with μ=50,000: $\mathcal{R}$= $\mathcal{R}_C$ + $\mathcal{R}_G$ = ℓ_{MP}/(μA)+g/A =12.54/(50,000*6.8) +0.015/6.8 = 0.0369*10^{-3} +2.2*10^{-3}. The gap reluctance is now 2.2/0.0369 = 60 times larger than the core's reluctance. 0.15mm is a small gap for audio chokes or transformers, with larger gaps the difference in reluctance would be even bigger.

Two lessons can be learned here. First, where an air gap is not needed (in power transformers, push-pull output transformers, and chokes with no DC current such as grid chokes), even a minute gap will profoundly affect the core's performance. This is why after strips are wound and cut, the joints of C-cores are precisely machined and polished. Nevertheless, a tiny "distributed" gap always exists, especially with EI laminations.

Secondly, in our calculations of chokes and transformers with a gap, we can simplify things by disregarding the core's reluctance since it is at least ten times smaller than the gap's reluctance. In other words, the gap defines the behavior of the magnetic core much more than the core itself.

This is good news. Often, we have no data for particular core or laminations on hand and thus no clue about its permeability.

As soon as a gap is introduced, the effective permeability is reduced; the formula that describes such deviation is μ_{EF}= μ_{INC}/(1+Kμ_{INC}). Obviously, for K=0 (no gap) the effective and incremental permeability are identical, μ_{EF}=μ_{IINC}. K is the ratio of gap and the length of the magnetic path of the core: **K=g/ℓ_{MP}**.

Transformer steel makers publish graphs showing the incremental inductance as a function of flux density (B), with H_0 (DC magnetization) as a parameter. One such graph for 3-4% non-grain-oriented cold rolled sheet steel (CRNGRO) is shown below.

Some lamination makers even publish graphs showing effective permeability as a function of incremental (AC) permeability, with gap ratio $K=g/\ell_{MP}$ as a parameter.

The graph below right is not for the same steel as the previous μ_{INC}-B_{MAX} graph, but it illustrates how with such two graphs for the same material, one can determine all important parameters of the magnetic circuit.

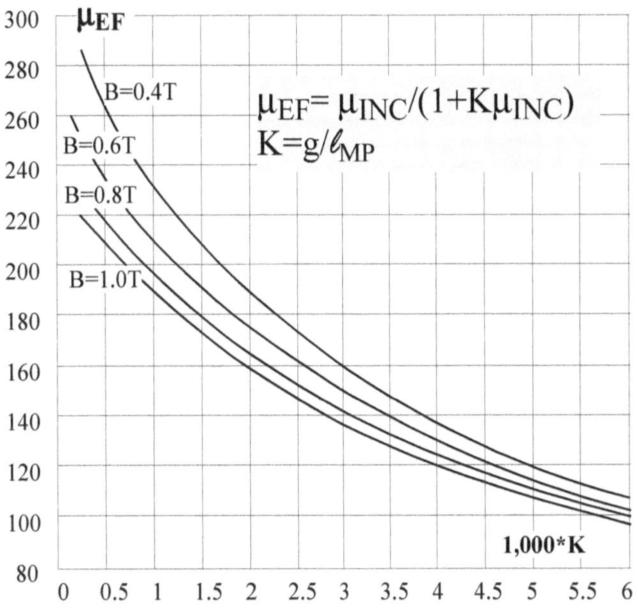

RIGHT: The μ_{EF} graph for 0.35mm 4% silicon steel EI laminations.

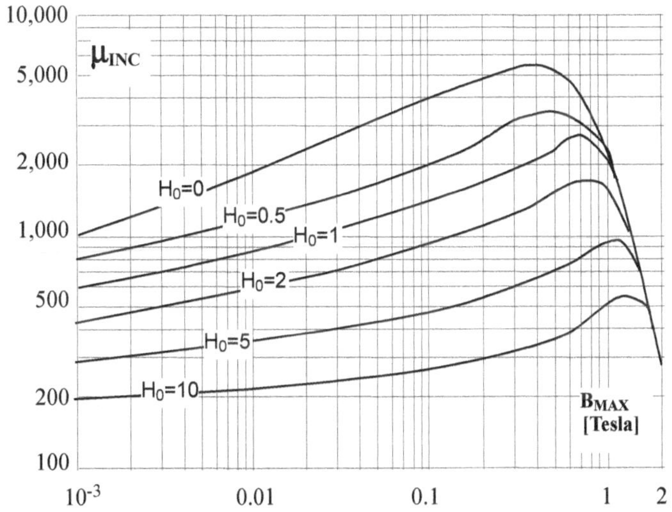

ABOVE: Incremental (AC) permeability as a function of flux density (B), with H_0 as parameter, for a typical 3-4% non-grain oriented cold rolled sheet steel (CRNGRO)

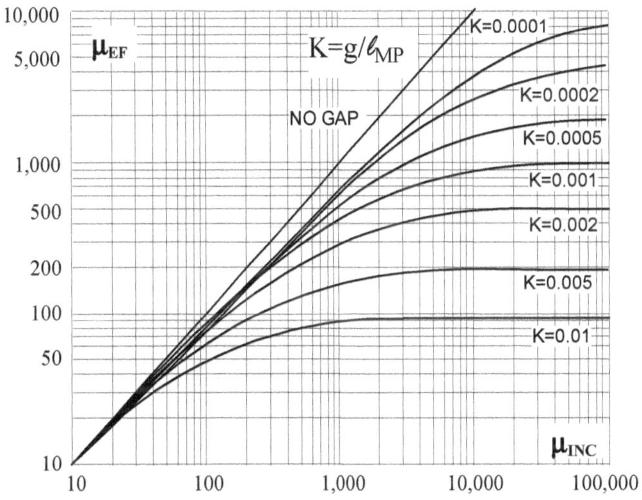

ABOVE: Effective permeability as a function of incremental (AC) permeability, with gap ratio $K=g/\ell_{MP}$ as parameter

EDDY CURRENTS

Ferromagnetic materials such as silicon steel are not just magnetic but also good electrical conductors. Eddy currents, also called Foucault currents after French physicist Léon Foucault (1819–1868) who is credited with having discovered them, are circulating currents that flow in the magnetic core, heat it up and produce their own magnetic flux that interferes with the flux produced by the transformer.

Power losses in a magnetic core due to eddy currents can be approximated (assuming a uniform magnetic field and no skin effect) by $P_{EC}=1.64B^2d^2f^2/(\rho D)$ [W/kg].

where B is the flux density, d is lamination thickness, f is the frequency of the signal (Hz), ρ is the electrical resistivity of the lamination material (Ωm), and D is the density of the lamination material (kg/m^3).

The characteristic eddy current frequency can be estimated as $f_{EC}=1.27\rho/(\mu_I d^2)$, where μ_I is the initial permeability (for H=0) of the lamination material. Above that frequency eddy current losses are deemed significant, below it "tolerable".

The table (next page) shows the limit of 14 kHz for low μ_I material and common thickness of 0.35mm, but only 1.4 kHz (10x lower limit) for laminations with 10x higher permeability.

Even significantly reducing lamination thickness to 0.1mm only increases the frequency limit of higher μ material (μ_I=3,000) to 17kHz, barely tolerable in audio applications. For high frequency converters and other HF equipment that would not be acceptable.

Reducing eddy currents

Since levels of B and the desired frequency range for tube amplifiers are pretty much fixed and determined, as are "d" and r (fixed for certain magnetic material), the only way to reduce eddy current losses is by using thinner laminations, i.e., by making "d" as small as possible.

Of course, laminations must be electrically insulated from one other. Vintage laminations had a thin sheet of paper glued to one side, but modern ones have a much thinner oxide layer or chemical coating.

Look carefully at a single lamination from both sides, and you will notice a slight difference in color and gloss because one side is coated or oxidized and the other one is not. Take the two probes of your multimeter (set to measure resistance) and press them against the same side of the lamination. You should get an open circuit (infinite resistance) on the insulated side and a short circuit on the non-insulated side.

The most common thickness values are 0.5mm (only non-oriented silicon steel) and 0.35mm (GOSS and non-oriented steel). GOSS laminations also come in 0.27, 0.23, 0.15, 0.10 and 0.05 mm thickness, but non-GOSS steel laminations are not made thinner than 0.35mm.

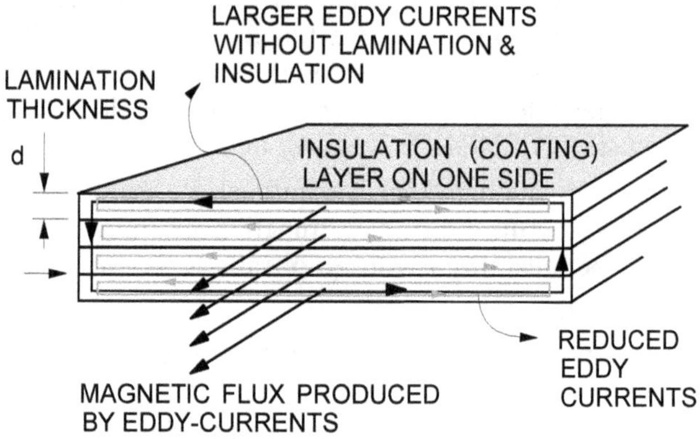

ABOVE: How eddy currents interact with the main flux of a transformer or choke

μ_I (initial permeability)	d=0.35mm	d=0.10mm	d=0.05mm
300	14 kHz	170 kHz	670 kHz
1,000	4 kHz	50 kHz	200 kHz
3,000	1.4 kHz	17 kHz	67 kHz

ABOVE: Characteristic eddy current frequency as a function of lamination thickness "d" and initial permeability μ_I

LAMINATION AND CORE TYPES

EI laminations

Named after the shape of the two pieces used, EI laminations are still the most common in audio transformers. Of all currently-produced types (C-cores, toroid, R-cores), they are the cheapest and inferior in most aspects. Toroid and R-cores require special winding machines and, as such, are not of interest to amplifier makers who want to wind their transformers. C-cores use the same bobbins as EI cores but require special mounting bases and steel-tensioning straps, so EI cores are the easiest to use and assemble.

Scrapless or wasteless laminations are EI-type stamped out in the manner illustrated on the next page. The lighter gray strips are the "I" pieces, so there is no wastage of magnetic material. Identifying their size is very easy. The longest dimension is 6U (units), which is the number in the name of the lamination. For EI114 size, 6U=114 mm, one unit is U=114/6= 19 mm, and the center leg is 2U or 2*19mm = 38mm. When the transformer is assembled, the other overall dimension is 5U, so the ratio of the two main dimensions is 6:5!

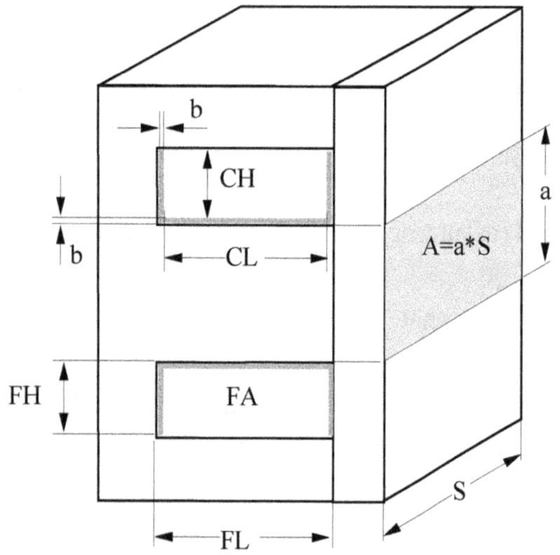

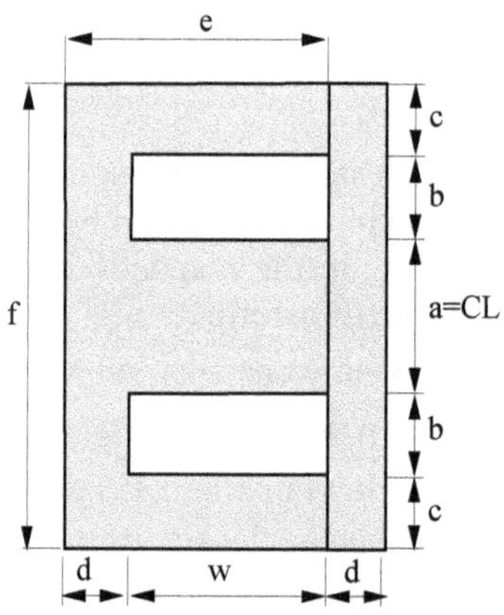

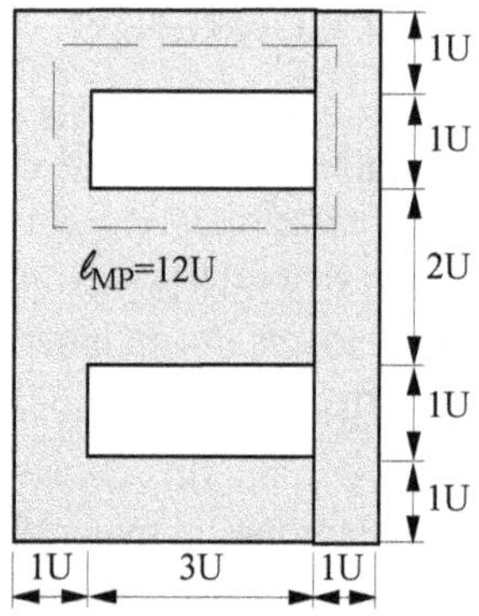

ABOVE LEFT: Dimensions and magnetic path length for all EI sizes, as referred to in the sizing table on the next page.

ABOVE RIGHT: General dimensions and magnetic path length for wastless (scrapless) EI sizes (gray shaded in tables) b = c = g and a = 2b = 2c = 2g

ABOVE: Definitions of various dimensional terms we will use in this book: a = center leg width, S = stack thickness, A = gross core area, FL= window length, CL-coil length, CL=FL-2b, FH= window height, CH= maximum coil height CH=FH-b, FA= window area, FA=FL*FH ("F" is from *Fenster,* the German word for window!)

While GOSS (grain-oriented silicon steel) materials improve the performance of transformers with EI-laminations, the superiority of GOSS material is not fully utilized in EI laminations because, in some sections of the magnetic path, the magnetic flux is perpendicular to the grain direction, cutting across the grain orientation. In those sections, the benefits of grain orientation are lost.

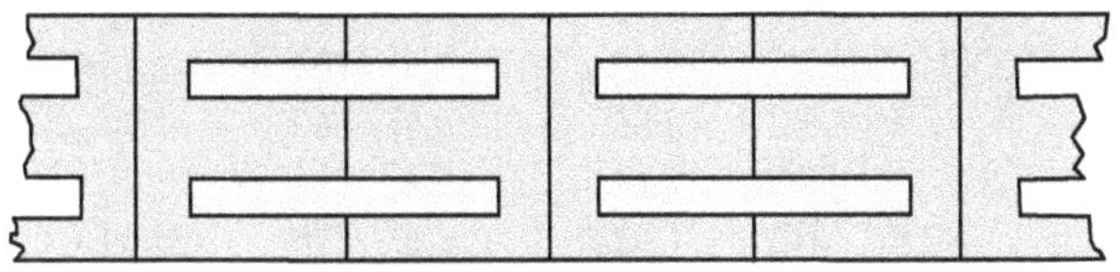

ABOVE: How scrapless (wastless) EI laminations are stamped out of a continuous strip

Magnetic path length (ℓ_{MP})

In the design calculations for audio transformers, we will often need to calculate the "length of the mean magnetic path" or ℓ_{MP}, so let's illustrate it on a "scrapless" lamination type.

Just as the actual physical gap for EI laminations is only one-half of the calculated or theoretical value, the length of the magnetic path is only considered in one-half of the core. Both are a consequence of the fact that flux splits into two paths in such cores.

Following the dotted line (previous page), $\ell_{MP} \approx 2(U/2+U+U/2)+2(U/2+3U+U/2)=12U$ Since 2U is CL (the width of the center leg), $\ell_{MP} \approx 6 * CL$! The exact formula is $\ell_{MP} = 2(b+e-d)+\pi c$.

TYPE	f	e	a	c	b	w	d
EI 19	19	12.5	5	2.5	4.5	10	2.5
EI 24	24	15	6	3	6	12	3
EI 25.4	25.4	16.1	6.4	3.1	6.4	13.1	3
EI 28	28	21	8	4	6	17	4
EI 35	35	24.5	9.6	5	7.7	19.5	5
EI 41	41	27	13	6	8	21	6
EI 48	48	32	16	8	8	24	8
EI 54	54	36	18	9	9	27	9
EI 57	57	38	19	9.5	9.5	28.5	9.5
EI 60	60	40	20	10	10	30	10
EI 66	66	44	22	11	11	33	11
EI 73	73	49	23	11.5	13.5	34.5	11.5
EI 74	74	51	23	11.5	14	34	14
EI 75	75	51.5	23	12	14	37.5	14
EI 76.2	76.2	50.8	25.4	12.7	12.7	38.1	12.7
EI 84	84	56	28	14	14	42	14
EI 85.8	85.8	57.2	28.6	14.3	14.3	42.9	14.3
EI 86	86	59	26	13	17	46	13
EI 88	88	60	29.4	13.3	16	44	16
EI 90	90	60	30	15	15	45	15
EI 96	96	64	32	16	16	48	16
EI 100	100	67	32	16	18	51	16
EI 105	105	70	35	17.5	17.5	52.5	17.5
EI 111	111	76	35	17.5	20.5	55.5	20.5
EI 114	114	76	38	19	19	57	19
EI 120	120	80	40	20	20	60	20
EI 133.2	133.2	88.8	44.4	22.2	22.2	66.6	22.2
EI 152.4	152.4	101.6	50.8	25.4	25.4	76.2	25.4
EI 181.2	181.2	120.8	60.4	30.2	30.2	90.6	30.2
EI 192	192	128	64	32	32	96	32

ABOVE: EI lamination sizes. Wastless sizes are highlighted in dark gray.

C-cores

C-cores (tape-wound cut-cores) are characterized by low noise, low magnetizing current, and fast assembly (no need to stack hundreds of EI laminations). Strips are cut from cold rolled grain oriented (CRGO) silicon steel along the grain direction and annealed in a high vacuum furnace. They are stacked, shaped, cut, and polished to ensure a good fit and minimum air gap (joints marked "X").

In terms of shape and performance, C-cores are in-between EI and toroids. Compared to EI-cores, they work at higher B levels and have a much higher power rating for the same cross-sectional area! The whole magnetic path is in line with grain orientation.

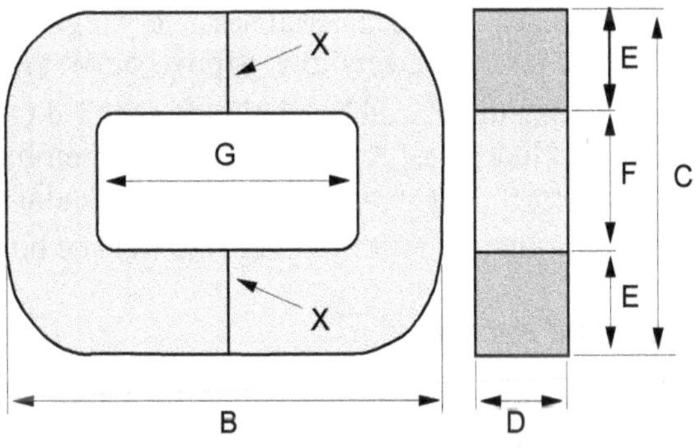

ABOVE: General dimensions of C-cores

Two C-pieces form one assembly or core. In this case, two bobbins are usually used, one on each longer limb (dimension "G"). This arrangement is shown in a). Also, two such assemblies (4 x C core) can be put together side-by-side, in which case only one coil is used around the center leg, as in arrangement b).

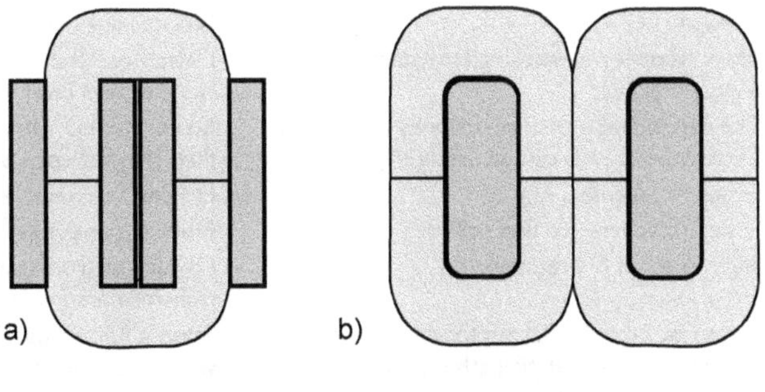

RIGHT: Two most common topologies of chokes and transformers made using C-cores:

a) Two C-cores and two identical coils

b) Four C-cores and a single coil on the center leg

TYPE	C [cm]	B [cm]	F_{MIN} [cm]	G_{MIN} [cm]	D_{MIN} [cm]	D_{MAX} [cm]	E_{MIN} [cm]	E_{MAX} [cm]	ℓ_{MP} [cm]	A [cm^2]	Weight [kg]	TPV	P_{OUT} [VA]	Power loss [W]
U/19	4.01	7.94	1.90	5.72	1.90	1.98	0.95	0.103	18.17	1.72	0.24	17.4	33	0.48
U/25					2.54	2.62				2.29	0.32	13.1	44	0.64
U/32					3.17	3.25				2.86	0.40	10.5	54	0.80
U/38					3.81	3.89				3.44	0.48	8.7	70	0.96
V/22	4.96	9.21	2.22	6.35	2.22	2.30	1.27	1.35	21.06	2.68	0.43	11.2	66	0.86
V/29					2.86	2.94				3.45	0.56	8.7	84	1.11
V/38					3.81	3.89				4.60	0.74	6.5	110	1.48
V/51					5.08	5.16				6.13	0.99	4.9	144	1.98
X/19	6.23	12.11	2.86	7.63	1.90	1.98	1.59	1.67	25.87	2.87	0.57	10.5	105	1.14
X/29					2.86	2.94				4.32	0.86	6.9	160	1.71
X/38					3.81	3.89				5.76	1.14	5.2	210	2.28
X/51					5.08	5.16				7.67	1.52	3.9	270	3.04
Z/25	7.50	13.02	3.50	8.9	2.54	2.62	1.90	1.98	30.68	4.59	1.08	6.5	215	2.15
Z/38					3.81	3.89				6.88	1.61	4.4	325	3.23
Z/51					5.08	5.16				9.17	2.15	3.3	425	4.3
Z/70					6.98	7.14				12.60	2.96	2.4	580	5.91

ABOVE: Data for standard metric-size GOSS C- cores, 0.3mm lamination thickness, measured at f=50Hz, B=1.7T

Source: http://www.aksteel.com/markets_products/electrical.aspx

Thickness (mm)	Grade JIS	Core Loss (W/kg)	B in Tesla at 800A/m	Stacking Factor (%)	Korean Posco	Japanese Nippon	Japanese Kawasaki	USA AISI	German DIN
0.23	23P095	0.95	1.87	94.5	23PH125	23ZH95	23RGH095		
0.23	23P100	1.00	1.87		23PH132	23ZH100	23RGH100		
0.27	27P100	1.00	1.88	95	27PH132	27ZH100	27RGH100		
0.27	27P110	1.10	1.88		27PH145	27ZH110	27RGH110		
0.30	30G120	1.20	1.80	95.5	-	30Z120	30RG120	M-3	VM97-30
0.30	30G130	1.30			30PG172	30Z130	30RG130	M-4	
0.30	30G140	1.40			30PG198	30Z140	30RG140	M-5	
0.35	35G135	1.35	1.80	96	-	35Z135	35RG135	-	VM111-35
0.35	35G145	1.40			35PG191	35Z145	35RG145	M-5	
0.35	35G155	1.55			35PG204	35Z155	35RG155	M-6	

ABOVE: GOSS EI laminations properties and part numbers of major suppliers

Amorphous Metglas® C-cores

Amorphous Metglas® C-cores use cobalt-based magnetic alloy 2714A. Although the saturation level is a low 0.57 Tesla, once annealed, the material has a very high DC Permeability (μ) of around 1,000,000 (as cast μ >80,000)! At 60Hz the core losses are very low, approx. 5mW/kg, almost 1,000 times lower than ordinary silicon steel!

The name seems to be an abbreviation for METal GLASs, and, indeed, the cores are so brittle that if hit or dropped, they break just like glass! Another negative is their high price, which may come down in the future. Of little benefit for mains transformers, these cores seem a good candidate for MC step-up and interstage transformers.

For more information, visit Metglas, Inc. website http://www.metglas.com/

Toroidal and R-cores

Toroidal cores are next to impossible to wind by hand; special winding machines are needed. Short lengths of winding wire can be wound on a spool and fed through the toroidal ring, but that is tedious and time-consuming.

EI Laminations	**C-Cores**	**Toriods & R-cores**
Cheaper (+)	No stacking needed (+)	Expensive (-)
More flexibility in stacking (any stack thickness is possible) (+)	Lower loss and smaller core area/size for the same power level (+)	Impossible to wind manually, without special winding machines (-)
The assembled transformer looks neater than C-cores, end bells can be used (+)	Lower number of turns needed for the same flux (Higher core permeability) (+)	Difficult to provide an air gap for SE transformers (-)
Easier to assemble (+)	Harder to assemble (steel straps needed) (-)	Need special mounting hardware (-)
Easier to source (to find sellers) (+)	More expensive (-)	Very sensitive to DC imbalance (-))
High levels of EM radiation (-)	Lower electromagnetic radiation than EI transformers (+)	The lowest EM radiation of all types
Does not take the full advantage of GOSS. A section of the magnetic path is perpendicular to the grain orientation (-)	Takes a full advantage of GOSS material, the whole magnetic path is along the grain orientation, none across (+)	Takes a full advantage of GOSS material, the whole magnetic path is along the grain orientation, none across (+)

Toroidal transformers are also very sensitive to even the smallest DC current through the coils (no air-gap), and their performance would suffer in push-pull amps with a DC imbalance in the output stage. We also had buzzing problems with toroidal power transformers due to the high DC voltage level on the Australian power grid.

Assuming a 30Ω DC resistance of the power transformer's primary winding, the measured $3V_{DC}$ superposed onto our $240V_{AC}$ mains voltage would cause a DC current of 3/30 = 100mA to flow through it, which is more than enough to take such a mains transformer into saturation!

Due to their small size and slim profile, low radiation, and high-volume production in China (meaning low price), R-cores are gaining popularity, especially as power transformers. They also require special winding machines.

The way it used to be: PU laminations

These 1950s Telefunken power transformers use the Philbert design. Each is made of two L-shaped cores, one corner is rounded, and the other has been cut at 45 degrees to achieve a more rounded magnetic path. There is no possibility of an error while assembling the stack - all insulated sides are automatically oriented properly.The two identical coils are connected in parallel (to double the current capacity).

The outermost 6.3 V heater winding is visible through the insulation. By counting its number of turns (41 in this case), one can easily determine the TPV figure (Turns-Per-Volt.) as 41/6.5= 6.3 TPV! 6.5 was the no-load voltage that should be used in TPV calculations.

After 60 years, the interconnecting colored wires looked brand new, no aging, fading, or cracking - truly remarkable!

For this particular size (P68), C=17mm and the stack thickness is 31mm, so $A=C*S=1.7*3.1=5.27\ cm^2$, enough for both power and output transformers in smaller Telefunken amplifiers.

The M6 type GOSS material used was called V111-35 and the power capability of the P68 core was 62VA, which is roughly double that for the same cross-section EI core (28VA).

These great transformers can be reused as they are in tube preamplifiers and low-powered amps, one per channel. Telefunken's larger 35 Watt push-pull V69B model used P96 core (C=24mm), good for up to 220VA.

PU-type transformers emit extremely low EM radiation. Sharp edges (where the magnetic flux lines tend to concentrate) were reduced. The low radiation is also due to 50% wider head sections ("1.5c" versus "c") and the fact that the two windings also act as electromagnetic shields themselves.

PU laminations have not been commercially available since the late 1980s, but many PU transformers are available on eBay and can be easily dismantled since they have not been glued or welded together.

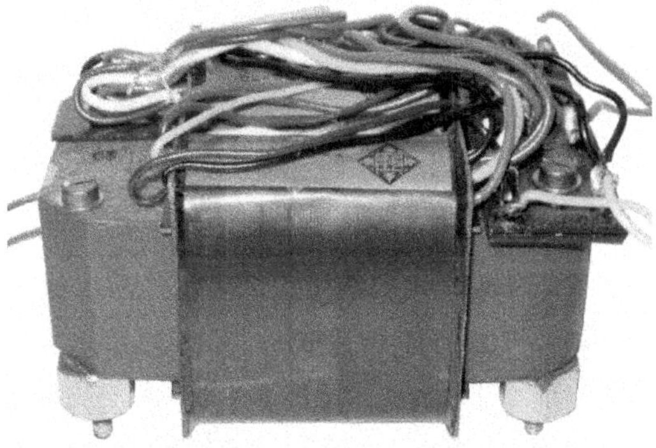

ABOVE: The P68 core used M6 type GOSS laminations (V111-35) and had the power rating of 62VA

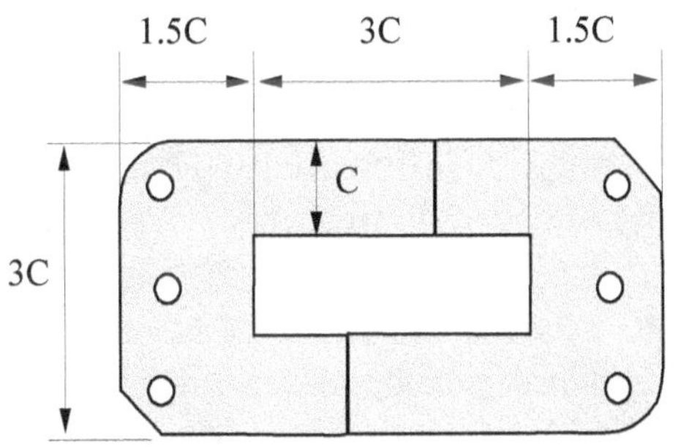

ABOVE: The profile and main dimensions of the Philbert design

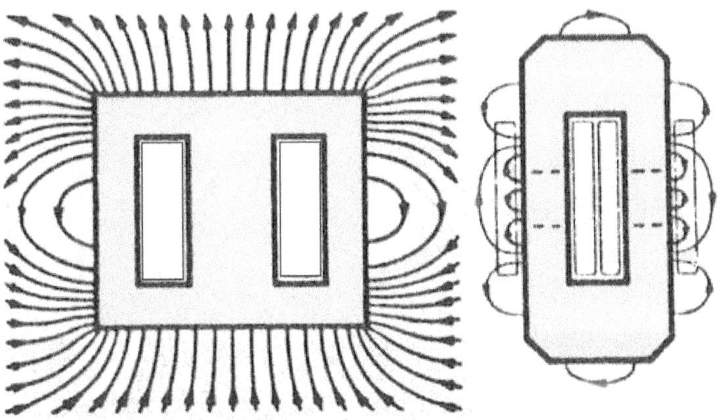

LEFT: This illustration from "Telefunken Laborbuch I" compares the magnetic field around an EI transformer and the low-radiation PU-type transformer.

The way it used to be: M laminations

Although the production of M-laminations ceased in the early 1980s, there are many power and output transformers available for sale on eBay and elsewhere, salvaged from European radios and amplifiers. They are easily recognizable with a square footprint (a=b) and slightly chamfered corners.

M-type are single-piece laminations with a slit in the middle leg. Smaller sizes have a fixed 0.3mm gap; the bigger ones come in three choices of the gap size plus the no-gap version. Although superior to EI-laminations, they are more expensive to make and more difficult to stack, so vintage transformer makers favored the EI type.

Chokes and transformers using M-lams can have a higher current density than EI- transformers, around 2.75 A/mm^2 for larger core cross-sectional areas (10-15 cm^2) and 3.5-4.0 A/mm^2 for smaller cores with A below 5cm^2!

Perhaps this comes from better heat dissipation. Relative to their size, M-lams have larger windows than EI-cores.

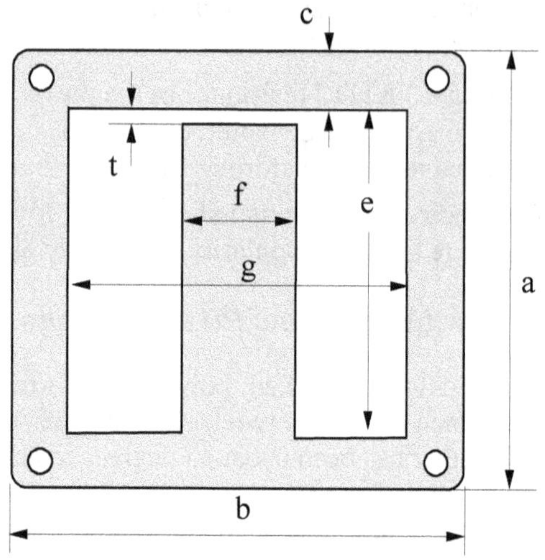

all in mm	a = b	c	e = g	f	t
M30	30	5	20	7	0.3
M42	42	6	30	12	0-0.3-0.5-1
M55	55	8.5	38	17	0-0.3-0.5-1
M65	65	10	45	20	0-0.3-0.5-1
M74	74	11.5	51	23	0-0.5-1-2
M85	85	14.5	56	29	0-0.5-1-2
M102	102	17	68	34	0-0.5-1-2

ABOVE: Dimensions of M-laminations. Notice that the air gap "t" comes in predefined sizes, 0 (no air gap), 0.3, 0.5, 1.0 and 2.0 mm! A transformer designer would have to determine the gap size before buying the laminations with such a gap!

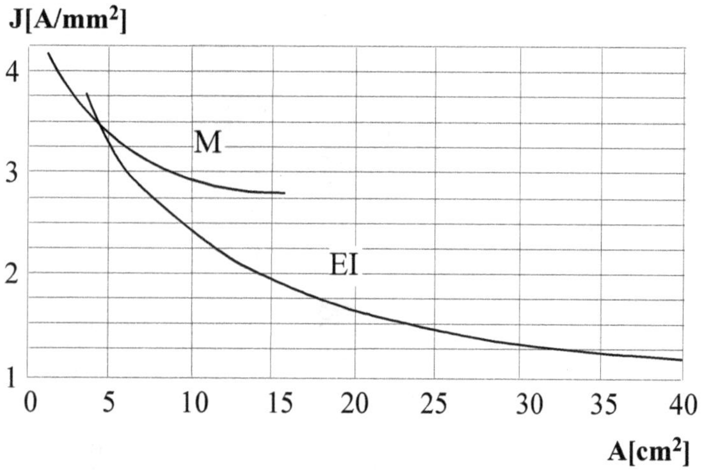

LEFT: Current density that can be used in transformers with M-laminations is higher compared to the EI laminations

THE OVERVIEW OF UNITS, SYMBOLS AND CONVERSION FACTORS

Magnetic and electric units and symbols

F = total field or MMF (magnetomotive force) [AT or AmpereTurns]

H = magnetic field strength or magnetizing force [AT/m]

Φ = magnetic flux [Weber]

B = flux density [Tesla=Wb/m^2]

μ = permeability μ_I = initial perm. μ_{INC} = incremental (AC) perm. μ_{EF} = effective perm.

f_L = lower -3dB frequency [Hz] or [cps] (cycles per second)

f_U = upper -3dB frequency [Hz] or [cps] (cycles per second)

Z_P = primary impedance [Ω - ohms]

R_I = internal impedance of the output or driver tube [Ω]

L_L = leakage inductance, L_P = primary inductance [H]

N_P or N_1 = primary number of turns N_S or N_2 = secondary number of turns

TR = turns ratio, IR = impedance ratio

R_P or R_1= primary resistance [Ω] R_S or R_2= secondary resistance [Ω]

P_C = power handling capability of the core [VA]

Mechanical (size-related) units and symbols

g = width of the air gap [mm]

ℓ_{MP} = mean length of the magnetic path [cm or mm]

ℓ_W = total winding length [m]

MTL = mean turn length [cm or mm]

A = gross cross-sectional area of the core [cm^2], A_{EF} = effective or net cross-sectional area of the core [cm^2]

WL = window length [mm], CL = coil length (winding length) [mm]

WH = window height [mm], CH = coil height (winding length) [mm]

F = gross window area, F_C = total copper area, F_P= primary copper area, F_S = secondary copper area [cm^2]

	SYMBOL	SI system unit	CGS unit	CONVERSION
FIELD STRENGTH OR MAGNETIZING FORCE	H	ATurn/m	Oersted	1 Oersted=1,000/(4*π) ATurn/m
TOTAL FIELD or MAGNETOMOTIVE FORCE	F or MMF	ATurn	Gilbert	1 Gilbert = 10/(4*π) ATurn
TOTAL FLUX	Φ	Weber (Wb)	Maxwell	1 Maxwell = 10^{-8} Weber
FLUX DENSITY	B	Tesla (T) =Weber/m^2	Gauss	1 Gauss = 10^{-4} T
AREA (core, window)	A	m^2	cm^2	1 cm^2 = 10^{-4} m^2
LENGTH (gap, magnetic path, etc.)	ℓ, g, LMP, MTL	meter (m)	cm	1 cm = 10^{-2} m
RELUCTANCE (F/Φ)	$\mathfrak{R}$	ATurn/Wb	Gilbert/Maxwell	$10^9/(4*\pi)$ ATurn/Wb
PERMEANCE (Φ/F) $\mathcal{P}$=1/$\mathfrak{R}$	$\mathcal{P}$	Wb/ATurn	Maxwell/Gilbert	4*$\pi 10^{-9}$ Wb/ATurn

FROM THEORETICAL BACKGROUND TO PRACTICAL DESIGNS

The black art of transformer design & construction

Although we have only scratched the surface, a few important issues should be obvious even to the most cursory reader. The first is that transformer design is the art of compromise, necessary to deal with often contradictory requirements. Optimizing one factor or performance aspect almost always worsens one or more other aspects of performance parameters. There is no such thing as a free lunch in electronics; you can not have your cake and eat it too, as the saying goes.

The second issue is that parameters often considered constant at the design stage are anything but in reality. For instance, we have already seen that the inductance of a choke or transformer's winding depends on relative permeability, which varies with quite a few factors, such as the level of DC current in the winding (DC bias) and the size of the air gap.

Consequently, in reality, what seemed to be a single equation with one unknown (solvable) is an equation with two or even three unknowns, which is not solvable unless we assume a conservative value for the second and third variables. Thus, transformer design is often a series of iterations, starting with one or more assumptions, working forward" to find the rest of design parameters, and then working "backward" to check if the initial assumptions were correct or reasonably accurate. If they weren't, a second and often a third iteration might be necessary, depending on what type of transformer designer you are, a sloppy, "that's good enough" type, or an anal-retentive or pedantic type which has to get all results to agree to two decimal points.

Finally, we have seen that the B-H characteristic of soft magnetic materials used for transformer laminations is highly nonlinear, and since the voltage induced in a choke or transformer winding is proportional to magnetic flux density (the fundamental transformer equation), and since the magnetizing force H is a product of the excitation current, it also follows that the V-I relationship is also nonlinear. This directly impacts the previous issue because what looked like a simple linear equation is a nonlinear relationship that cannot be solved using linear algebra.

Or, in the words of one of my university electronics professors, "if the theory does not agree with reality, disregard reality"!

Theory versus practice

The good news is that although the information in this chapter was necessary from an educational and theoretical point of view, it is seldom needed in practice. If you stick to the few basic rules of thumb, all that mathematical exhibitionism becomes academic.

I still remember the time spent in one of our local transformer winding shops. While waiting for the small batch of power transformers to be packed, I tried to get some insider information from the two owners of the business, but they could not answer any of my questions.

It turned out that they didn't have much knowledge of transformers at all. They simply typed all the required parameters (voltages and currents of a power transformer's windings, the lamination size and stack thickness used, etc.) into some kind of computer program and a detailed winding diagram would be printed out without any further input from them.

However, once we installed their transformers in our tube amplifiers, they buzzed and vibrated, rendering them useless for our purpose. The transformer winding shop "specialized" in control transformers for mining industry, where they would sit in a switchboard and the buzz would not even be noticed. For hi-end audio applications, the calculations of that software program were obviously wrong.

In the case of those two "experts", the gap between their theoretical knowledge (almost none) and their practical knowledge was too large, and even their practical know-how was questionable. I mention this story so you don't underestimate your own knowledge and skills and/or overestimate the capabilities of the so called "experts". To make matters even worse, most transformer shops only deal with power transformers and know nothing about audio applications.

Somebody once remarked that the only man who truly, intuitively and completely understood transformers died in 1943. There may be some truth in that claim. Here is a short summary of his remarkable life.

Nikola Tesla

Nikola Tesla was born to Serbian parents in 1856 in Smiljan, a village in Austro- Hungarian empire, later Yugoslavia, today's Croatia. After emigrating to the USA at age 28, he worked for a short while at the Edison Machine Works before setting up his own laboratories in New York, with the financial backing of George Westinghouse, a rival of Thomas Alva Edison.

His most notable invention is the brushless alternating current (AC) induction motor and related polyphase transformers and AC patents, which he licensed to Westinghouse Electric.

In 1891, he invented the Tesla coil, an air-cored, dual-tuned resonant transformer capable of producing very high voltages at high frequencies.

Despite Edison's huge lobbying and "marketing" efforts, which often degenerated into outright lies, Tesla's AC power generation, transmission & distribution system (still in use today) won over Edison's troublesome and inefficient DC (direct current) system.

Tesla then worked on a wide range of projects, such as electrical discharge tubes, X-ray imaging, wireless communications, and wireless electric power distribution. He died in New York City on 7 January 1943.

In 1960, the General Conference on Weights and Measures named the SI unit of magnetic flux density "Tesla" in his honor. The international airport in Belgrade, the capital of former Yugoslavia (today's Serbia), was also named in his honor, not to mention quite a few companies around the world, most notably "Tesla" in the USA (electric cars and battery systems).

In the field of electrical engineering, he is considered the greatest inventor of all time.

ABOVE: Nikola Tesla (1856 - 1943), the greatest inventor of all times in the field of electrical engineering

FILTERING CHOKES (INDUCTORS WITH DC CURRENT)

- EVALUATING VINTAGE & COMMERCIAL CHOKES
- SWINGING CHOKES
- CHOKE DESIGN METHODS
- HOW TO DETERMINE THE SIZE OF AIR GAP
- DIY CHOKE DESIGNS

3

EVALUATING VINTAGE & COMMERCIAL CHOKES

Commercial benchmark: vintage USA chokes

These two hermetically sealed USA-made vintage chokes have the same declared inductance of 4H. The larger one has a DC resistance of 85Ω and can pass 270 mA; the smaller one has a DC resistance of 115Ω and is good for up to 180 mA of DC current.

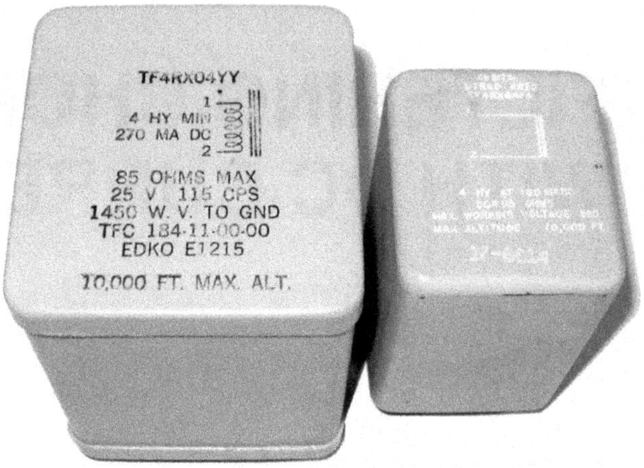

Depending on the secondary voltage of the power transformer you have available, the choke's DC voltage drop may become an issue. The higher the voltage drop on the rectifier and the choke, the lower the final DC voltage on the anodes of output tubes and the lower the amplifier's maximum power.

At its maximum current, the voltage drop on the larger choke will be $\Delta V=0.27*85=23V$ or $\Delta V=0.18*115=20.7V$ on the smaller choke.

The voltage rating of the larger choke is also higher, $1{,}450V_{DC}$, versus $590V_{DC}$ for the smaller one.

The money and sanity saver: Cheap & ready chokes for tube preamplifiers and screen grid supplies

These Taiwanese-made baby mains transformers with 6.3 or 12.6 V secondaries (below) make ideal HV filtering chokes for tube preamps and phono stages. You may find them in surplus stores or clearance sales. If rated at 5VA and 240 V_{AC} mains, their primary current, blue & brown wires (1), is around I = P/V= 5/240 = 21 mA, which means the primary wire can also take 20mA DC current, and can be used as a 20mA choke, plenty of current capacity for both channels.

Depending on the quality of the laminations and the actual current, you will get 8-15H of inductance. With primary rated for 115V as in the USA, the primary current will be 5VA/115V = 44 mA, even better! Use one choke for each channel, or connect them in a chain for an even better filtration (CLC - CLC filtering).

These baby trafos were not stacked to have an air gap, but from our experience, they don't saturate easily (there is always a residual gap present). There is no need to restack the laminations for preamps with two or even four duo-triodes. The secondary is left open circuit, so just chop those three leads off (2), and your preamp choke is ready.

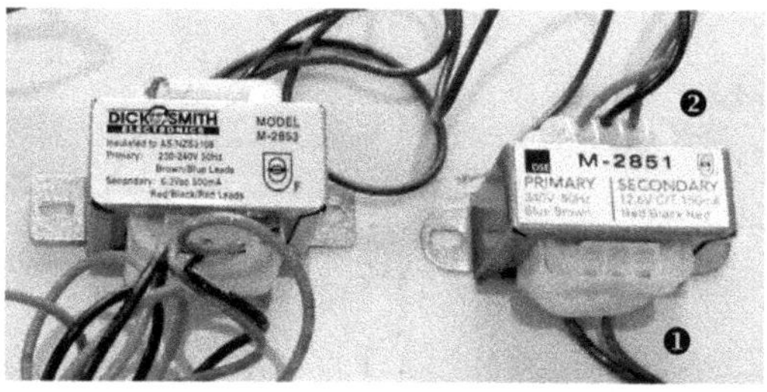

The primary windings of small low-voltage mains transformers (ABOVE) and salvaged SE output transformers from guitar amps or vintage radios (BELOW) can be reused as they are (without restacking) as low current filtering chokes for preamps and screen grid supplies.

This vintage output transformer from a 5W single-ended German EL84 pentode amp has core dimensions 54 x 44 x 18.25 mm, so the laminations are EI54 size with center-leg cross-sectional area $A=1.8*1.8=3.24\ cm^2$.

The primary DC resistance is $R_P= 772\ \Omega$, and the primary inductance is $L_P=22.5$ H, a great result for such a small transformer, indicating quality laminations.

Tested by simply connecting its primary to the mains voltage (in our case 256 V_{AC} at 50Hz) the secondary voltage was 6.91V, so $TR=256/6.91=37$ and $IR=TR^2=1{,}369$. With 4Ω load $Z_P=1{,}369*4 = 5{,}476\ \Omega$, or 5.5 kΩ nominal primary impedance.

These can be reused as output transformers in guitar amps (you wouldn't get any bass below 100Hz or so, but for guitar amps, that does not matter a bit), but since EL84's DC current is 35-50 mA, you can use their primaries as chokes for preamplifiers, or as chokes for power amp's screen grid supplies and preamp sections. 22H is one mighty choke.

Assuming a current draw of 20mA for both channels, the DC voltage drop on primary resistance of $R_P= 772\Omega$ would be 15.5V, easily accommodated by most power supplies.

Vintage Motorola filtering chokes

An American eBay seller, electronic surplus parts company in Virginia, was selling half-a-dozen NOS chokes with "Motorola" printed on one and a number code on the other end-bell. However, they only showed a photo of the box and wrapping and not the actual chokes. No technical details were listed, not even the dimensions, so buyers were not interested.

We asked only one question - how many would fit in a Medium Flat Rate USPS Priority Mail International box. The seller could ship four chokes in that box. We could have asked for dimensions and a photo of one choke, some sellers would even be willing to perform basic ohmmeter tests, but the fact that only four would fit in a relatively large box with 20 lb maximum weight was enough to give us a pretty good idea about their size and weight.

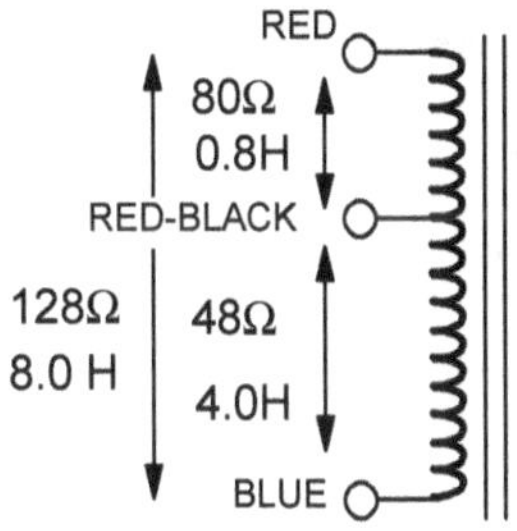

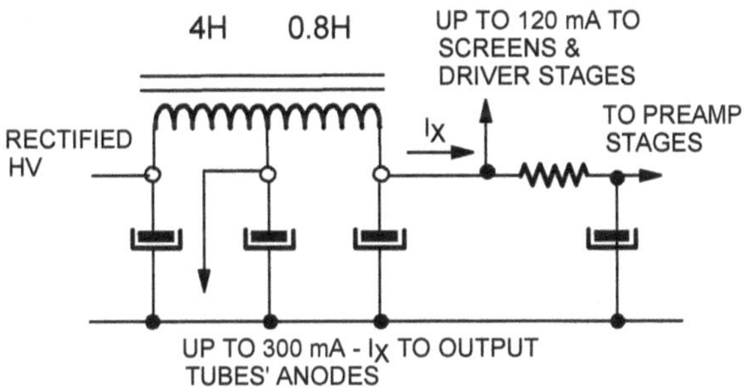

FAR LEFT: The results of the basic resistance and inductance measurements.

LEFT: A typical application when both chokes are used to their maximum current-carrying capacity.

The laminations are 76mm x 63mm with 50mm stack thickness (12.7cm^2 center leg area), meaning that at 50Hz, the magnetic cores are rated at 160 VA. The DC resistances are straightforward; however, the inductances cannot simply be added up. The two sections measure 0.8H and 4.0H, but the total winding measures 8.0 H!

When used as an 8H choke, the maximum current is limited by the section wound with a thinner wire. In this case, that is the high DC resistance section that measures 0.8H. Its winding wire was 0.22mm in diameter, which is good for up to 120 mA.

The topology illustrated here can be used to better utilize the 4H section, which is wound with 0.35mm wire and can pass up to 300mA. The filtered high voltage for the output stages is taken after the 4H choke section, while the DC supply voltage for output tubes' screens (if used as pentodes) and driver and preamp stages go through another filtering stage (another CLC filter).

How the inductance of a filtering choke varies with the magnitude of its DC current

These vintage Halldorson chokes are rated at 8H @ 150mA, with 145Ω DC resistance. The printing on their boxes provides some valuable clues. Notice how the inductance increases by 50% to 12H at a lower DC current of 100mA (point A) and reduces to half of the nominal inductance, only 4H at 200mA (point C).

Most chokes, especially ones made by reputable manufacturers, will tolerate higher DC current than that specified as nominal. However, the winding may get a bit warmer than normal (due to increased losses in copper), and the effective inductance will be reduced.

Likewise, the lower the load current below the nominal, the higher the inductance and the better the filtering action of the LC filter, resulting in reduced ripple!

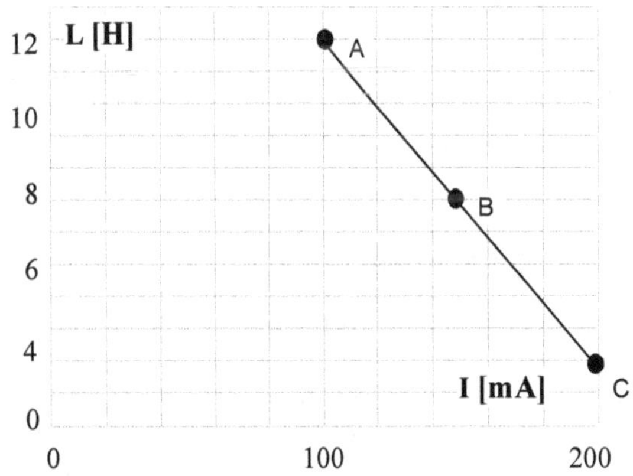

ABOVE: Between 100 and 200 mA the choke's L-I dependency is almost linear.

LEFT: A pair of HV power supply filtering chokes.

SWINGING CHOKES

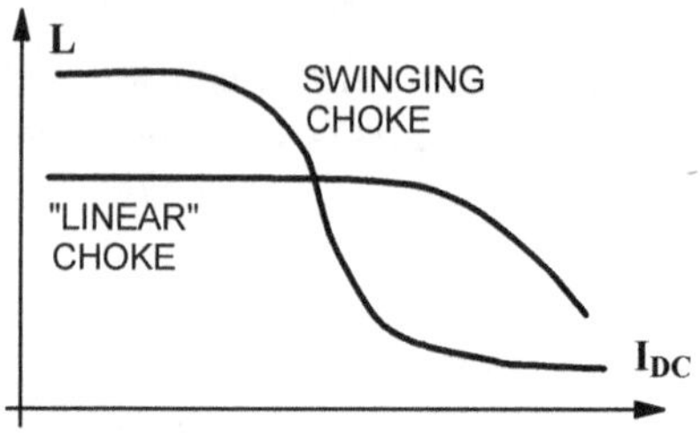

The inductance of a conventional or "linear" choke has a fixed value in the normal operating range and only drops as the magnetic core approaches saturation. A swinging choke is an inductor whose inductance drops as the DC current flowing through it increases.

A swinging choke is a simple voltage regulator. As the current increases and the inductance decreases, the voltage on the filter capacitor increases. The filter changes from predominately inductive at low currents (large inductance of the swinging choke) to largely a capacitive input filter at a higher load (small inductance of the swinging choke).

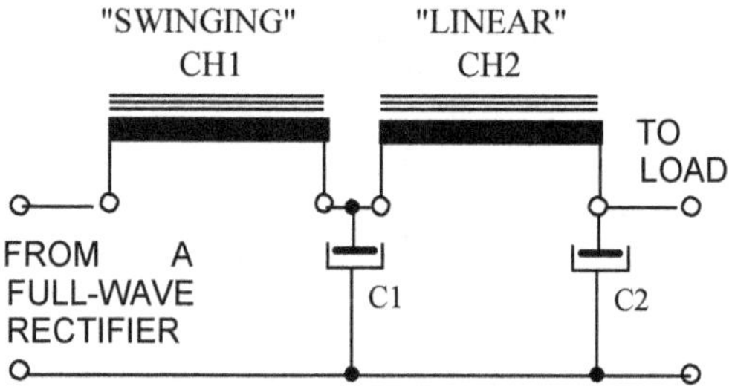

Apart from applications such as class B transmitters, audio swinging chokes were only occasionally used in vintage high power class AB amplifiers, where the load current raises to high levels at high volumes. It was the first choke after the rectifier and was usually followed by a "linear" or standard choke CH2. Swinging chokes make no sense in class A amplifiers where the current through the power supply is practically constant!

CURRENT CAPACITY, INDUCTANCE AND PHYSICAL SIZE OF A DC CHOKE

$A = 27.5\, I\sqrt{L}$

A = cross-sectional area of the center leg [cm^2]

L = inductance [H]

I = maximum allowed DC current [A]

Reverse engineering a swinging choke

Stancor's 20-4 H choke (model C-2307), with a current range of 30-300 mA, has an inductance of 4H @ 300 mA and 20H @ 30 mA. Its lamination size is 4 3/4" x 4" (approx. 12x10cm) or E120. We don't know the stack thickness, but can estimate that **$A = 27.5\, I\sqrt{L}$** $= 27.5*0.3*\sqrt{4} = 16.56\ cm^2$. Since EI120 laminations have a center leg width of 4.0cm, the stack would need to be S= 16.56/4 = 4.14 or 1 3/5 " thick.

From the R_{DC} of 80 Ω and 300 mA current rating, we can guess that d=0.4mm wire was used.

The Mean Length of Turn is MLT=2*(a+S+4B) + π*CT where B=bobbin thickness, CT=coil thickness, CT= WH - B - top insulation - top clearance = WH - 2.5mm, bobbin thickness of B=2 mm is assumed.

For EI120, MLT = 2*(40+40+4*2) + 3.14*17.5 = 231mm = 0.231m. Since $R = 2.126*\ell_W/d^2*10^{-2}$ ohms (if d is in mm), we can calculate total winding length $\ell_W = 100*R/2.16*d^2 = 602$ m, and finally we can estimate the number of turns by dividing total winding length with a mean length of a turn: $N = \ell_W/MLT = 602/0.231 = 2{,}606$ turns.

It is possible to combine laminations made of different materials, such as ordinary silicon steel and GOSS laminations. We had done that in a few cases when we didn't have enough laminations to fill a certain bobbin.

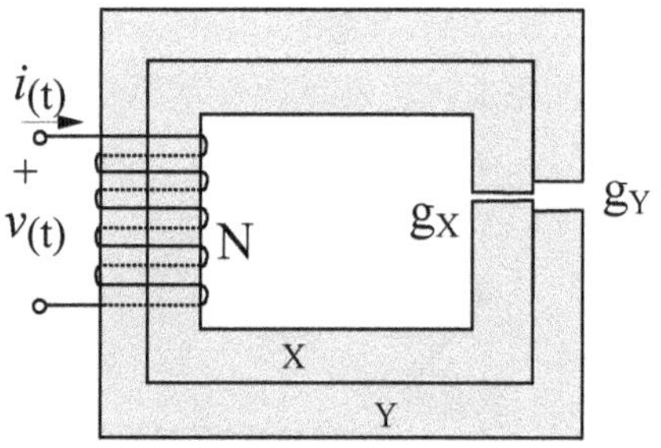

A core can be assembled from identical laminations but with a different size gap. That structure (right) is used in swinging chokes. With a small gap gX, Core X has a high effective permeability μ_{EFFX} while the DC current through the choke is at low values. With a much larger gap g_Y, Core Y reaches its optimum μ_{EFFY} at high DC current levels.

Although the stacks as drown here are of equal width and cross-sectional area, in practice, the low current stack X is usually only around 30% of the total stack width, and Y comprises the bulk of the stack at 70%!

LEFT: The double gapped magnetic lamination stack structure of a swinging choke

CHOKE DESIGN METHODS

The maximum turns method

There are two ways to design a choke. The first is a "brute-force" approach. Once you know the rated DC current, choose a suitable wire size. Then simply wind as many turns as you can fit into the window of the lamination stack you have on hand and want to use. Assemble the laminations and then adjust the air gap to get a desired value of L, or adjust the gap so that the maximal possible L is achieved without taking the core into saturation.

With inductors, forget about layers and insulation between them - life is too short for that. Use the so-called "bulk winding" method, wind the whole coil at once, in one go. Wind it in a left-to-right fashion and then back right-to-left neatly and consistently, so you get the uniform thickness and thus maximize the number of fitted turns.

The simplified graphical method

You can start with these graphs if you have no idea what size of laminations to use and how thick the stack should be for a certain current and inductance. First, calculate LI^2, then estimate the cross-sectional area A of the core from the LI^2 ⇨A graph. As an example (see Design #1 that follows), for L=7H and I=0.12A we get LI^2=0.1 (1).

The LI^2 ⇨A graph says that cross-sectional area of the center leg A should be between 6.5 and 7.5 cm^2 (2). Since the core in Design #1 is 6.8 cm^2, let's choose this value to see how close this approach will be to our brute force method. Now we read the NI or Ampereturns product from the A ⇨NI graph. For A=6.8 cm^2 (2) we get NI=350 (3), or N=350/0.12= 2,916 turns, very close to our actual result of 3,000 turns.

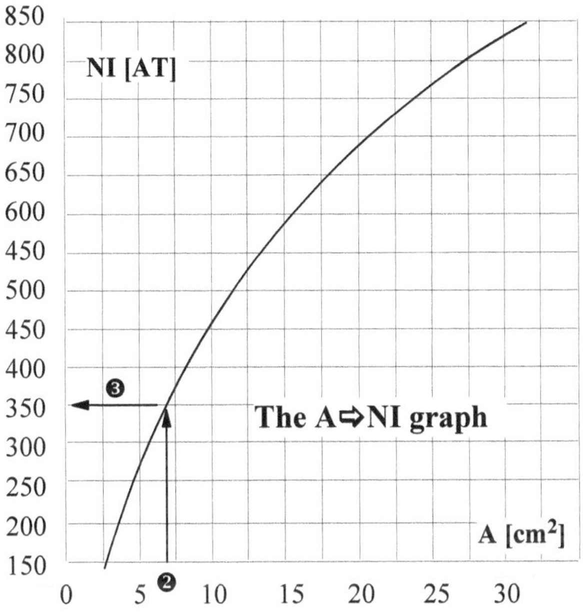

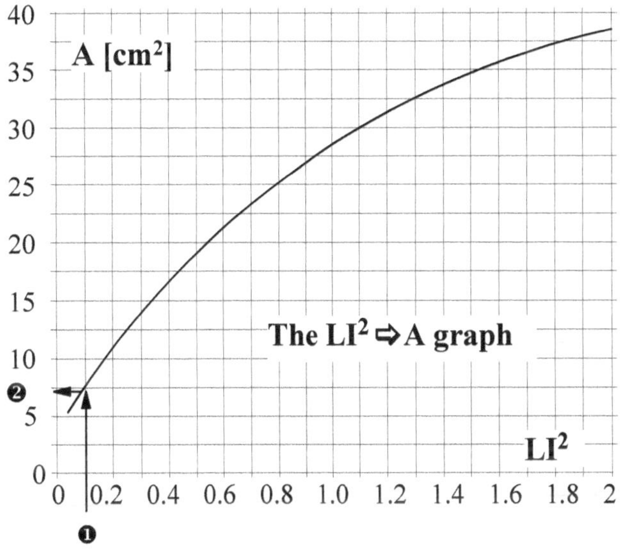

HOW TO DETERMINE THE SIZE OF AIR GAP

The air gap needs to be adjusted to achieve a maximal inductance L while still retaining the gap's saturation prevention effect. There are complex yet not entirely accurate formulas, so a practical adjustment is much easier and yields better results.

In his book "Practical Transformer Design Handbook," Eric Lowdon summarizes the whole approach quite well: "So we must not expect to achieve high levels of accuracy using such equations. The design of inductors, like that of transformers, is a cut-and-try affair based to a great extent on empirical data and bench adjustments."

Gap size as a function of DC current

The same principle of maximizing the primary winding's inductance applies to single-ended interstage and output transformers, not just chokes.

In the example illustrated on the right, five different inductance curves are shown for various levels of magnetizing DC current flowing through the primary.

Notice how the optimal gap (for L_{MAX}) rises with the rise in DC current and how the maximal inductance falls. Also, the peak of the curves gets flatter and flatter with increasing current, making it less critical to find the exact value of the gap.

For instance, for 60 mA, the optimal gap is somewhere in the 0.12 - 0.17 mm range. The range for 80 mA is even wider.

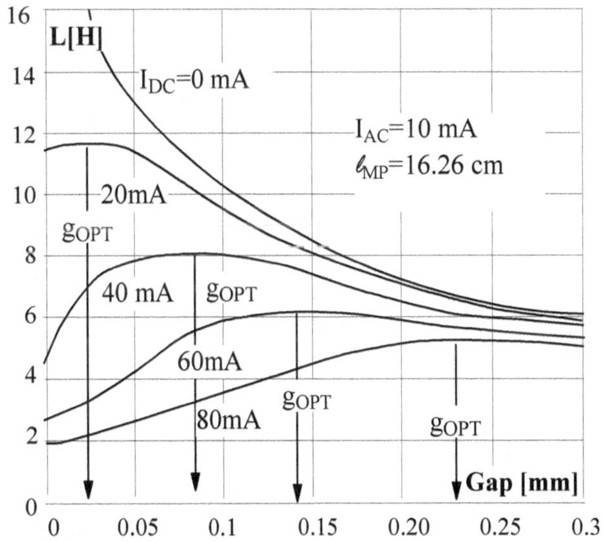

Determining the size of the air gap experimentally

To determine the size of the gap for our chokes, we perform a simple experiment, measuring the ripple on the high voltage supply line to one channel of class AB amplifier, as a function of the increasing load current I_{DC} and for various gaps in the filtering choke for that channel. The results of one such trial are on the next page.

Without any gap, the core went into saturation at 25-30 mA. Saturation is recognized by the sharp increase in the ripple, resulting from a drop in inductance, which itself is due to reduced permeability.

With a 0.005 mm physical gap, the same choke started to saturate at about 80mA and was deep in saturation at 110 mA. At 120 mA and 0.1 mm gap, the choke was far from saturation.This particular choke's optimal physical gap range was just above 0.1 mm.

That was the case for most small chokes. To save you time and effort, don't bother with smaller or larger gaps for commonly used cores and standard amplifier power supplies (up to 200 mA per channel).

Notice how the ripple with a 0.2 mm gap was almost double the ripple with a 0.1 mm gap (at higher currents), so for this choke, the optimal gap would be closer to 0.1mm.

The gap width in this experiment is the actual measured or physical gap between lamination E and I profiles; the "theoretical" gap is double that figure (since the flux crosses the physical gap twice).

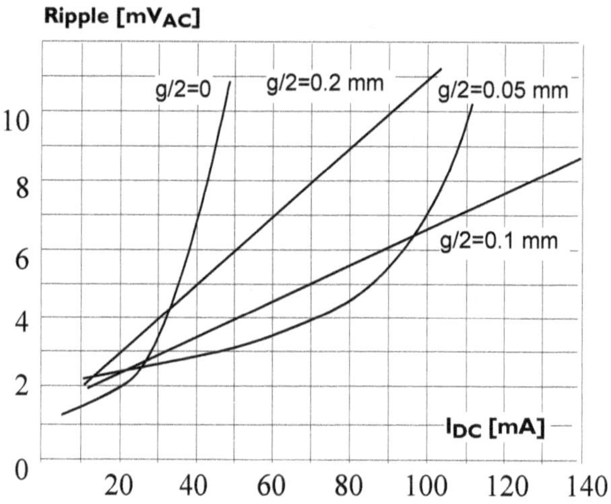

ABOVE: The optimal gap can be determined by measuring AC ripple after the choke for various values of DC load current. Use a power rheostat as a variable load)!

The graphical way to determine the optimal air gap

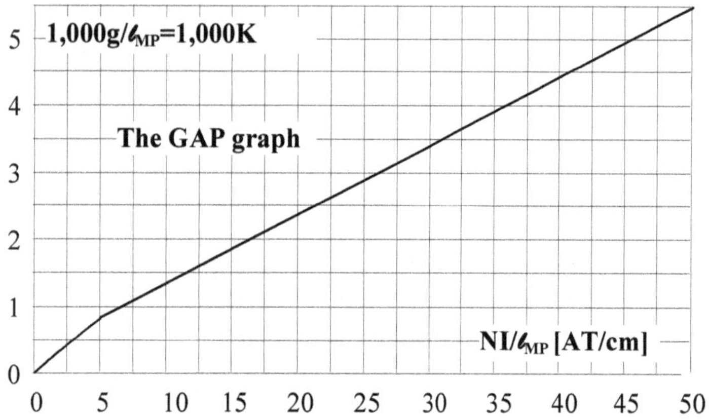

Remember that $H=NI/\ell_{MP}$? Once we determine the H, we can use this graph to determine gap g. The values on the Y-axis are in 1,000*K, and K is the gap ratio g/ℓ_{MP}.

To get the value of g, we have to divide the reading by 1,000 and multiply it by ℓ_{MP}.

For L=7H, I=0.12A and EI66 laminations with ℓ_{MP}=6*CL=13.2cm, we calculate N*I/ℓ_{MP}= 3,000*0.12/13.2 = 27.27 AT/cm.

From the gap graph we get 1000*g/ℓ_{MP}=3.1, so g=3.1*132/1,000=0.4mm, double what we determined experimentally (g=2*0.1= 0.2mm).

DIY CHOKE DESIGNS

Design #1: 120 mA choke

We had a few surplus 30VA (30V, 1A) mains transformers, suitable for conversion into guitar output transformers and filtering chokes. The core is EI66, 22mm center leg, with S=31 mm stack thickness. A=aS=2.2*3.1= 6.8 cm^2, so P = A^2= 6.8*6.8 = 46VA, meaning their power rating is 50% higher than the declared 30VA, which is great (usually manufacturers use underrated cores)! The bobbin is divided vertically in half. The primary was wound on one half, and the secondary on the other.

With 240V primary and 30V secondary, the voltage ratio is TR=240/30=8. With 1A secondary current, the primary current is $I_P=I_S/TR$=1/8=125 mA, meaning the primary wire was already sized for our load current of 120 mA.

Once we removed the laminations, we kept the primary winding, unwound the secondary (its wire was too thick for this application), and wound as many turns as we could on the empty half of the bobbin.

Finally, we joined the two windings in series and stacked the laminations back together, only this time we had to include an air gap, so all E pieces were together on one side, and all the I pieces were on the other side.

ABOVE: 30VA-rated mains transformer rewound and restacked with an air gap as a filtering choke.

The cross-section S of a wire with diameter d is $S = d^2\pi/4$ and $S = I/J$ (current/current density). For 120 mA current rating of the choke we need a wire $d = \sqrt{(1.27*0.12/2.5)} = 0.247$ mm, so we used wire d=0.25mm.

The issue with this "brute-force" approach is that you never know what will be the inductance of the finished choke. Once you know the window size and the wire diameter, you can estimate the number of turns you can fit in, but you still don't know the laminations' permeability. Don't despair; it doesn't matter if you get 5H or 6.6H.

In this case, we got L=7.1H and DC coil resistance of 145Ω. Remember that inductance was measured by a digital LCR meter without any DC current! In a power supply, such DC load current will reduce the choke's effective inductance value to around 5.5H.

Remember that $L=N^2\mathcal{P}$, so permeance of the core in this case was $\mathcal{P}=L/N^2 = 7.1/3{,}000^2 = 7.9*10^{-7}$ ATurn/Weber and reluctance was $\mathcal{R}=1/\mathcal{P}= 1{,}267{,}606$ Weber/ATurn!

Since $L= N^2\mathcal{P}$ and permeance is $\mathcal{P}=1/\mathcal{R}= \mu_0\mu_R A/\ell_{MP}$, we get $L= N^2\mu_0\mu_R A/\ell_{MP}$ and can find out the relative permeability $\mu_R = L\ell_{MP}/(N^2\mu_0 A) = 115.7$ (substitute $\mu_0 = 4*\pi*10^{-7} = 1.257*10^{-8}$, $A=6.8$ cm^2, $\ell_{MP}=12.54$ cm, N=3,000, L=7.1H). That is the effective permeability μ_{EF} and since $K=g/\ell_{MP} = 0.015/12.54 = 1.2*10^{-3}$, from $\mu_{EF}= \mu_{INC}/(1+K\mu_{INC})$ we can express $\mu_{INC}= \mu_{EFF}/(1-K\mu_{EF}) = 115.7/(1-115.7*0.0012) = 115.7/0.861 = 134$. Such a low incremental permeability indicates a low quality ordinary silicon steel laminations.

Design #2: 300 mA choke

To get a lower DC resistance and higher DC current capability use thicker wire, but you will fit fewer turns and get lower inductance. Say you don't have 0.25mm wire, so you decide to use 0.44mm wire. You start winding but can only fit 1,050 turns. How much inductance should you expect?

We got 1.2 H. On the plus side, this choke is good for at least 300mA, but the air gap had to be increased for such a high current. The increase in wire size and a larger air gap resulted in approximately 3X lower number of turns, 7X lower coil resistance and 6X lower inductance!

DESIGN #1:	MEASURED RESULTS:
1. EI66 core, 31mm stack	1. N=3,000 turns
2. Current rating: 120mA	2. Gap g=0.15mm
3. Wire dia: 0.25mm	3. R_{DC}=145Ω, L_{120}=7.1H

DESIGN #2:	MEASURED RESULTS:
1. EI66 core, 31mm stack	1. N=1,050 turns
2. Current rating: 300mA	2. Gap g=0.25mm
3. Wire dia: 0.44mm	3. R_{DC}=19Ω, L_{120}=1.2H

Designs #3 and #4: 120mA/6H & 120mA/3.3H chokes

We also had some mains transformers with EI60 laminations, 20x25mm cross-sectional area (A=2*2.5=5.0 cm^2). When rewound as chokes, we got an inductance of around 6 Henrys. The previous design used 6.8 cm^2 cores with a bigger window area and 20% more turns for 7H.

The fact that we managed to get 6H of inductance indicates that the permeability of these cores was higher than those in Design #1.

Small power transformers rated at 15VA by their manufacturer were also tried. The core was EI57 with a 19mm stack. The actual power rating was $P = A^2 = (1.9x1.9)^2 = 13$VA; the core was undersized, lower than its declared VA size.

Due to smaller windows and a lower number of turns, we only got half of the inductance compared to design #3, and, as already indicated, we suspected a lower quality core as another factor.

DESIGN #3:	MEASURED RESULTS:
1. EI60 core, 25mm stack	1. N=2,500 turns
2. Current rating: 120mA	2. Gap g=0.1mm
3. Wire dia: 0.25mm	3. R_{DC}=100Ω, L_{120}=6H

DESIGN #4:	MEASURED RESULTS:
1. EI57 core, 25mm stack	1. N=2,000 turns
2. Current rating: 120mA	2. Gap g=0.1mm
3. Wire dia: 0.25mm	3. R_{DC}=82 Ω, L_{120}=3.3H

Estimating the number of turns and DC resistance

Before you start winding on the chosen bobbin size (to suit the magnetic core), you can estimate the number of turns that can be fitted from the lamination stack's dimensions and the DC resistance of the finished choke. Calculations for each of the previous four designs are in the middle (unshaded) section of the table on the next page.

NOTE: All dimensions are in mm or mm^2. WL= Window Length, b = Window Height, CT= Coil Thickness, calculated as CT= b - BT- top insulation- top clearance, where BT is the thickness of bobbin's walls, assumed to be 2 mm, and the top insulation and top clearance are 0.5mm, so CT=b-2.5 mm.

CL=Coil Length, calculated ass CL=WL-2BT, so CL=WL-4 mm.

WA= Gross Window Area, WA= WL*b and NWA=Nett Winding Area, NWA=CT*CL.

There are three figures for the number of windings. N is the theoretically maximal number of turns that can fit into the Window Area WA, and N_N is the number of turns that can fit in the Nett Winding Area NWA.

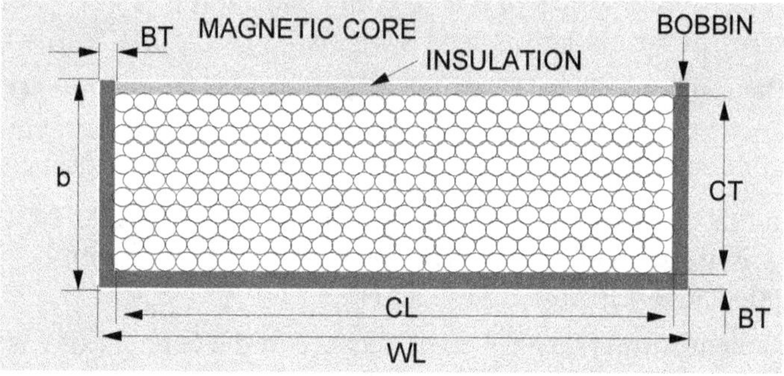

DESIGN	CORE	a	S	MLT	WL	b	WA mm^2	CT	CL	NWA mm^2	N	N_N	N_R	d	GA mm^2	ℓ (m)	S wire (mm^2)	R_C Ω	R_M Ω	Error %
1	EI57	19	19	114	27.5	9.5	261	7	23.5	165	4,180	2,257	2,031	0.25	0.073	231	0.049	78.6	82	-4.1
2	EI60	20	25	129	30.0	10.0	300	7.5	26.0	195	4,800	2,675	2,407	0.25	0.073	312	0.049	106	100	+6
3	EI66	22	31	148	33.0	11.0	363	8.5	29.0	247	5,808	3,381	3,043	0.25	0.073	452	0.049	153	145	+5.5
4	EI66	22	31	148	33.0	11.0	363	8.5	29.0	247	1,875	1,165	1,048	0.44	0.212	156	0.152	17	19	-10

Finally, $N_R=0.9*N_N$ is the practically achievable (realistic) number of turns. The HFF of 0.9 was used, which is reasonably ambitious and assumes a skilled winder.

The shaded columns on the left are parameters related to the lamination and stack size, the next three columns are the results for the number of turns, and the shaded columns on the right are related to the wire size.

The Gross Area (GA) is the area of a square with a side equal to the gross wire diameter (with insulation) d_I, and ℓ is the total length of the winding. Once we know the cross sectional area of the wire, the MLT (Mean Length of Turn) and the number of turns, we can calculate the estimated DC resistance R_C of the winding. The same principle and the same formulas apply to mains and audio transformers as well.

To calculate the length of the winding wire, multiply the Mean Length of Turn (MLT) with a realistically achievable number of turns N_R: ℓ=MLT *N_R. $R_C=\rho\ell/S$, where ℓ=length of wire, S = cross-sectional area of the wire, and ρ (rho) is specific resistance or resistivity of the material, for copper $\rho=1.67\times10^{-8}$ Ωm.

Since $S=r^2\pi=d^2/4\pi$ we have $R_C=4\rho/\pi\ \ell/d^2=2.126\ell/d^2*10^{-2}$ Ω (ℓ in meters and d in mm). The calculated values R_C are close to the measured values R_M, the error (last column) ranging from -4 to +6%, except for the low value of resistances such as in Design #4, where a measuring error of only 1 ohm (cannot be avoided if digital multimeters are used) would cause a discrepancy of 6% or more (19Ω instead of 18Ω is a relative error of 5.5%).

Calculating MLT (Mean Length of Turn)

The Mean Length of Turn is MLT=2(a+S+4B)+πWT where B=bobbin thickness, WT=winding thickness for transformers (primary or secondary winding or each section) or total CT (Coil Thickness) for chokes, which is CT= b - B - top insulation - top clearance = b - 2.5mm (since there is only one winding). Bobbin thickness of B=2 mm was used.

The drawing (RIGHT) shows a transformer with one primary (thickness W1) and one secondary winding of thickness W2, but the principle behind the calculation is the same for any number of windings and chokes with only one winding.

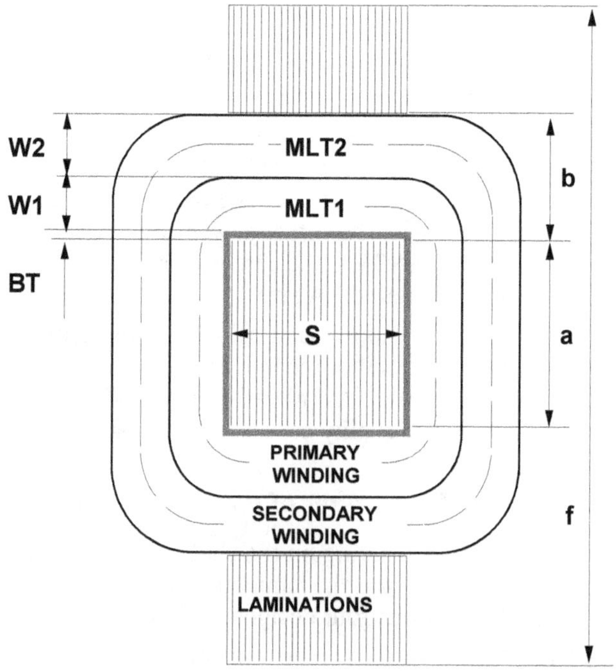

These are MLT figures for the three core sizes used in the four choke designs just discussed:

EI57: MLT = 2*(19+19+4*2) + 3.14*7.0 = 114 mm

EI60: MLT = 2*(20+25+4*2) + 3.14*7.5 = 129 mm

EI66: MLT = 2*(22+31+4*2) + 3.14*8.5 = 148 mm

TRANSFORMER MATERIALS, CONSTRUCTION METHODS AND ISSUES

- TRANSFORMER MATERIALS
- WINDING TECHNIQUES AND ISSUES
- CONSTRUCTION ASPECTS OF TRANSFORMER MAKING
- SHIELDING OF TRANSFORMERS AND CHOKES

4

TRANSFORMER MATERIALS

Reusing old power transformer C-cores and EI laminations in your output transformers

For practical purposes, magnetic materials don't age, so reusing them is perfectly fine, and you can save yourself lots of money. Most older laminations are superior to some currently produced overpriced rubbish. To add insult to injury, in many countries buying retail (small) quantities of transformer laminations is very difficult if not impossible, even in relatively developed countries such as Australia or New Zealand, since very few transformers are made in such places. The supply situation is slightly better in USA and UK.

You may see old TVs dumped during your neighborhood's verge pickup days. I keep a toolkit in my car and, in the past 25 years, have "liberated" hundreds of transformers and chokes destined for the rubbish tip. Almost all happily perform today in somebody's tube amp. Many power transformer cores even made great audio transformers.

However, few TVs with cathode ray tubes are being dumped these days. Flat-screen TVs (LCD, LED, etc.), which use switch-mode power supplies, are no use from a transformer salvage point-of-view.

Laminations from microwave ovens

Transformers in microwave ovens generate a high voltage needed for the magnetron and cannot be reused as they are since the secondary voltage is way too high for our needs in tube amps, typically around 1.1 to 1.5kV. You could reuse the laminations, but in most cases, the lamination stack is welded to prevent buzzing, so it cannot be taken apart and restacked.

Laminations from line-matching audio transformers

Line transformers convert the output of any standard 4 or 8Ω amplifier for use with 70.7V or 100V public address sound distribution systems. They belong to the family of audio transformers and can be converted into interstage transformers, plate chokes, grid or even output transformers. They may be cheaper than buying laminations by weight if you find them on sale or as surplus stock. Plus, you get all the trimmings - bobbin, mounting frame, terminals, even wire can be reused. They almost always use high-quality M6 grain-oriented steel, 0.35 mm thick.

Magnet wire

In trade circles, transformer winding wire is called "magnet wire." Not all magnet wires were created equal. Apart from the type and purity of copper used, the main differences are in the wire grade, which indicates how many coats of insulation were applied and the type of insulation used. Wires from different manufacturers and specifications also have different mechanical stiffness and feel different during the winding process.

Older materials such as "Formvar" (the main type used in the 1950s and 60s) had a relatively low temperature rating, only 105°C, compared to around 200-220°C for modern polyamide-imide and aromatic polyimide materials.

Insulation type	Thermal rating
Polyvinyl Acetal (Formvar)	105°C (220°F)
Polyurethane	155°C (310°F)
Polyurethane with a Polyamide top coat	155°C (310°F)
Polyurethane	180°C (355°F)
Polyester-imide	180°C (355°F)
Polyester-imide with a Polyamide top coat	180°C (355°F)
THEIC Polyester	200°C (390°F)
THEIC Polyester with a Polyamide top coat	200°C (390°F)
Polyamide-imide	220°C (430°F)
Aromatic Polyimide (ML)	220°C (430°F)

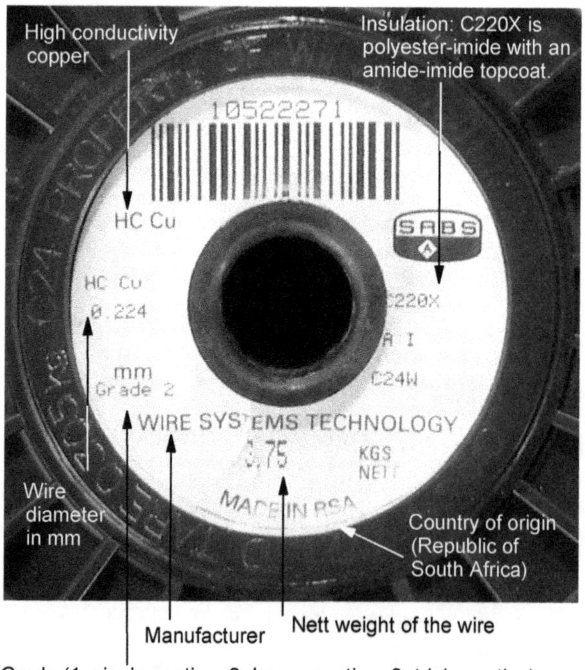

ABOVE: The label on South African wire spool tells you all you need to know, the best wire of all we've used so far.

LEFT: Insulation types and their thermal ratings

PUR wire is insulated with cross-linked polyurethane, either with one or two coatings (PUR1 and PUR2). Temperature class 130°C (Class B) PES wire is manufactured similarly to PUR; however, a small amount of polyimide is added to the enamel to improve the temperature index without reducing the solderability. PEI is a polyester-imide coated wire, most commonly used today, 155°C (Class F).

PEI-AI has an additional overcoat of polyamide-imide enamel, making it suitable for continuous operation at temperatures of up to 200°C. PAI is a polyamide-imide enamel wire suitable for use in 220°C applications.

We haven't conducted any serious listening experiments, but anecdotal evidence suggests that magnet wire of the same diameter and nominal specifications but by different makers and using different insulation types sound different when used in audio transformers.

So, even if you plan to do lots of transformer winding, don't rush and buy large 10kg spools of wire. Try small quantities of the same wire from various manufacturers to see which ones are easier to work with. Unfortunately, in most countries, you will not have such a luxury; the supply is very limited.

Insulation materials

There are five internationally recognized insulation temperature classes, which specify either the maximum allowed temperature or the temperature rise. Class C is not universally recognized. Most commercial transformers are made to class A standards.

Presspahn paper is traditionally used in electric machines, capacitors, and oil-cooled transformers. Its insulation class is A (105 °C). Available in sheets and rolls, thickness from 5 mil (0.125 mm) to 20 mil (0.50 mm) and various grades. Older bobbins and coil formers were also made of Presspahn.

Electrical grade insulating craft paper and diamond dotted paper are also used. The diamond dotted paper has diamond-shaped epoxy resin dotted on both sides. For data sheets, visit https://www.pucaro.com.

Mylar® polyester film was invented in the early 1950s and the name is often used to generically refer to plastic film. Mylar® is a registered trademark of Dupont Tejjin Films and refers to the specific family of products made from the resin Polyethylene Terephthalate (PET).

Class	Max. Temp.	Temp. Rise	Typ. insulation
A	105 °C	60 °C	Presspahn or polyester foil
E	120 °C	75 °C	Polyester foil
B	130 °C	80 °C	Polyester foil (Mylar®, Hostaphan®),
F	155 °C	100 °C	Aramid paper (Nomex®) or Polyimid foil (Kapton®)
H	180 °C	125 °C	Same as F
C	220 °C	160 °C	Same as H

ABOVE: Major insulation temperature classes

AC Dielectric Strength for Mylar® 92 EL at 25°C (77°F) is 7.0 kV/mil, while DC Dielectric Strength is 11.0 kV/mil. Dielectric Constant at 25° C and 1kHz is 3.2, melting Point 254 °C.

Kapton® (Polyimide) Tape is an insulation tape with silicon-based adhesive to withstand high temperatures up to 180° or 260° C. 3M calls it "Electrical Tape 92" (Polyimide Film Tape with Thermosetting Silicone Adhesive). Kapton® is a trademark of E.I. du Pont de Nemours and Company.

WINDING TECHNIQUES AND ISSUES

The bowing or bulging effect

No matter how tightly we wind the coil, the winding will always deviate somewhat from the ideal profile. This effect is called bowing or bulging out. The height of the bulge is greatest in the middle of the longer sides of the bobbin.

Bulging or bowing reduces the maximum winding thickness that can be fitted into the core's window and has to be taken into account when checking the fit. The rule of thumb is to add about 15% to the winding height, but the coils wound by beginners are often much looser and higher, so adding 20-25% is not unreasonable.

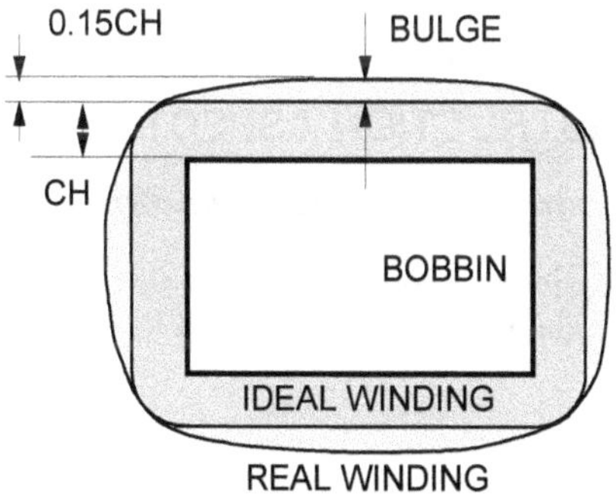

ABOVE: The less experienced you are in transformer winding, the more bulging allowance you should include in your calculations!

Power transformer winding order

Usually, the primary winding is wound first, followed by insulation and the ES (electrostatic) shield. Instead of one layer of wire, a piece of copper foil can be used, providing the two ends are not joined together, so it does not form a closed loop. Such a loop would create a "blind" or shorted turn and render the transformer unusable.

All primary connections are taken out on one side of the coil and all secondaries on the other side. If you have many secondaries, you may run out of bobbin slots and may need to take some out on the same side as the primary.

The secondary windings are usually wound in the descending order of power dissipation, higher VA windings first, closer to the primary. That usually means that the high voltage secondary is wound first after the ES shield.

The final winding should be of a larger diameter wire to give the coil strength and make the whole packet tighter. This is usually one of the higher current heater windings, and here we may deviate from the descending power order rule.

Vertical and horizontal sectionalizing

Depending on the transformer's purpose, the designer has to decide on the optimal winding arrangement, in order to minimizing leakage inductance L_L and shunt capacitance C_S. These are only an issue in audio transformers, power transformers operate at low frequencies 50 or 60Hz, where these effects are of no importance.

The leakage inductance can be reduced by sectionalizing, or interleaving the primary and secondary windings in horizontal layers, one on top of the others.

However, the more layers we have, the more paralleled capacitors we create, and the higher the shunt capacitance gets. To reduce the stray capacitance we can sectionalize the secondary windings (or even the whole bobbin, including the primary) vertically.

Alas, vertical dividing increases the leakage inductance, requires more complex bobbins and slows down the winding process, so it significantly increases cost.

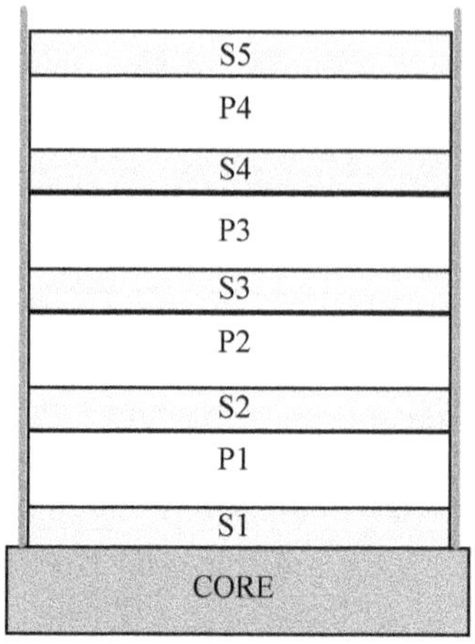

Horizontal sectionalizing: L_L reduced, C_S increased

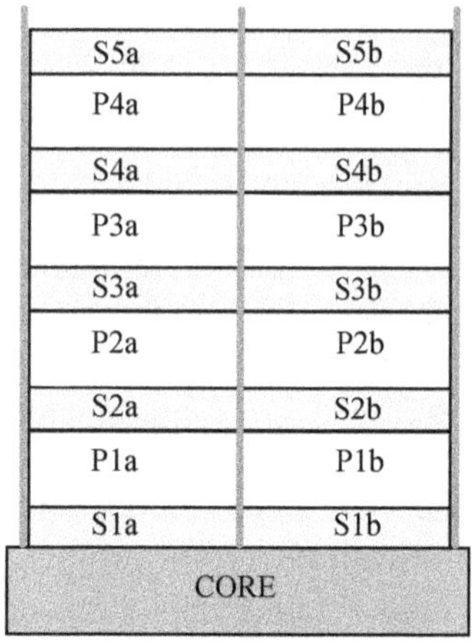

Vertical sectionalizing: C_S reduced, L_L increased

Bifilar and multifilar winding

If perfectly balanced windings (the same number of turns and identical DC resistance) are needed (for instance, around a CT - center tap for power transformers or push-pull output transformers), the bifilar winding is the solution. The two halves are connected out-of-phase with respect to the CT and have identical AC parameters and DC resistances, which would not be the case if they were wound one on top of the other.

Bifilar winding means that two thinner wires (from two spools) are fed and wound in parallel instead of one thicker wire. Smaller diameter wires are easier to wind and achieve a higher window filling factor than large wires.

Parallel connected bifilar windings (or multifilar windings, trifilar with three wires, and so on) can be used to reduce skin effect by increasing the relative amount of surface area compared with using one larger wire. This usage of bifilar does not affect the volts per turn or the required insulation strength.

Primary/secondary combined in a multifilar winding is sometimes used to get the tightest possible magnetic coupling (minimum leakage inductance). This comes at the expense of increased capacitive coupling (poor HF isolation) and reduced dielectric isolation between primary and secondary (compared with having more physically separated winding layers).

CONSTRUCTION ASPECTS OF TRANSFORMER MAKING

The simplest possible winding machine

Professional winding machines are physically large, operationally complex and very expensive. However, making your own simple winder is not hard even for most mechanically challenged. You can use new or recycled parts.

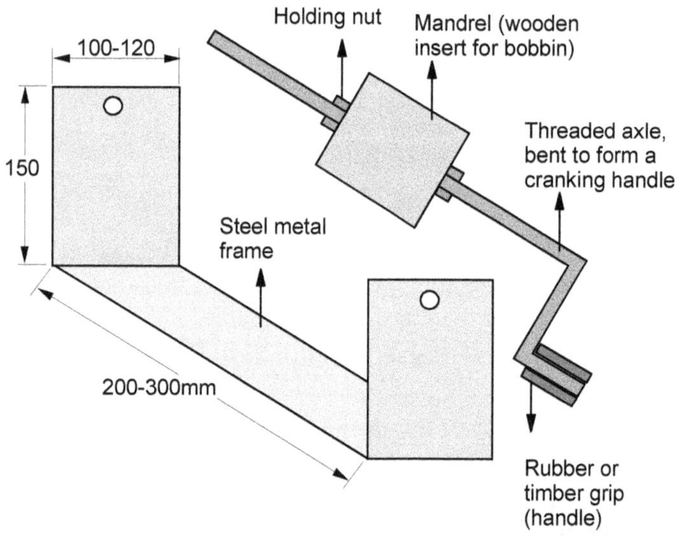

There are only two main components: one stationary frame (usually metal, steel, or thick aluminium, but even timber or plastic can be used) bolted or clamped onto a desk. Through its two holes, a threaded handle is fed. It has to be threaded because a timber mandrel will sit in its middle, and the mandrel needs to be secured onto the rod.

The easiest way to do that is to use a threaded rod, bent into a handle shape (as illustrated), and a pair of wing nuts, which will hold the mandrel in place.

You will need a different mandrel for each bobbin size (plastic winding frame) you intend to use. Each mandrel needs to be drilled through at its center, the hole should be only slightly larger than the diameter of the rod, so the rod fits snugly into the mandrel, which should not be able to rotate around the rod.

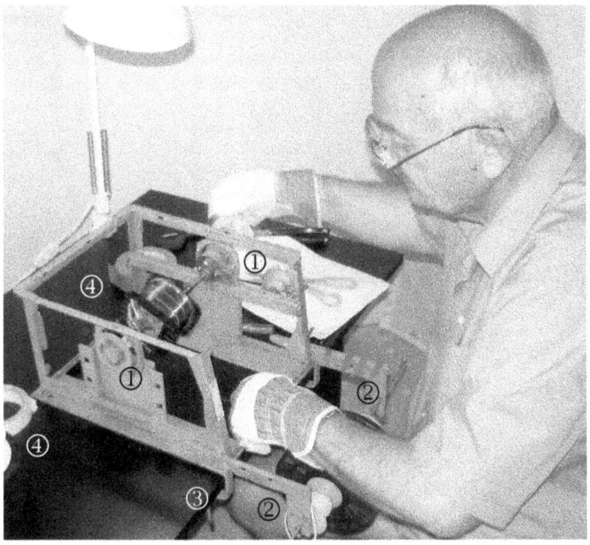

Dr. Bob made his winding machine from a steel frame salvaged from God-knows-what, onto which he welded a pair of ball bearings (1), a wire spool holder arms (2), and adjustable clamps (3).

It looks hideous, but it works well. Many hundreds of power and audio transformers were wound on this contraption.

In the photo, he's using gardening leather gloves since tensioning the wire with your bare hands starts damaging your skin after a while.

Apart from the wire and flexible sleeving, you will need scissors and a few insulating paper and plastic tapes (4).

Practical winding tips

Since a picture is worth a thousand words, let these snaps from the winding and assembly of a power transformer for a 300B amplifier give you a few tips and tricks about transformer and choke winding.

Insert the mandrel (1) into the bobbin (2), feed the threaded rod through it, and tighten nuts (3) on both sides.Feed the magnet wire through a flexible plastic sleeve (4), approx. 5-6 cm long, and through the first slot (or hole) on the left side of the mandrel (5). That is the start of your transformer's primary winding.

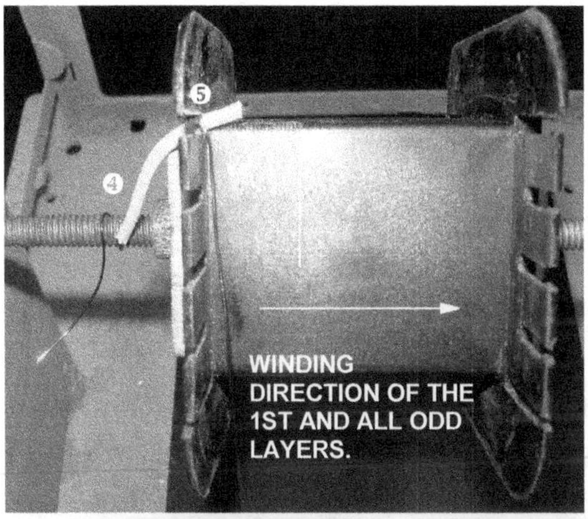

Wind the first layer slowly and steadily by turning the handle with your right hand, tensioning the wire, and moving slightly towards the right with your left hand.

Try not to leave any space between the turns. Some gaps are inevitable; it happens even to best coil winders, as at (7). In any case, better to have slight gaps than for wires to cross; that should never happen!

When you reach the left end of the bobbin, start moving your left hand (with wire) in the opposite direction, L to R, "filling" the second layer of the bulk winding, etc.

Once you finish the whole primary (in this case) or the first section of a sectionalized transformer, cut the wire a bit longer (10-20 cm), feed it through another piece of flexible tubing (sleeve), and the first slot on the right side of the bobbin frame (6). Gently wrap the wire around the threaded rod (3), and you have just finished the primary winding.

To prevent the wire from moving and often even partially unwinding itself if you hadn't tensioned the winding properly, tape the end down using ordinary masking paper tape. That tape, sold in hardware shops and even some supermarkets and craft shops, has one dry and one glue-covered (sticky) side and is ideal for this job.

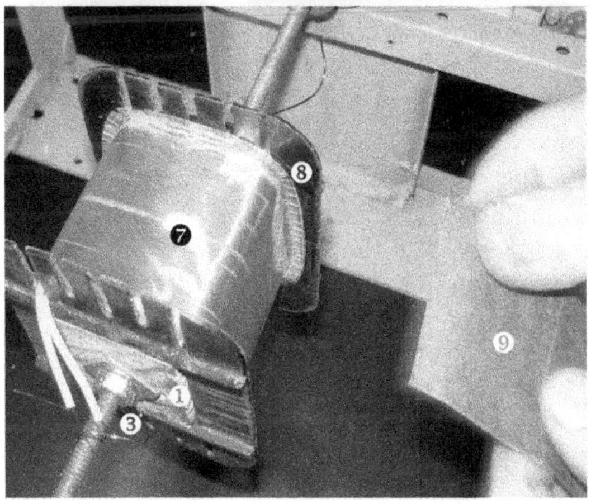

Next, cut a piece of plastic or paper insulation long enough to encircle the whole winding. The strip (that you precut earlier) should be a few (7-8) mm wider than the bobbin width.

With small scissors, make a series of small cuts (2-3mm) along the whole length of the strip (8). This will form a "lip" whose job is to prevent turns from subsequent windings (that will follow on top of the first winding) from dropping down amongst the first winding's turns.

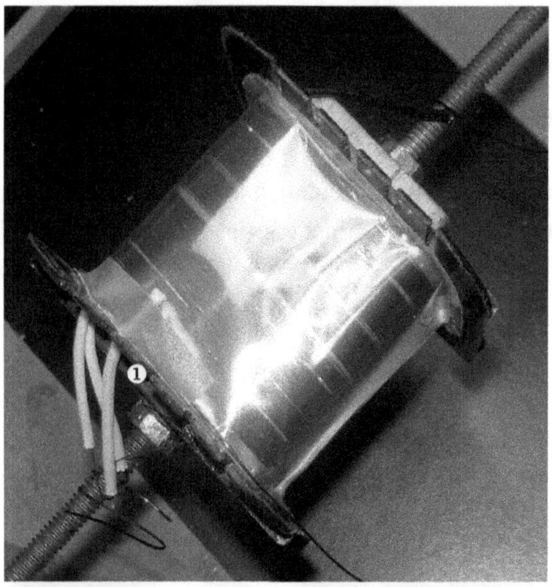

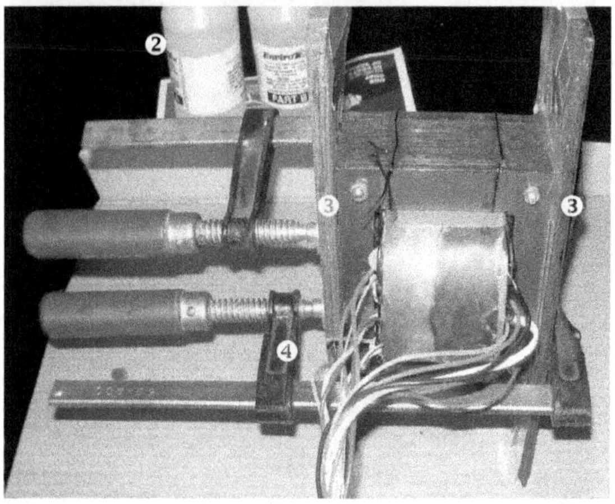

Since the insulation between the primary and secondary of a power transformer is critical, Dr. Bob is applying another layer of insulation on top of the Mylar film (previous page), this time craft paper (9). Besides increasing electrical insulation, that paper layer also makes the firmer base for the next winding (electrostatic shield).

Only the start of the E/S shield is to be grounded, so we only need that end of the winding (1); at the other end, the wire at the edge of the bobbin is simply cut off and taped down with masking tape.

The photo on the right shows the Mylar insulation on top of the E/S shield layer of wire before the first secondary winding is wound.

Once the winding stage is finished and the final (topmost or "outer") insulation layer is applied, before stacking the laminations, always test the finished coil with an ohmmeter meter, just in case something went wrong during the winding process. You should be able to measure the DC resistance of each winding (except the E/S shield), and there should be infinite resistance between different windings.

Once the lamination stack is assembled, bolt it temporarily together (without the end bells). Paste all four sides of the core with 2-part pour-on polymer varnish (2) and clamp them together overnight. The two thin pieces of timber (3) are there to prevent the steel clamps (4) from damaging the lamination edges.

Also, insert a strip of Mylar or similar plastic film between the timber strip and the lamination stack, or you will glue them together forever!

If you use a vacuum chamber to impregnate the whole transformer, this last step (the gluing part) is not required.

Burr (lamination deformity)

Their edges will be bent if laminations are stamped using worn-out (blunt) dies. Many poor-quality currently produced laminations suffer from this problem. Before you commit to buying a large quantity of EI laminations, ask for a few samples. Run your finger over the edges and try to identify the "rough" side. With good quality stampings, it will be difficult to tell which side is which.

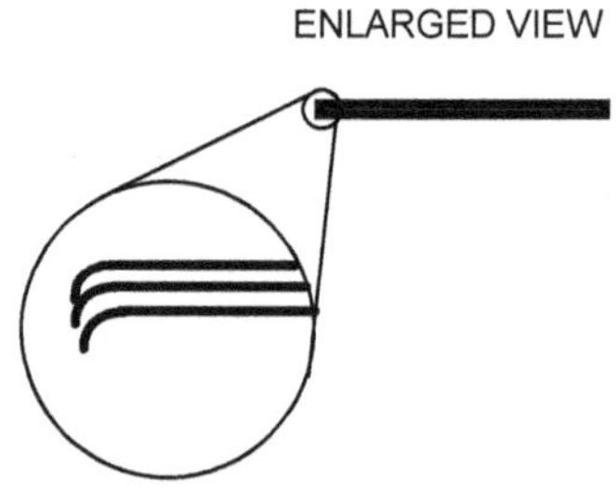

If the laminations you have are deformed, stack them all facing the same way (as pictured); otherwise, the stack will have gaps between laminations, and the stacking factor will be poor, reducing the effective cross-sectional area A_{EF}!

The final assembly

The photo on the next page shows major parts of a single-ended output transformer for a 300B amplifier. The finished coil is in the bottom-left corner, all the I-pieces in one stack, all the E pieces in another, plus the insulating paper that will act as the "airgap" between the two stacks. Vertically-mounted end-bells are used; the only part missing is the bolts' insulating sleeves.

Ideally, the bolts holding a transformer together should be made of non-ferromagnetic material, such as brass or aluminium, so they don't shunt the magnetic flux through them! Hardware stores sell threaded brass lengths of various diameters. They should not be in contact with laminations; otherwise, they will act as a magnetic short circuit. This requires fiber washers and insulating plastic or Teflon® (PTFE) sleeves.

Transformer buzz

When subjected to varying magnetic fields produced by alternating currents, transformer laminations expand and contract at twice the frequency of the AC field. This effect is called magnetostriction and is the main cause of transformer buzz. Another source of vibration and buzz is transformer windings. If not tightly wound and properly secured, the coil can also vibrate and produce a buzzing sound.

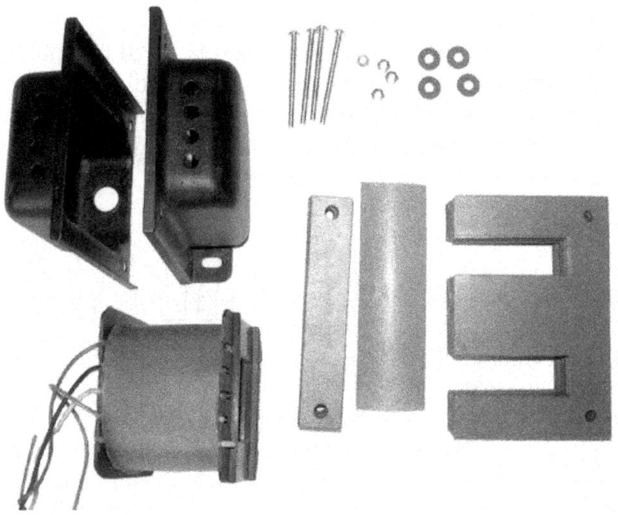

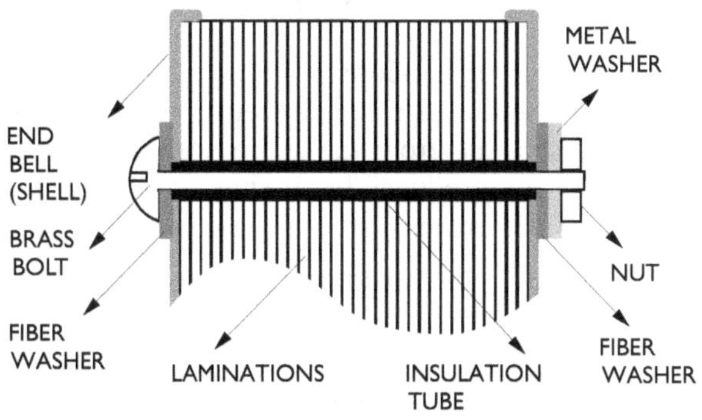

ABOVE: Cross-section of a finished transformer showing the mechanical assembly details and parts.

LEFT: All components of a 300B SET output transformer, ready for the final assembly, except the bolt insulating sleeves (tubes).

Improperly designed transformers will also buzz due to a very high level of B (density of magnetic flux), approaching saturation, or due to overload, with high power losses and overheating.

Audio transformers will buzz in accordance with the music signal, and such slight buzz will not be noticeable during listening. You will only hear that buzz when you test an amplifier using a dummy load. The buzz from mains transformers is constant and can be heard, so it must be addressed.

To impregnate or not, that is the question!

After assembly, many (but by no means all) commercially made transformers are impregnated with varnish. This is done to seal the windings so they are not exposed to the oxidizing atmosphere, or, for industrial transformers, the atmosphere could be corrosive, humidity could be an issue, and so on. The second reason is to prevent the movement and thus buzzing of windings and laminations. The varnish "glues" or bonds the assembly together.

However, there are a few issues with impregnation in vacuum chambers we amateurs must deal with. Firstly, air must be evacuated from spaces between transformer laminations and windings using a vacuum chamber. Then, varnish has to be injected under pressure to fill all those nooks and crevices. Specialized (expensive) equipment is needed unless you are a mechanical whizz-kid and can design and build your own.

Secondly, the varnish is expensive, and large quantities are needed. And thirdly, in the case of audio transformers, the presence of varnish inside the windings will have a major (mostly negative) impact on their frequency range and sound. Generally, the bandwidth will be reduced, probably due to increased parasitic capacitances. Remember, transformer winding layers with insulation between them form one giant capacitor; the varnish will change the dielectric constant of such a capacitor and thus its capacitance

!If not by impregnating them in a vacuum chamber, how can you prevent or reduce buzzing in your mains and audio transformers? Here are the tips to minimize the buzz caused by loose windings or by the laminations:

1. Use laminations of the smallest thickness you can get. 0.35mm thick laminations are better than 0.5mm!
2. Use laminations of the best mechanical (no burrs caused by blunt die-stampings!) and electrical quality.
3. Use low-loss GOSS laminations even for mains transformers.
4. Design transformers using as low levels of magnetic flux density (B) as possible.
5. Design and wind the windings so all layers are fully filled, using proper wire tensioning. A firm, tight winding will not buzz!
6. Bolt the laminations tightly together and coat the surface of the stacked transformer by a coating of contact adhesive or an acrylic pour-on finish.

EnviroTex Lite pour-on acrylic finish is achieved by mixing a bottle of resin with a bottle of hardener (catalyst) which then cures into a thick, clear glossy coating in about eight hours at normal room temperatures.

Due to its gel form, Selley's *Kwik Grip Gel* contact adhesive is easy to brush on and ideal for use on transformer's vertical surfaces. On contact, it sets almost immediately and reaches its maximum strength in 24 hours. One liter covers up to five square meters, enough for quite a few transformers.

Once cured, the adhesive is resistant to acids and alkalis and heat resistant to 130°C, high enough for most, if not all, transformers. If the surface temperature of your finished power or audio transformer or choke reaches such high temperatures, something is seriously wrong.

SHIELDING OF TRANSFORMERS AND CHOKES

C-cores, toroids, and R-cores feature low magnetic radiation, but transformers with EI-laminations, most commonly used in audio, radiate strong magnetic fields. For that reason, the amplifier's power supply section should ideally be completely separate from the audio section, on its own chassis, and well shielded. That way, the magnetic and electrostatic flux produced by the power transformers and filtering chokes wouldn't reach the interstage and output transformers. However, that is an expensive and often impractical solution to hum and buzz problems.

Simply enclosing the offending magnetic components in a metal box (shield) usually achieves adequate results.

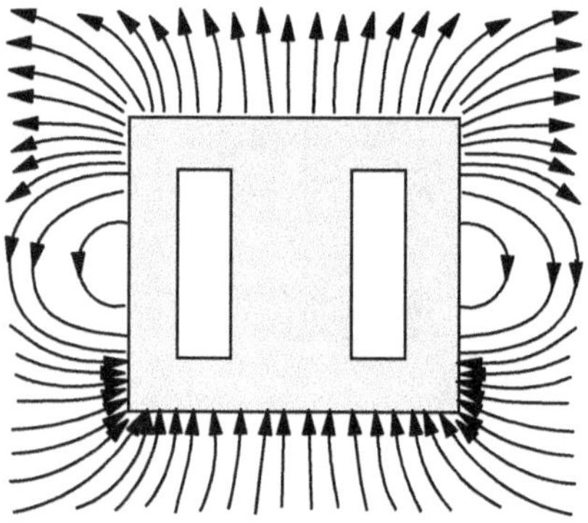

The electromagnetic field has magnetic and electrostatic components. The removal of or shielding from either component prevents the other one from radiating since they must travel (propagate) together.Two different magnetic shielding mechanisms are the diversion through high μ shield and the generation of opposite flux (also called the "shorted turn method").

The first uses high m ferromagnetic shields or transformer covers made of such material, which present a low reluctance path for the radiated magnetic field, so very little of it escapes outside the enclosure. In shields made of aluminium, copper, and other non-ferromagnetic metals, the generation of eddy currents in the shield opposes the original flux radiated by the transformer and contains the radiation within the enclosure.

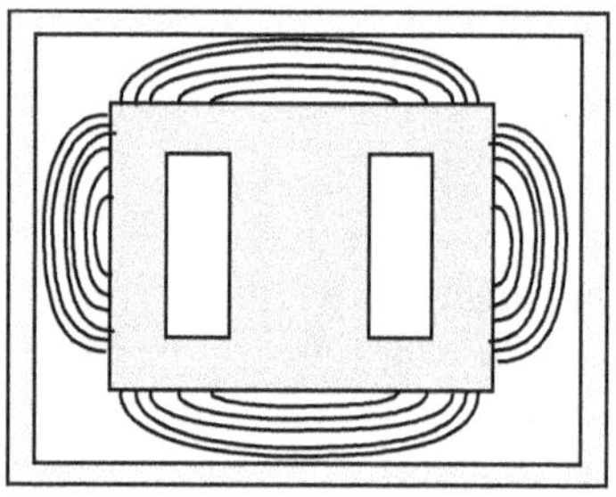

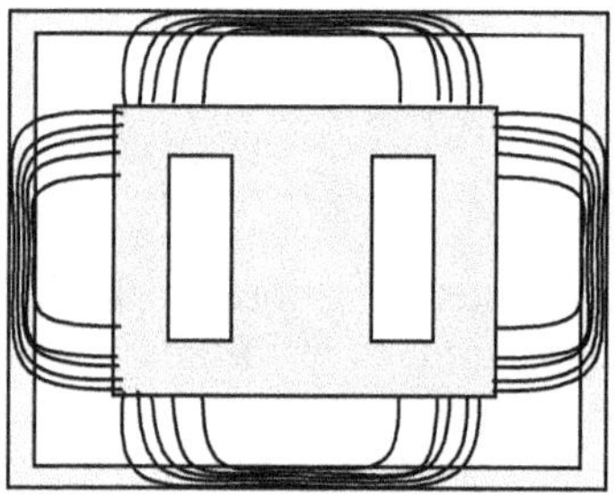

ABOVE: The magnetic field lines around a typical EI transformer or choke.

The shielding principle behind diamagnetic and paramagnetic shields such as aluminium or copper (FAR LEFT) works through the generation of eddy currents in the shield, which oppose the original flux. The high μ ferromagnetic shields, such as steel or mumetal (LEFT) channel the flux through their walls (path of least magnetic resistance).

High or radio frequency shielding

The best shields for high-or radio-frequencies are made of a material with high electrical conductivity, such as copper or aluminium. The HF magnetic flux induces voltages in the shield, which causes eddy or circulating currents to flow in the shield, and these eddy currents oppose the action of the flux and reduce its penetration through the shield. So, the aim here is to actively "encourage" the formation of eddy currents. As we have seen, the situation with transformers is the opposite; we try to minimize eddy currents by using thin laminations.

The attenuation in dB is approximately A=8.78a/d [dB], where a is the thickness of the shield, and d is "skin depth," which is defined as $d=5{,}033\sqrt{(\rho/\mu f)}$. ρ is the resistivity of the conductor (shield) in Ω/cm^3, f is the radio signal frequency (interference), and μ is the magnetic permeability of the shield material. For copper at 20°C the skin depth is $d=6.62/\sqrt{f}$.

Low or audio frequency shielding

Notably, nonmagnetic (diamagnetic or paramagnetic) but conductive shields, such as copper or aluminium cases, still shield against the magnetic fields. Because of their high "reluctance," they repel the magnetic flux lines and thereby contain the flux within its confines. They also provide electrostatic shielding.

Due to their lower "reluctance" compared to free air, the ferromagnetic shields, made of materials such as steel or Mumetal, "attract" magnetic flux, so the flux lines close through the shield and don't escape into the outside air. While power transformers and chokes are shielded, so they don't radiate outwards, signal transformers, such as input-, MC-, interstage- and output-transformers, are shielded, so they don't pick up external radiation. The protective principles, however, are the same.

The shielding action is based on the magnetic material short-circuiting the magnetic flux and eddy currents that make the shield behave as a conducting shield.

By completely encasing their transformer in a high-permeability metal casing (Mumetal®, Permalloy®, HyMu®, and Co-Netic® are the most commonly used for that purpose), Stanley Transformer Company managed to achieve -70 dBm attenuation of external electromagnetic fields (RIGHT).

Since A(dB)=20logA, so A=INVlog(-70/20) = $0.316*10^{-3}$, or 1/A=3,162! Any external interference would be attenuated more than 3,000 times.

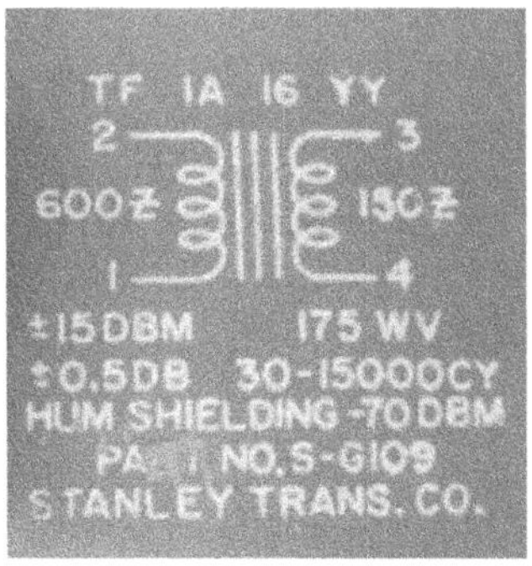

Such extensive shielding is not normally required for power amplifiers but is always a good idea on power transformers for preamps and phono stages, and sometimes with interstage and preamp output transformers. Due to extremely low voltages, double- or triple-shielding for MC (moving coil) step-up transformers is mandatory. 3mm thick steel reduces pickup by about 12 dB, compared to a 30 dB reduction for a Mumetal® can. Double and triple Mumetal® cans provide 60 and 90 dB of shielding.

Multiple shields

A strong magnetic field could easily saturate Mumetal® and other shielding materials with high μ but low maximum B (magnetic field density). That would significantly lower their magnetic permeability μ and reduce their shielding effectiveness. For that reason, the outer shields are often made of steel or similar material of lower μ, but higher saturation levels of B. The outer shield will reduce the external magnetic field to a level not capable of saturating the inner shield.

The discussion above applies to shielding a sensitive "receiver" such as the MC step-up transformer. To shield a source or emitter of magnetic interference such as a choke or power transformer, the situation would be reversed; the inner shield would be of low m material, and the outer shield would be made of Mumetal®.

If you have a magnetic shield made of high permeability alloy such as Permalloy®, don't drill it, cut it or modify it in any way. Such materials are adversely affected by mechanical strains such as drilling, punching or bending and must be heat-treated after mechanical fabrication to be effective, and any subsequent stress will render them useless!

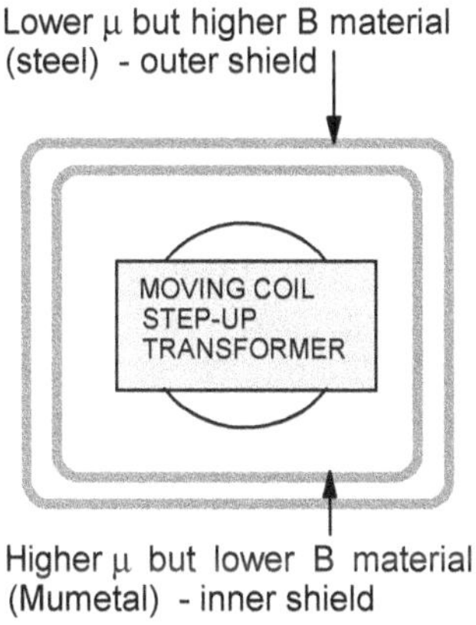

Transformer positioning

When open construction transformers are used (not fully enclosed in metal boxes), to minimize the unwanted magnetic coupling, position the power and output transformers at 90° angle. If that is not possible or due to aesthetic reasons, slant them at an angle. The same applies to interstage transformers, if used. Even if full metal enclosures are used, it is still a good practice to keep transformers inside them at 90° with respect to one another.

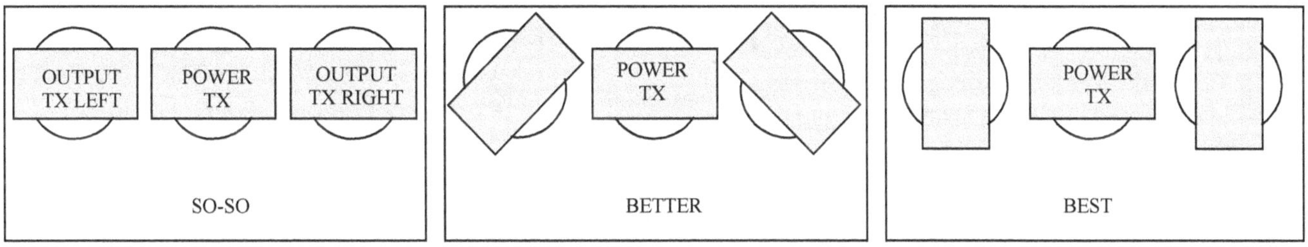

Electrostatic shielding

Two types of E/S shielding of transformers are used. The first is a full layer of wire between the primary and secondary, which is then earthed at one end. The diameter does not matter; use whatever wire you have at hand, but you should wind those turns as tight and as neatly as possible to not leave any gaps between turns and fill the full length of the bobbin.

Alternatively, a copper foil turn can be used, but its overlapping ends must be insulated from one another; otherwise, the shield would form a short-circuited turn.

As illustrated in the photograph (left), the second type of shielding also uses a thin copper strip, but it's wrapped around the whole transformer once it's assembled (1). As per the photo, since it is outside the magnetic circuit, its ends are soldered together, and it does not need to be earthed.

LEFT: power transformer with an external copper shield (1) and welded laminations (2).

BY THE SAME AUTHOR:

Sound Improvement Secrets For Audiophiles: Get Better Sound Without Spending Big

Publisher: Career Professionals
Year published: 2021
Language: English
Paperback: 328 pages
ISBN: 978-0648298205

Avoid the hit-and-miss approach and stop wasting money on overpriced high-end products in the blind hope of sonic improvement. Achieve the ultimate audio synergy and get more enjoyment from your audio system by making it as good sounding as possible.

"Sound Improvement Secrets for Audiophiles" will teach you how things work, why some circuits, designs, and technologies sound the way they do, and how to make them sound even better through simple modifications and improvements. It is like having an audio and acoustic consultant by your side to guide you through optimizing and voicing your audio system and your listening room.

While relatively technical and in-depth, this practical manual goes way beyond "a dozen quick tips" and the simplistic advice you read elsewhere. Instead, the focus is on dozens of DIY projects, case studies, and examples of commercial audio components – turntables, preamplifiers, amplifiers, loudspeakers, power supplies, and acoustic treatments.

With over 400 photographs, diagrams, and illustrations, "Sound Improvement Secrets for Audiophiles" makes it easy for you to understand and comprehend complex technical concepts and issues.

The author does not shy away from many controversial and hotly debated topics. Tubes vs transistors, objectivists vs subjectivists, measurements vs listening, and digital vs analog: all of these are discussed in detail.

The money invested in this book would not even buy you a budget-priced pair of cables: it will prove to be one of the best financial investments you ever make. Even if you implement only a few improvements from the hundreds described within its pages – you will never look back!

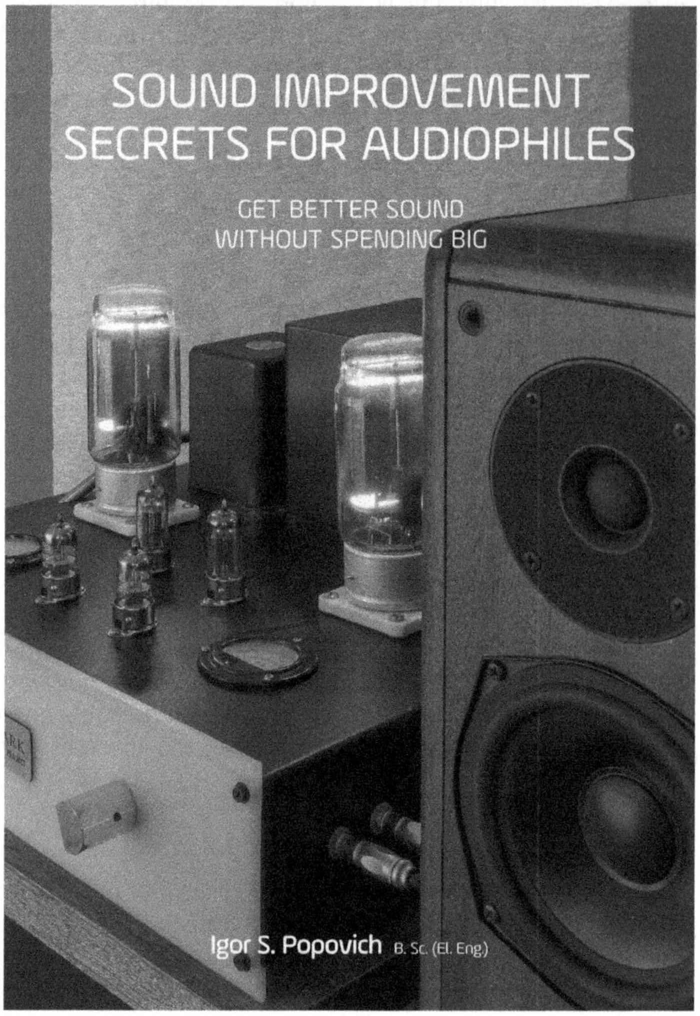

BOOK CONTENTS:

1. WHY YOU SHOULD READ THIS BOOK AND HOW YOU WILL BENEFIT FROM IT
2. BEFORE YOU BUY AN AUDIO SYSTEM OR COMPONENT - THINGS TO DO & MISTAKES TO AVOID
3. WHAT DO WE LISTEN FOR AND WHAT DO WE ACTUALLY HEAR?
4. CLEANING UP THE POWER SUPPLY TO REDUCE NOISE, HUM, AND INTERFERENCE
5. CABLES, FUSES, CONTACTS, AND CONNECTIONS
6. UPGRADING & FINE-TUNING THE SOURCES: OPEN REEL RECORDERS, TURNTABLES, PHONO STAGES AND CD PLAYERS
7. AUDIO AMPLIFIERS - HOW THEY WORK AND HOW TO IMPROVE THEIR SOUND
8. HEADPHONES AND HEADPHONE AMPLIFIERS
9. LOUDSPEAKER TYPES, TESTS, AND IMPROVEMENTS
10. COMPONENT MATCHING AND AUDIO SYSTEM INTEGRATION ISSUES
11. LOUDSPEAKER POSITIONING
12. OPTIMIZING THE ACOUSTIC PERFORMANCE OF YOUR LISTENING ROOM
13. ACOUSTIC TREATMENTS
14. MINIMIZING UNWANTED VIBRATIONS & OSCILLATIONS
15. TROUBLESHOOTING YOUR AUDIO SYSTEM

MAINS (POWER) TRANSFORMERS

- POWER SUPPLIES, RECTIFIER AND FILTERING CONFIGURATIONS
- AUTO-TRANSFORMERS
- COMMERCIAL POWER TRANSFORMERS - EVALUATION & ANALYSIS
- LESSONS LEARNED: VINTAGE GERMAN POWER TRANSFORMER
- DESIGN EXAMPLE: HEATER TRANSFORMER FOR A HIGH-POWER AMPLIFIER
- DIY DESIGN: EI POWER TRANSFORMER FOR A 300B SE STEREO AMPLIFIER
- DIY DESIGN: C-CORE POWER TRANSFORMERS FOR MONOBLOCK PP AMPLIFIERS
- MAINS TRANSFORMER DESIGN SEQUENCE

5

POWER SUPPLIES, RECTIFIER AND FILTERING CONFIGURATIONS

The five aspects of a power supply: voltage, current, energy storage, voltage regulation and ripple

A power supply inside audio equipment is invariably a DC (direct current) supply. It performs AC to DC conversion, converting mains AC voltage of a certain frequency (50 or 60Hz) to a steady DC voltage of the required level. This could be a low voltage used for tube heaters, typically 6.3 or 12.6V, a negative bias voltage, typ. -10 to -60V,and a high anode voltage, typ. 200-500V.

Apart from the desired voltage, a power supply must have a required current capacity; for instance, it must supply $400V_{DC}$ anode voltage at 160mA, which would be the current drawn by the load (the audio section of the amplifier).

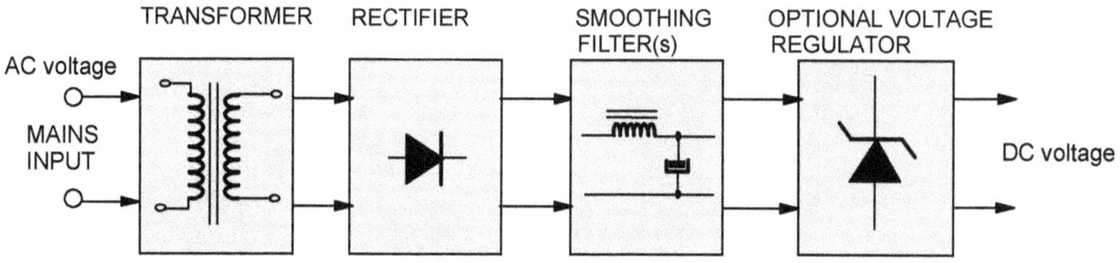

LEFT: Block diagram of a typical power supply in a tube guitar or hi-fi amplifier

In Class A amplifiers, the current draw of the audio section is constant, so the audio section behaves like a fixed resistor; in this case, its load resistance is R_L=400V/0.160A= 2,500Ω.

A power supply must be efficient in its energy conversion from an AC to DC supply, i.e., its internal losses should be as low as possible. Low losses mean less heat dissipation, less vibration, no buzzing of transformers, and a cooler and quieter running amplifier.

Internal losses are modeled by the internal resistance of a power supply, which should be as low as possible (zero in an ideal case).

One consequence of the internal resistance of a power supply is the voltage drop that load current will cause on such a resistance, thus reducing the output voltage (voltage divider effect). So, the output of a real power supply will drop (sometimes called a voltage "droop" or "sag") with increasing load current.

An ideal DC supply provides a constant output voltage regardless of the load and its current demands (current draw). The factor that describes how close a real power supply comes to an ideal one in that regard is called voltage regulation.

Since the output DC voltage is obtained by rectification and filtration from an AC waveform, some remnants of that AC, called *ripple*, remain.

While an ideal DC supply would provide a perfectly horizontal line, the output voltage of real power supplies fluctuates above and below some steady DC value V_0. Thus, the smaller that "ripple" or AC component, the lower the hum on amplifier's output.

For instance, if the voltage of a push-pull amp drops from 350V at no load to 300V at full load, its regulation factor is 16.67%.

Once suitable AC voltages are present at power transformer's secondary windings, all except heater voltages must be rectified and filtered into DC voltages needed for various amplification stages.

Due to its unidirectionality, a diode (a solid state or vacuum diode) is a rectifier. Since it allows the current to flow in one direction only, it "rectifies" an AC signal at its input so only positive peaks are passed through, thus making it a signal with a DC component.

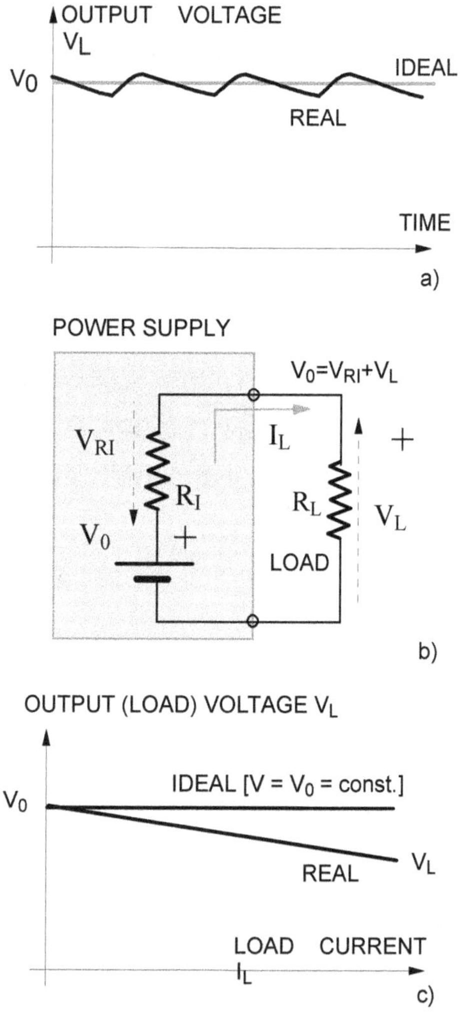

ABOVE:

a) The output voltage waveform of an ideal (no ripple) and real DC power supply

b) The model of a real DC voltage source with internal resistance R_I

c) The output voltage versus load current characteristics for the ideal power supply, with (constant output regardless of the load (V_L=V_0) and a real, unregulated power supply (drooping voltage V_L)

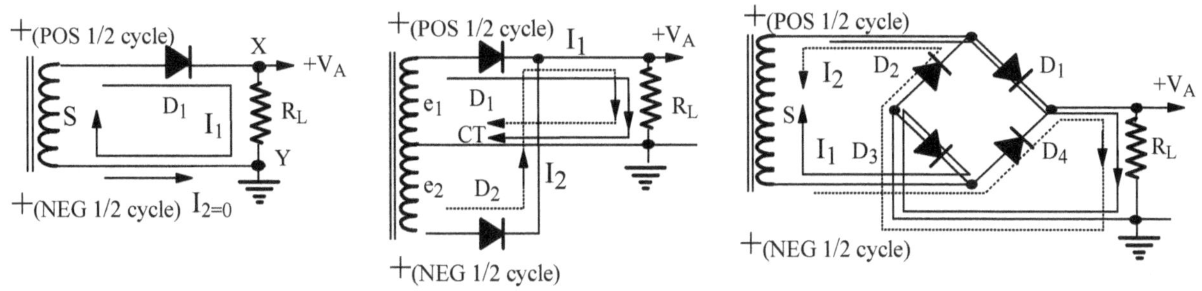

Three basic rectifier circuits (L-R): half-wave, full wave (center-tap) and full-wave bridge. Current flows through diodes for the positive (I_1, full line) and negative (I_2, dotted line) halves of the secondary voltages are indicated.

Half-wave rectifier

Half-wave rectification suffers from three main problems: 1. high ripple 2. unbalanced DC current flowing through the secondary can saturate the transformer's core 3. low efficiency of 40% (η=DC power out/AC power in). Therefore, it is seldom used, except for negative bias supplies in Class A_1 and AB_1 amplifiers, where it supplies a practically infinite load impedance (no grid current flowing). To get the negative bias voltage, point X should be grounded and the negative voltage should be taken from point Y (diagram above left).

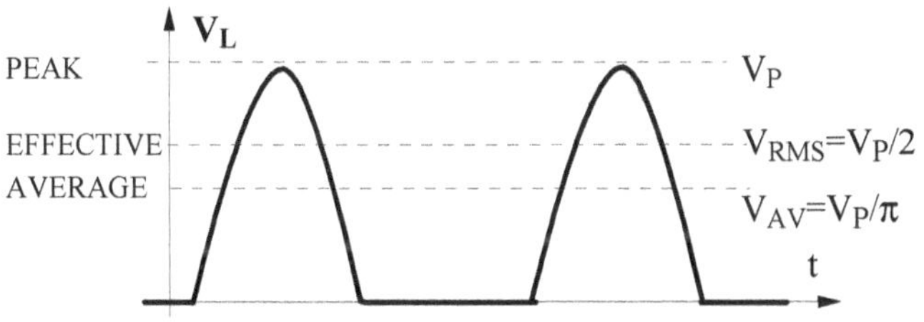

Half-wave rectified voltage (resistive load without any filtering)

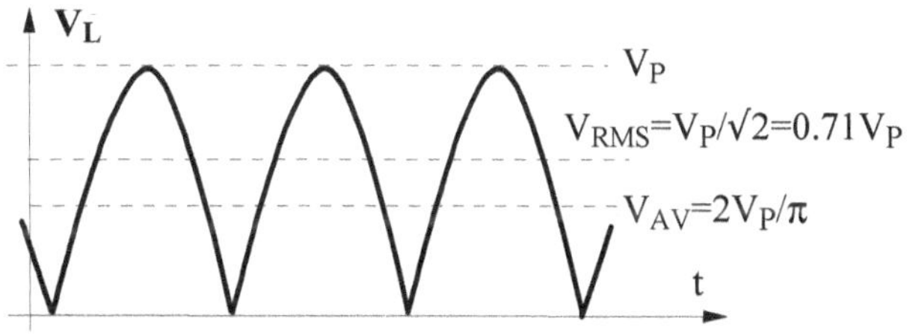

Full-wave rectified voltage (resistive load without any filtering)

Full-wave center-tap rectifier

This was the most common circuit with tube rectifiers in the 1950s and 60s due to the large size and high cost of rectifier tubes. Once silicon rectifiers took over, this topology was abandoned in favor of the solid-state bridge rectifiers. The bridge rectifier with tubes would need two separate heater power supplies (or secondary windings), while only one was needed for the center-tap arrangement.

The two DC currents (I1 and I2) through the secondary winding cancel out, so there is no DC magnetization of the core. The efficiency is much higher than that of the half-wave rectifier; however, two secondary windings are needed, doubling the number of secondary turns.

Full-wave bridge rectifier

As with the center-tapped circuit, there is no DC magnetization of the transformer's core. Higher efficiency than the center-tapped arrangement, plus half the secondary turns required. The price is the need for four rectifiers instead of two, but with small and very cheap silicon diodes, that is not an issue.

In this case, the effective value of the output (load) voltage V_{RMS} is 41% higher than the half-wave circuit ($0.71V_P$ versus $00.5V_P$), the same as for the center-tapped circuit.

PEAK, AVERAGE AND RMS RECTIFIER VOLTAGES

HALF-WAVE:	FULL-WAVE:
$V_{AV}=V_P/\pi$	$V_{AV}=2V_P/\pi$
$V_{RMS}=V_P/2=0.5V_P$	$V_{RMS}=V_P/\sqrt{2}=0.71V_P$
RIPPLE FACTOR γ= 1.21	RIPPLE FACTOR γ= 0.48

Voltage doubler

If the window area in the power transformer is at premium and you cannot fit in enough winding turns to get the voltage high enough for a full-wave center-tapped arrangement, and there isn't enough space even for a bridge topology, all is not lost. Providing you can fit half of the required turns, you can always use the voltage-doubling circuit where $V_{DC}=2V_P$, where V_P is the peak value of the secondary voltage.

There is no DC magnetization of the core. However, notice how the secondary winding must be floating; you aren't allowed to ground it due to the specific topology of the circuit. That point can be used as an additional DC voltage source at half of the main V_{DC}.

When point X is positive, current I_1 flows through D_1 and charges capacitor C_1. D_2 does not conduct. When point Y is positive, during the next half-period of the sinusoidal secondary voltage, I_2 flows through D_2 and charges C_2. The voltages on C_1 and C_2 are in series and add up!

The doubler is also less affected by the variations in the AC input than the other full-wave supply circuits. Its topology prevents spikes and interference from being fed from the amplifier back to the transformer's secondary and primary.

The output voltage V_{DC} drops as the load current increases, meaning that voltage doublers have a poorer regulation. The ripple factor also worsens in that case. The maximum reverse voltage on diodes is approximately equal to double the AC voltage amplitude on the secondary.

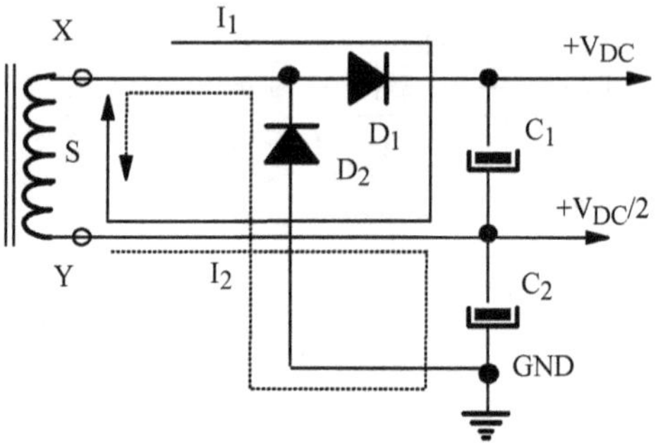

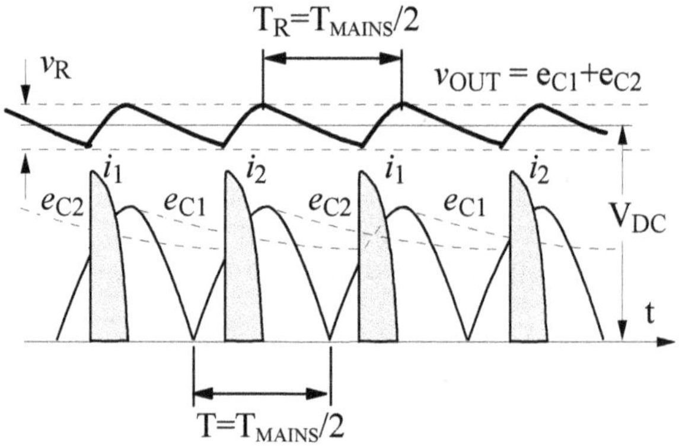

Capacitive filtering

We have already seen capacitive filtering in action in the voltage doubling circuit. A capacitive filter is inherent in this circuit; voltage doubling would not be possible without the capacitive action!

A full-wave rectifier using a vacuum rectifier tube is illustrated on the right. For a moment, disregard the choke L and the second capacitor C_2 and assume that only C_1 is connected, a simple capacitive filter.

At point A the output voltage (on the capacitor) follows the sinusoidal waveform. The capacitor is charged by a current supplied by the transformer's secondary winding and the rectifier tube. As soon as the voltage reaches point B, the peak, the voltage on the capacitor becomes higher than the input voltage pulse from the rectifier tube, and the charging current iS stops flowing. The capacitor now provides the load current, and due to the discharging effect of the load current, the voltage on the capacitor starts to drop.

With capacitive filtering, the current through the transformer secondary flows in very short high current pulses, whose spectrum contains many high order harmonics. These high-frequency components heat the transformer core and the winding (hysteresis and copper losses). In some commercial tube amps we had on our test bench, the mains (power) transformers got so hot that one could barely touch them after an hour or so of operation. These transformers will have a short and stressful life.

Half-wave rectification is used here to better illustrate the more pronounced voltage drop between the pulses (right), but the discussion also applies to full-wave rectification. The lower the required or desired ripple, the higher (sharper) the current pulses must be (to charge the capacitor in a shorter and shorter time), and the more nasty harmonics injected back into the power transformer's secondary!

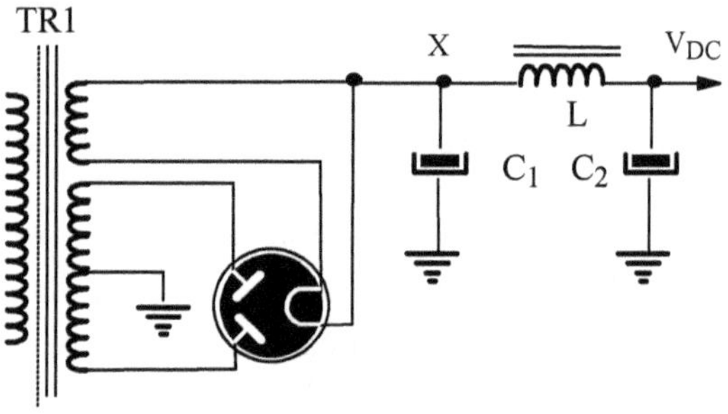

ABOVE: Typical HV power supply using a rectifier tube and a CLC-filter

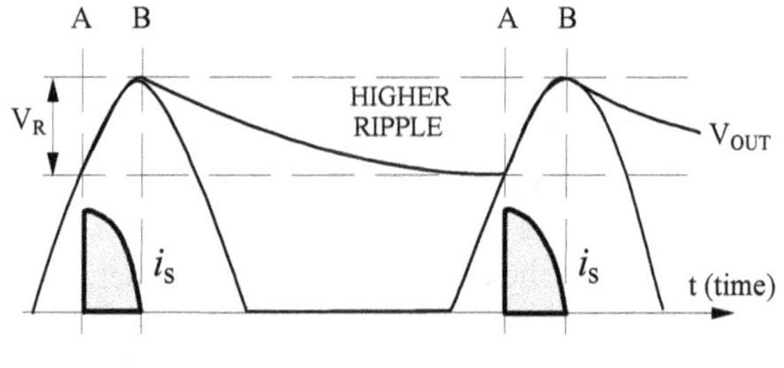

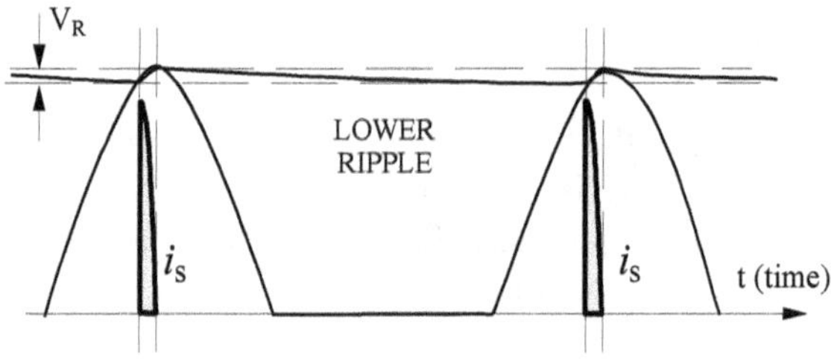

Inductive (LC) filters

L-filter (also called LC-filter) features a choke connected to the rectifier output without any filtering capacitor between that point (marked X) and the ground (unlike CLC or PI-filter).

The currents through two halves of the high voltage secondary winding are almost constant, in contrast to currents in transformers feeding CLC filters, short-duration high amplitude spikes. The inductor represents a high reactance to the ripple voltage, while the shunt capacitor has low reactance and shunts or bypasses any AC ripple remaining after the choke smoothing to the ground.

The ripple factor γ is the ratio of the AC (ripple) current and DC load current, or the RMS value of the ripple voltage divided by the DC component of the load voltage:

$\gamma = I_{RMS}/I_{DC} = V_{RMS}/V_{DC}$ and for LC filter $\gamma = \sqrt{2/3} * X_C/X_L = \sqrt{2/3} * 1/2\omega C * 1/2\omega L$. With L in [H] and C in [μF], $\gamma = 0.83/LC$ for 60Hz mains and $\gamma = 1.2/LC$ for 50Hz mains.

Ideally, above I_X, the minimum load current, the output voltage V_L voltage stays constant ($V_L = V_0$) as per curve A). The load current is constant and is comprised of two approximately square-wave components i_1 and i_2. In reality, due to DC resistance and AC reactance of transformer's secondary winding, the rectifier tube's internal resistance and choke's resistance, the voltage drops with increased load current (curve B), since the voltage drop on all those resistances increases and subtracts from the ideal level V_0.

Secondary transformer voltage (for each half of the HV winding) is $V_S = V_M \sin\omega t$, its RMS value is V_{RMS} and the output DC voltage is $V_L = 2V_M/\pi = 0.637V_M = 0.9V_{RMS}$.

The only advantage of LC filters is improved sonics; that is why many audiophiles prefer hi-fi amps with LC filtering. Nobody knows why amplifiers with LC filtered power supplies sound better. The most likely reason is the elimination of sharp current pulses through the power transformer. These pulses contain high energy harmonics which propagate through both the mains circuit and the audio circuitry of amplifiers and "smudge" or defocus their sonic presentation.

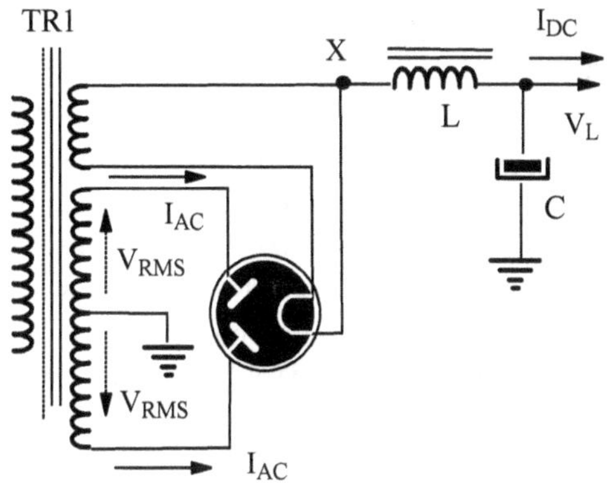

ABOVE: Typical HV power supply using a directly heated rectifier tube (5U4GB) and LC-filter

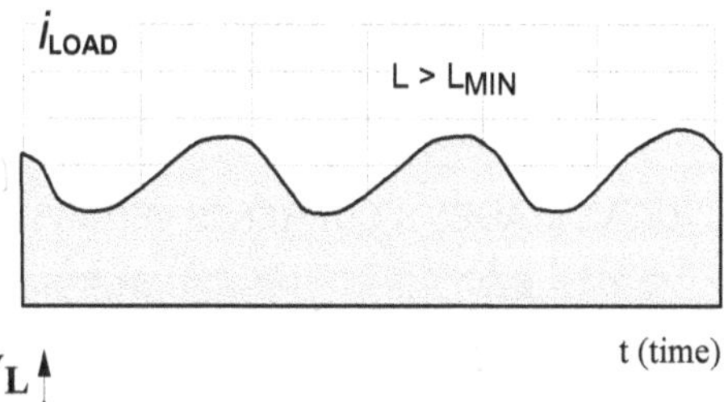

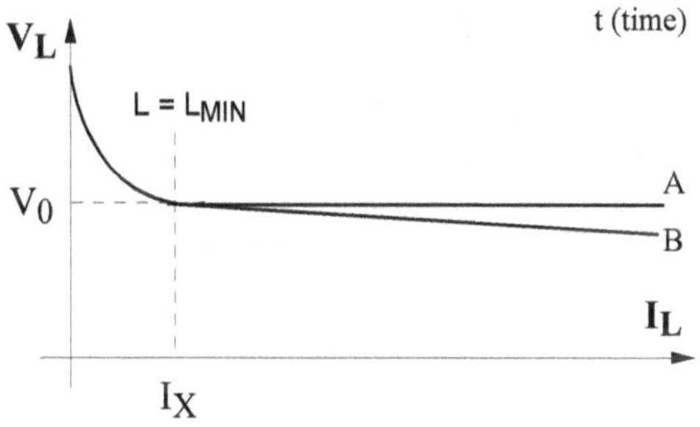

ABOVE RIGHT: Load current in a full-wave rectifier with an LC filter for $L > L_{MIN}$, where L_{MIN} is the critical inductance of the filtering choke

RIGHT: The regulation curve of an L-filter, ideal (A) and real (B).

Tube versus solid state rectifiers

Before selenium, germanium, and silicon rectifiers were invented, tube rectifiers ruled the field. Compared to their solid-state descendants, tube rectifiers and rectifying circuits suffer from at least seven drawbacks.

The mains transformer requires an additional winding (1) to provide heating voltage for the rectifier tube, typically 5V, 2A (10 Watts) or 5V, 3A (15 watts). Furthermore, a high voltage winding with a center tap (CT) is needed (2), and such a secondary has double the number of turns compared to the secondary designed for bridge rectification.

The value of the first filtering capacitor is limited, typically from 10 to 47 μF (3). Each tube rectifier has the specified maximum value of capacitance it can tolerate. With higher value capacitors, the rectifier tube will start arcing when the amp is powered on, due to high capacitor charging currents, and its life will be drastically shortened.

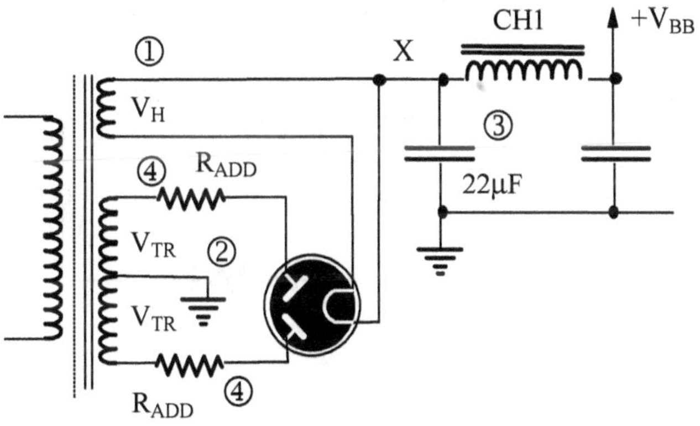

ABOVE: A typical high voltage power supply using a tube rectifier

RIGHT: Common tube rectifiers and their main operating parameters, heater voltage/current, maximum secondary AC voltage on their anodes (plates), minimum required transformer secondary resistance, maximum capacitance of the 1st filtering capacitor (3), and maximum DC current they can provide

TUBE	V_H (V)	I_H (A)	V_{TR} (V)	R_{MIN} (Ω)	C_{MAX} (μF)	I_{MAX} (mA)
5R4GY	5	2	750	250	4	250
5R4WGY	5	2	850	500	4	150
5U4G	5	3	300	170	32	245
5U4WG	5	3	450	75	32	225
5U4GB	5	3	300	21	40	300
5U4GB	5	3	450	67	40	275
5V4	5	2	375	100	10	175
5W4	5	1.5	350	50	4	100
5X4G	5	3	300	170	32	245
6X4	6.3	0.6	325	525	10	70
EZ80	6.3	0.6	350	175	50	90
6CA4	6.3	1	350	230	50	150
5Y3	5	2	350	50	20	125

The total energy reserve of a tube power supply is much lower than the energy reserve of the same voltage solid-state supply. $E=CV^2$, so the energy stored in 32 μF filtering cap used with a tube rectifier producing $500V_{DC}$ is 8.0 Joules, while that of the 680 μF with SS rectifier at the same voltage is 170 Joules. That is 21.25 times more energy, which means better bass response and dynamics.

Tube rectifiers have high internal resistance, so their voltage drops are 20-50 times higher than silicon diodes, typically 15-40V, compared to only 0.5-0.7V for silicon rectifiers. For the same transformer secondary, the voltage at point X on the diagram will be much lower than the same circuit using SS diodes.

Since the voltage drop on tube rectifiers increases with load current, their voltage regulation is worse than SS power supplies. With the increased amplifier power draw, more and more power is wasted as heat, plus the anode voltage will sag or droop, limiting the output power. All other aspects being equal, an identical amplifier with a SS rectifier can produce higher output power than the same amplifier with tube rectification.

Another limitation in the table is that the minimum resistance R_{MIN} on the AC or anode side tube rectifiers must be connected. If a transformer primary and secondary were wound with a large diameter wire, the total secondary transformer winding resistance might be below the minimum needed, so series resistors may need to be added (4).

Both audiophiles and guitar players claim that amps with tube rectifiers sound better than solid-state diodes despite these limitations and drawbacks. By better, they mean smoother, softer, more "natural," and "emotionally engaging."

DC VOLTAGES & CURRENTS VERSUS AC VOLTAGES & CURRENTS

RECTIFIER	FILTER	DC VOLTAGE	DC CURRENT
HALF-WAVE	CAPACITIVE	$V_{DC}= 0.45*V_{AC}$	$I_{DC}= I_{AC}$
FULL WAVE - CT	CAPACITIVE	$V_{DC}= 1.41*V_{AC}$	$I_{DC}=2I_{AC}$
FULL WAVE - CT	INDUCTIVE (LC)	$V_{DC}= 0.9*V_{AC}$	$I_{DC}=2I_{AC}$
FULL WAVE - BRIDGE	CAPACITIVE	$V_{DC}= 1.41*V_{AC}$	$I_{DC}=I_{AC}$
FULL WAVE - BRIDGE	INDUCTIVE (LC)	$V_{DC}= 0.9*V_{AC}$	$I_{DC}=I_{AC}$

NOTE 1: V_{AC} and I_{AC} are the effective (RMS) values of transformer secondary's voltage and currents

NOTE 2: V_{AC} for CT secondaries is 1/2 of the total voltage (between CT and one output) and I_{AC} for CT secondaries is 1/2 of the total current the winding can supply.

NOTE 3: Ideal diodes and no losses of any kind are assumed.

AUTO-TRANSFORMERS

Principle of operation

An autotransformer does not have two separate, galvanically isolated windings. There is only one winding that has a tap to which either the secondary is connected (the load), if the transformer is a step-down type, or if the transformer is of the step-up type, the primary voltage is connected to the tap, and the secondary load is connected to the full winding (see diagrams below). Of course, multiple taps are also possible. We will focus on the step-down type, but a similar analysis can be done for the step-up type.

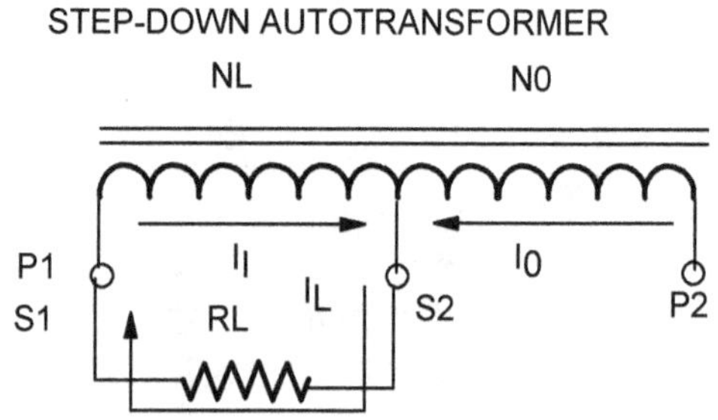

N_0= total no. of turns (P1-P2)

N_L= load turns (S1-S2)

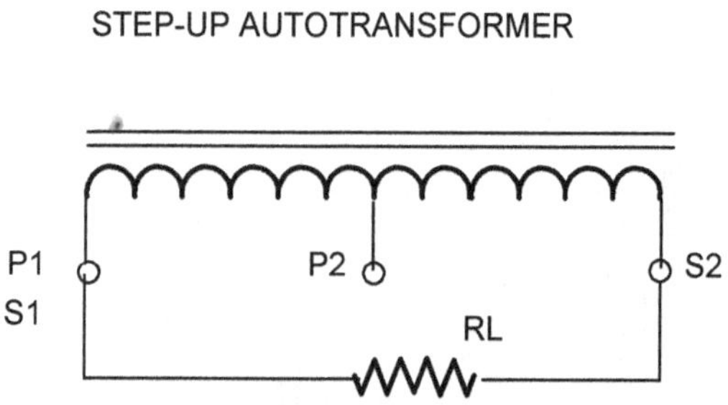

Voltages are divided as per the ratio of turns $V_L/N_L=V_0/N_0$

AT (Ampereturns) must be the same, so $I_0N_0=I_LN_L$ and $I_0(N_0-N_L)= I_IN_L$

The Kirchoff's law says the sum total of currents entering a node equals the sum of currents exiting the same node, so for the point S2: $I_L=I_0+I_I$ or $I_I=I_L-I_0$

Combining the two equations (eliminate I_0) we get: $I_I = I_L(1-N_L/N_0)$

Load power is $P_L=V_LI_L$ and the power in the "secondary" or load part of the coil is $P_{SEC}=V_LI_I = V_LI_L(1-N_L/N_0) = P_L(1-N_L/N_0)$.

The power the secondary part of the winding N_L needs to supply depends on the ratio of turns N_L/N_0; we can call the $(1-N_L/N_0)$ term "Reduction Factor" or RF.

The power rating of the autotransformer is $\mathbf{P=P_{SEC}=P_L*(1-N_L/N_0) = P_L*RF}$.

The closer the step-down voltage is to the input voltage, the lower the needed power rating of an autotransformer!

QUICK EXAMPLE: Your mains voltage is 240V, and you need a 240⇨220V autotransformer for a tube amp designed for 220V, with 200VA consumption. Since the two voltages are so close (less than 10% difference), the power of the core needed is only 200*(1-220/240) = 200*(1-0.917)= 200*0.0833 = 16.7 VA.

DESIGN EXAMPLE: Step-down autotransformer 240V⇨117V

We need a step-down autotransformer for a vintage tube amplifier. "117V, 60Hz, 175W" was printed at the back of the amp. Since the load is slightly inductive, the apparent power of the load is around 180VA. We are in Australia, where the nominal mains voltage is $240V_{AC}$, but in our area, the voltage is around $247V_{AC}$.

We need a step-down ratio 247/117 = 2.11 but also need to allow for losses, so we will raise the secondary voltage by 5%. The practical voltage ratio is VR= 247/(117*1.05) = 247/122.85 = 2.01 = 2.

The reduction factor is $RF=(V_1-V_2)/V_2$ = (247-123)/247= 124/247 = 0.502 = 0.5, meaning the power rating of the core will be half of the load power: $P_I = RF*P_L$ = 0.5*180 = 90VA, from which we can determine the required core size: $A=\sqrt{P_I} = \sqrt{90} = 9.5\ cm^2$.

We have EI84 laminations with a=2.8 cm, so we need a stack thickness of S=A/a = 9.5/2.8 = 3.4 cm, meaning the standard 35mm bobbin will be perfect. We will get A= 2.8*3.5 = 9.8, which, taking the stacking factor into consideration, would give us A_{EF} of around 9.5 cm^2.

$TPV=40/A_{EF}$ = 40/9.5 = 4.2

N_1= 4.2*247= 1,038 T and $N_2= N_1*VR$= 1,038*0.5= 519 T

Load currents: I_L=180VA/117V = 1.54 A and $I_0=P_L/V_1$ =180/247 = 0.73A, $I_I=I_L RF$ = 1.54*0.5= 0.77A

RF is close to 1/2, so the two currents are practically equal, meaning both parts of the winding can be wound with the same wire thickness of $d= 0.71*\sqrt{I} = 0.71*\sqrt{0.77}$ = 0.71*0.88=0.62 mm (current density J=2.5 A/mm^2).

We can go up to 3A/mm^2, so a 0.56 mm wire would be fine. The amp's power consumption was specified for maximum output power; a push-pull Class AB amp will never work on that 100% power level for long, if ever.

CL=42-4=38 mm, so maximum turns-per-layer TPL_{MAX}=38/0.56= 68.

1,038/68 = 15.3 layers, so we will have 8 layers, a secondary tap, followed by another 8 layers, a total of 16 layers with TPL=1,038/16 ≈ 64.

COMMERCIAL EXAMPLE: Step-down autotransformer 240V⇨110V

These Chinese-made step-up or step-down transformers (some models have a switch at the back panel) use EI76.2 laminations with a=2.54 cm, with 40mm stack, A=a*S= 10.16 cm^2, meaning the core is rated at $P=A^2$ = 103 VA. This model is step-down only and is sold as a 200 VA transformer under various brand names.

The 240V primary (whole winding) has a DC resistance of 20.6Ω, and the 110V portion of the winding measured 11.2Ω. It is an autotransformer, so there is no galvanic isolation between the mains and the load.

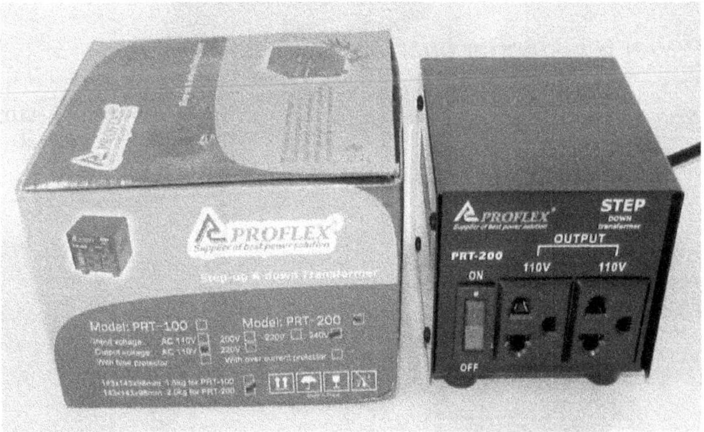

Available under various brand names, this series of China-made autotransformers comes in 100VA, 200VA, 500VA, 1kVA and 2kVA power ratings. 200VA model pictured.

COMMERCIAL POWER TRANSFORMERS - EVALUATION & ANALYSIS

It is important to know how to determine if a certain transformer is suitable for use in your design, properly sized for the load, both in magnetic (the core size) and electric terms (the wire sizes), so it can deliver the voltages and currents needed by the audio section of the amplifier.

How to identify windings of an unknown power transformer

The rules for identifying transformer windings are "the lower the voltage, the higher the current" and "the thicker the wire, the lower the winding resistance." The heater winding will have the thickest wire and the lowest winding resistance (say 0.3-3Ω). The primary winding will have an order of magnitude higher resistance (say 10-30Ω), and the high voltage winding an order of magnitude higher again (1:10:100Ω).

Quick estimation of EI transformer's power rating

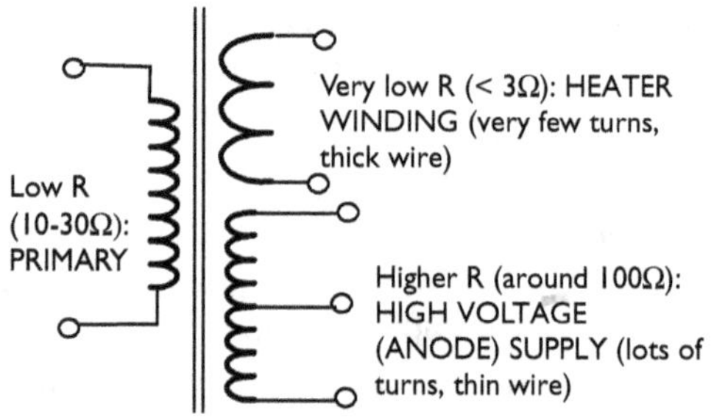

EI power transformers are still the most common in tube amplifiers. The following discussion applies only to transformers that use standard (wasteless) EI lamination, not toroidal or C-core transformers.

The power rating of the core is the area of the center leg's cross section in cm^2 (CL times the thickness of the lamination stack S) squared. The width of the Center Leg is double the width of the I-section. From the drawing, you can see that you can determine U (one unit or the width of I) by measuring the outside dimensions. Ensure you measure the laminations, not the overall dimensions, including the thickness of end-bells or covers.

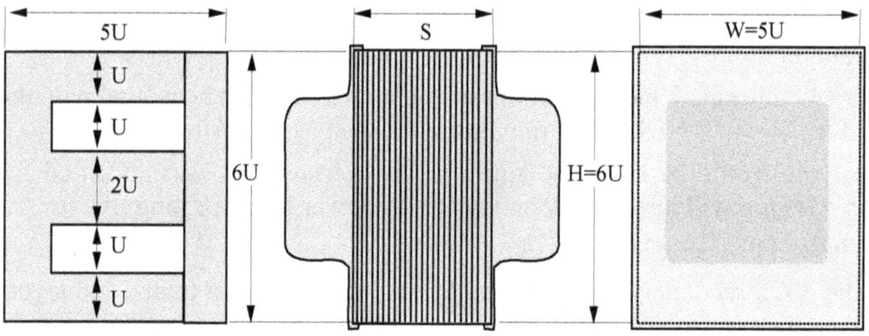

ABOVE: Typical DC resistance values of mains transformer windings

LEFT: By measuring external dimensions of wasteless EI-laminations on finished transformers, their type and internal dimensions can easily be determined.

S is the thickness of the lamination stack, H is the height (the larger of the two outside dimensions).

For example, transformers using EI96 laminations have H= 96mm, so I=96/6 = 1.6 cm. To get the cross-sectional area, divide the height by three and multiply by the stack thickness S. H/3 = 3.2 cm and if the stack thickness is 4.5 cm (one of the standard bobbin sizes), A is 14.5 cm^2, and the power rating of that transformer core is $P=A^2 = 14.4^2 = 207$ VA.

Measure the largest dimension (height H) and stack thickness S in centimeters, multiply them, divide by three and square, and you have a quick estimate of the EI core's power capability!

MAINS TRANSFORMER CORE SIZING

EI cores @ 50 Hz $A_{EF} \approx \sqrt{P}$

EI cores @ 60 Hz $A_{EF} \approx 0.9\sqrt{P}$

P in VA, A_{EF} in cm^2

C-cores @ 50 Hz $A_{EF} \approx 0.5\sqrt{P}$

C-cores @ 60 Hz $A_{EF} \approx 0.45\sqrt{P}$

Power transformer buck or boost connection

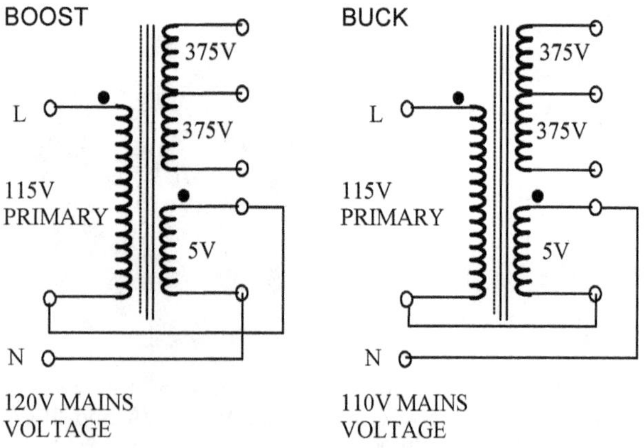

To use a 110V or 115V transformer on 120V (or higher) mains, connect its unused 5V or 6.3V secondary winding in series with the primary. Likewise, if the primary was designed for 230V and your mains voltage is nominally 240V but in reality closer to 250V, you can fix that problem with a 12.6V secondary winding or make it less troublesome with a 5 or 6.3V winding!

BOOST: Connect the primary and the rectifier heater winding in phase, so their voltages add up. Instead of 115V primary you now have a 120V primary.

BUCK: The windings are in series, but this time out-of-phase. The winding ends without a dot are connected together. Instead of 115V primary you now have a 110V primary.

CAN POWER TRANSFORMERS DESIGNED FOR 60Hz BE USED ON 50Hz MAINS, AND VICE VERSA?

From the fundamental TX equation, we can express turns per volt $TPV = N_1/V_1 = 10^4/(4.44fBA_{EF})$.

Since TPV and A_{EF} are constant for a certain transformer, when a 60Hz unit works on 50Hz mains, the frequency will go down 16.67%, so B (magnetic flux density) will have to go up 20% (1.2 times) to make up for it. Thus, it all depends on the level of B the transformer was designed to operate at 60Hz.

If the transformer was operating at high levels of B, 1.5-1.6T, it might saturate at 50Hz (B up to 1.8T)! Many American mains transformers from vintage tube amps were made to a strict budget with a penny-pinching attitude and would be marginal at 50Hz at best or totally in saturation at worst.

For 50Hz transformers working on 60Hz mains, the frequency f will go up, and the flux density B will go down, making them run cooler and quieter.

CASE STUDY: Triad N-68X isolation transformer

Triad N-68X is made in China and sold by Allied Electronics. Its claimed power rating is 50VA, which should be enough for small single-ended guitar amps. Its bigger brother, model N-77U is rated at 100VA.

The two identical primary windings can be connected in parallel, making it a 1:1 isolation transformer (115V-115V), or in series, making it a step-down isolation transformer (230V-115V).

The datasheet claims that primary and secondary windings were not wound concentrically, on top of one another (separated by a thin layer of insulation), but on separate arbors (bobbins), so a short circuit between primary and secondary winding (resulting from overheating or isolation breakdown) would not be possible.

The idle current draw of a transformer is a good indication of its quality. While 64mA for a 100VA unit was quite high, by coincidence the idle draw of N-68X was also 64mA!

Such a high idle current for such a tiny transformer was the first warning sign, indicating that poor quality (high loss) laminations were used.

The magnetic core of N-68X is only rated at around 30VA, so proclaiming it to be capable of 50VA is fraudulent. When loaded with a 50 Watt load N-68X started buzzing, and after five minutes, it got so hot that one could smell the melting varnish and the metal frame was far too hot to touch!

The 100VA N-77U has a 118 VA core, so a 100VA rating is accurate once losses of approx. 18VA are deducted.

Once we took one N-68X transformer apart, the claim about separate arbors also proved to be a total lie! The 680 turns of the secondary wire were wound first, followed by insulation and a copper foil as E/S shield, then another layer of insulation. 600 primary turns were wound next, followed by plastic insulation and another 600 primary turns. The bulk winding method was used (no careful layering).

Since $d = 0.71*\sqrt{I}$ we can find the nominal secondary current as $I = (1.41*d)^2 = (1.41*0.33)^2 = 217mA$ (current density $J = 2.5\ A/mm^2$).

The secondary is supposed to be rated at 435mA, meaning the designer assumed or allowed a current density of exactly double (435/217=2), $5A/mm^2$, which is way too high!

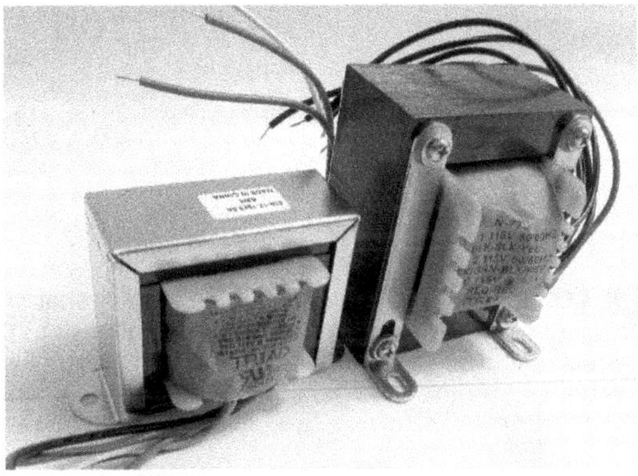

ABOVE: Triad isolation transformers: N-68X (left) and N-77U (right)

TRIAD N-68X MEASURED RESULTS:

- No load primary current: 64mA
- EI66 laminations (a=22mm)
- S=25mm stack thickness
- Core power rating: $P=(a*S)^2 = (2.2*2.5)^2 = 30.25$ VA

TRIAD N-77U MEASURED RESULTS:

- No load primary current: 64mA
- EI85.8 laminations (a=28.6mm)
- S=38mm stack
- Core power rating: $P=(a*S)^2 =$ 118 VA

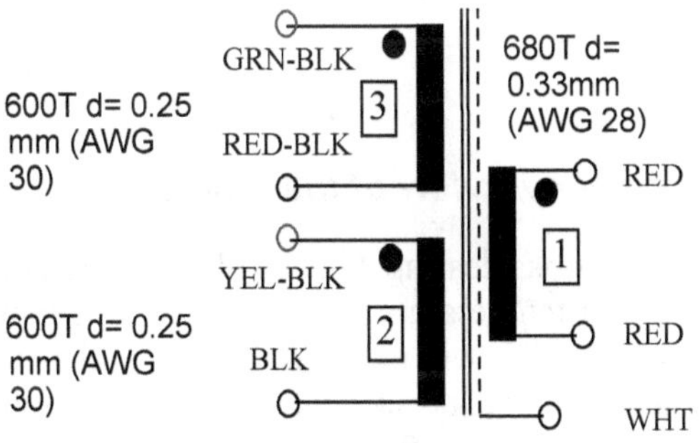

ABOVE: Triad N-68X winding diagram

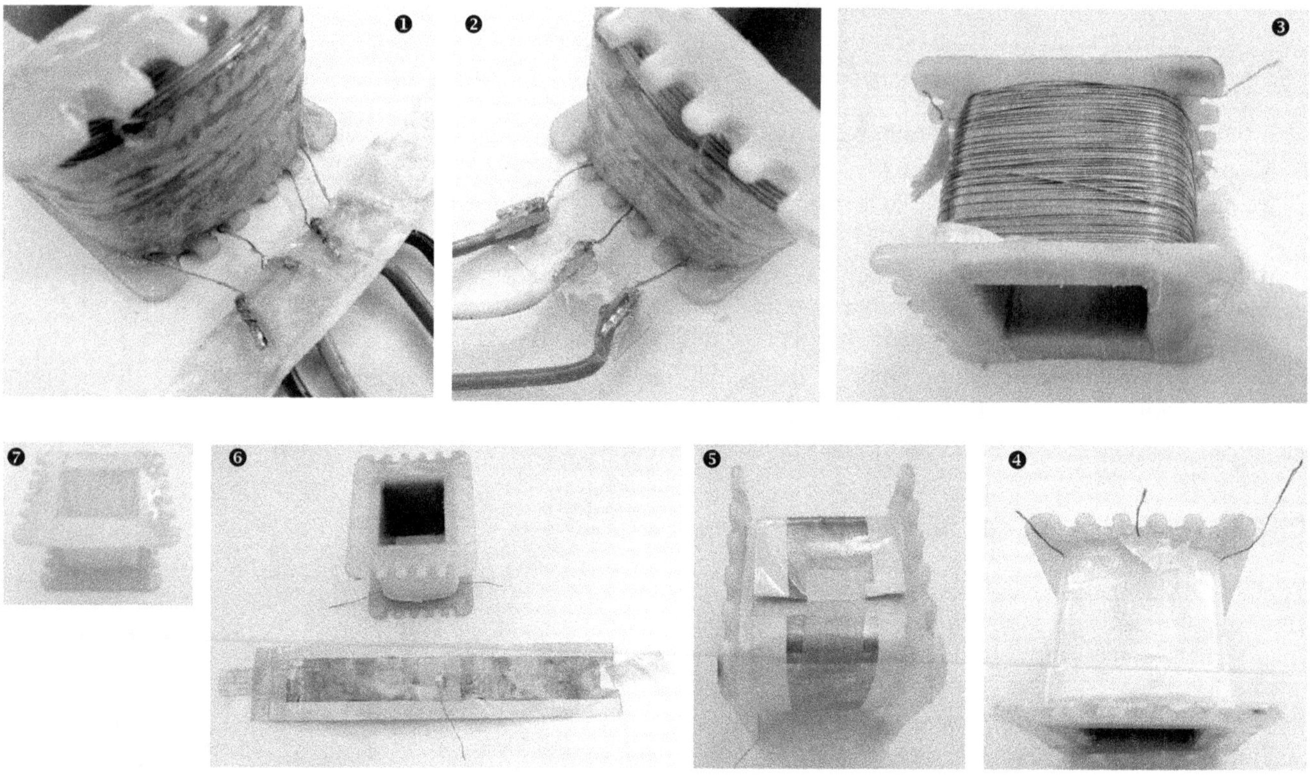

The deconstruction of Triad N-68X isolation transformer in 7 steps: 1) Primary connections - terminations 2) Secondary connections - terminations 3) Top insulation removed, outer (top) primary winding exposed, ready for unwinding 4) Primary unwound 5) Insulation between the inner primary winding and E/S shield removed, showing the electrostatic shield 6) E/S shield (one turn of copper foil) removed 7) Secondary unwound, bare bobbin left

CASE STUDY: Power transformer

Let's practice the basic transformer deciphering skills. Please study the sticker on this power transformer and draw its wiring diagram. Then, add up the total nominal secondary power draws and compare the total secondary power draw with the rule-of-thumb estimate of the magnetic core's power rating.

EI76x40 means EI76 laminations with a 40mm wide stack. EI76 laminations have a=25.3 mm, the width of their center leg. Thus, the center leg cross-section Thus, the center leg cross section is A= aS= 10.1 cm^2. The core is rated at $P=A^2= 10.1^2= 102$ VA.

Since P=VI, the total power rating of the three secondary windings is $P=P_1+P_2+P_3 = V_1I_1+V_2I_2+V_3I_3 =$ $6.3*3.2 + 40*0.18 +350*0.25 =$ $20.16 + 7.2 + 87.5 = 114.86$ VA.

This agrees with the declared power rating of 115V, but exceeds our estimation by 115/102 = 1.127 or 12.7%. Unless very high quality (low loss) laminations were used, if fully loaded this transformer will get very hot.

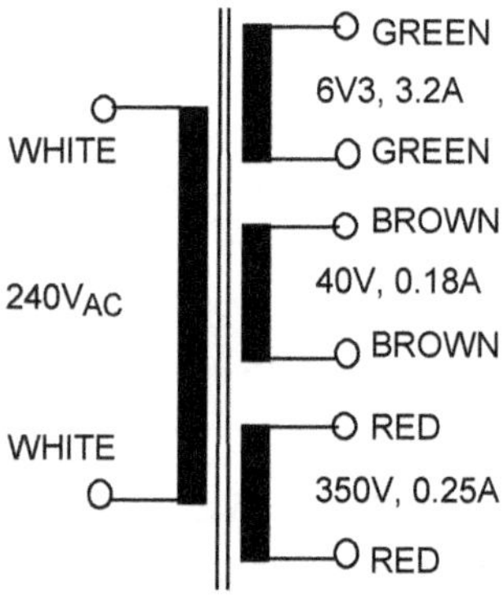

A brief overview of thermal issues

Some books dealing with small power transformers, such as those used in tube amplifiers, devote a whole chapter to the issue of temperature rise due to various power losses, going through dozens of complex formulas that help estimate such rise. We will not go down that path here. First, I am not a sadist to expect you to understand all that mathematical masturbation, and second, if you use the right rule-of-thumb for estimating the power capability of a transformer's core or lamination stack (as we do in this book), the temperature rise will be very mild anyway, and there is no need whatsoever to double-check that such a temperature increase will be acceptable. It will be!

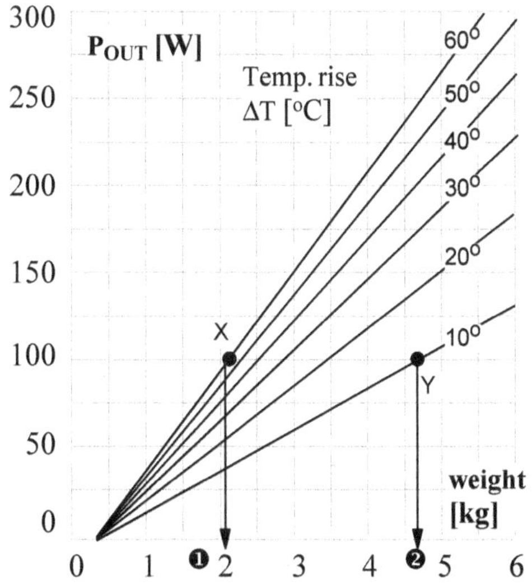

LEFT: The power rating of the core as a function core weight and allowed temperature rise as a parameter.

RIGHT: The required weight as a function of the allowed temperature rise for a 100 W 60 Hz power transformer

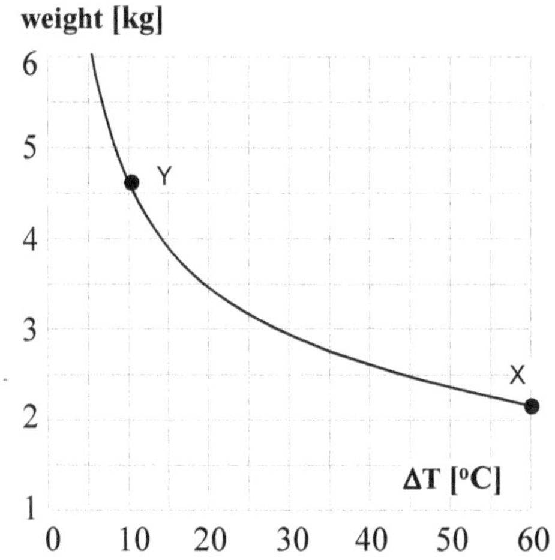

The graph above left shows that the required weight of the core increases linearly with its power rating. The slope of that rise depends on the allowable temperature rise. The lower the temperature rise, the faster the weight increases with power rating.

Using a 100W transformer as an example, a 2.1kg core (1) would result in a 60°C temp. rise (point X), but a 4.6kg core would warm up only by 10°C (point Y) under the same load.

If a weight-vs.-temperature rise curve is drawn for this transformer, it will look like the graph above right. The two points mentioned, X and Y, are indicated as references.

CASE STUDY: Power transformer

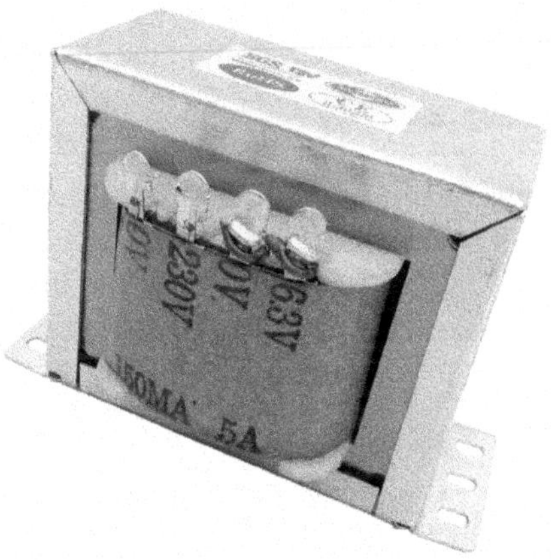

Bought on ebay in 2017 for US$20.50 and US$37 airmail and made in Taiwan by SanLang Co., the primary of this transformer has three taps, 120V, 230V and 240V.

With a 130VA magnetic core, it can supply 230V@150mA and 6.3V@5A, or a total of $P=230*0.15+6.3*5 = 66VA$.

So, its electrical power rating is very conservative for such a large core (only 50%), always a good sign, meaning it will run quietly in an amplifier and will stay cool even under full load.

With its HV secondary loaded with a 1,500Ω 50W rheostat (150mA of AC current) for 15 minutes, it stayed cool, so it seems that its HV secondary winding could supply 200mA or more intermittently.

The voltage regulation of the HV winding can be calculated from the voltage drop under full rated load (150mA) as $VR=(V_L-V_0)/V_0 = (219-236)/236 = -0.072$, or -7.2%, a good result.

The very low idle current draw of only 22.2mA indicates that high quality (low loss) laminations were used.

MEASURED RESULTS:

- Idle primary current (no load): 22.2mA
- EI85.8 (a=28.6mm), S=40mm stack
- Core power rating: $P=S^2=(a*S)^2 = 130$ VA
- HV sec: 236V (no load), 219V (150 mA)
- Heater sec: 6.66V (no load)
- Weight: 1.90 kg

"What were they thinking??" Another Taiwanese power transformer

We considered this power transformer, also made in Taiwan by SanLang Co. for one of DIY projects in my book "Tube Guitar Amplifiers". Most similar China- and Taiwan-made units only come with a 110 or 220V primary and are unsuitable for use with our 240-250V mains voltage. This model has three primary taps, 120V, 230V, and 240V; so far, so good.

EI96 laminations were used, but the stack thickness wasn't specified. The power rating at 50Hz was specified as "79.5VA max.". The transformer weighs just under 3kg.

There are three secondary windings, 0-6.3V@5A, for audio tubes' heaters, 0-5V@3A, presumably for the heater of a rectifier tube, and a high voltage (HV) winding 0-275V@120mA.

The winding is neat, and winding wires protrude between the bobbin and the terminal lugs onto which they were soldered. Thus, their thickness can easily be measured with a micrometer.

The current and power ratings for 60Hz operation were also given: 6.3V @6A, 5V@3.6Aand 275V@144mA, for a total of 95.4 VA. A quick check: 60Hz/50Hz=1.2, $P_{60Hz}=P_{50Hz}*1.2 = 79.5*1.2 = 95.4$VA! So, the voltages stay the same, but current ratings are 20% higher when a transformer designed for 50Hz operation works on 60Hz mains.

So far, so good. However, go back and look at the secondaries and their current ratings. Firstly, a 5A heater winding would make this transformer suitable to power the heaters of four 6L6 and three 12AX7, for example. That would be a 100 Watt amplifier. Can this transformer power such an amp?

Notice the very low HV current rating of 120mA. Remember, that is an AC current rating, meaning that when rectified, we will not get more than 80mA of DC current! That isn't enough even for two 6L6 in push-pull.

So, the heater winding is oversized (the wire used is too thick), while the HV winding is undersized, its wire is too thin, and as a result, its current capacity is too low!

This transformer is only suitable for a single-ended amplifier with one 6L6, EL34, etc. It will be marginal even in a small PP amp with two 6V6, EL84, or similar low power output tubes.

Finally, the most mind-boggling issue. The HV winding is not center-tapped, so a rectifier tube cannot be used, it requires a solid-state rectifier bridge or a voltage doubler, yet there is a 5V, 3A winding for a tube rectifier?

This case is another illustration of a shocking lack of critical (or any!) thinking on behalf of transformer designers in particular and amplifier designers in general.

LESSONS LEARNED: VINTAGE GERMAN POWER TRANSFORMER

This power transformer by Loewe Opta uses M-laminations, as do most of the smaller German and European power transformers of the 1950s and 60s. The normal marking practice has been to specify nominal voltages and full load currents for each winding except the primary, where the current is usually not specified.

In this case, the manufacturer specified the voltages under full load but not the maximum allowed current levels. Instead, the number of turns and the wire diameters are marked.

The transformer weighs 1.4kg, and its core measures 74.3 x 74.3 x 34.7 mm, which indicates M74 laminations. The published dimensions (table below) specify a 2.3cm wide center leg, and with a 3.4cm stack the effective cross-sectional area is approximately 90% of the gross area, or $A_{EF}=0.9A = 0.9*2.3*3.4 = 7.0\ cm^2$.

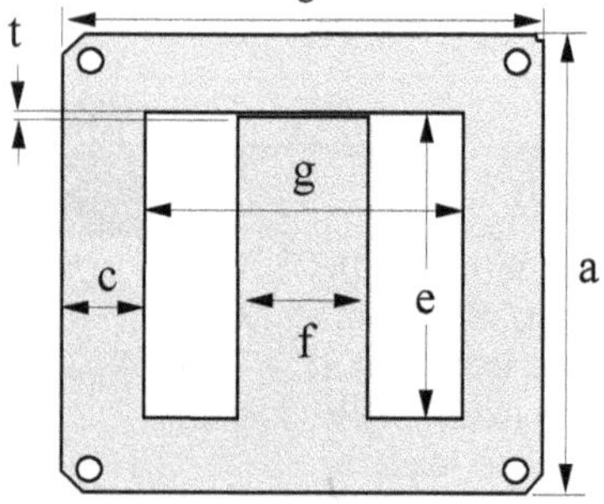

CORE	a = b	c	e = g	f
M30	30	5	20	7
M42	42	6	30	12
M55	55	8.5	38	17
M65	65	10	45	20
M74	**74**	**11.5**	**51**	**23**
M85	85	14.5	56	29
M102	102	17	68	34

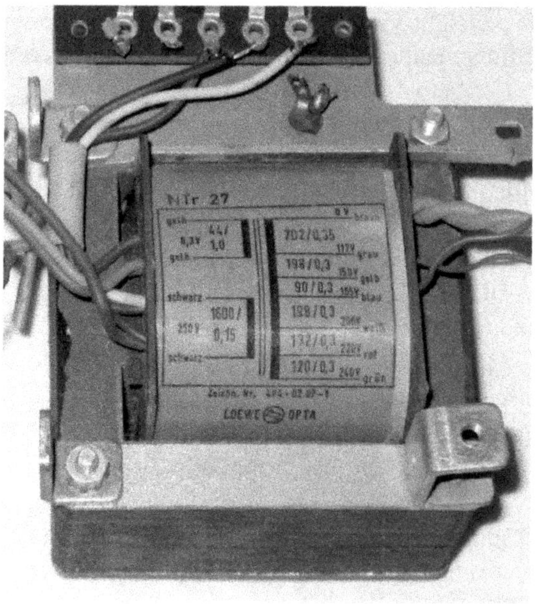

ABOVE: Power transformer from Loewe Opta tube console radio

LEFT: The significant dimensions of "M-schnitten" and the table with standard laminations sizes, from the smallest (M30) to the largest (M102).

Determining the TPV (Turns-Per-Volt) figure

If you add up the individual number of turns in all the primary sections, up to the maximum of 240 Volts, you'll get 1440 turns, so the Turns-Per-Volt figure is 1,440/240 = 6.0 TPV. The figure is slightly higher for the high voltage secondary winding (250V), 1,600/250 = 6.4. With TPV=6, the secondary number of turns would be 250*6=1,500, meaning 100 turns were added, or 100/1,500 = 6.67%!

The 6.3V secondary that supplies tube heaters is wound with TPV=44/6.3 = 7.0! With TPV=6, as for the primary, the secondary number of turns would be 6.38*6=38, so six turns were added (44-38=6) or 6/38 = 15.8%.

The explanation lies in the winding order. In small power transformers, the primary is wound first, closest to the magnetic core, followed by the HV secondary, while secondaries supplying the heaters are wound last, furthest away from the core. This order is usually also the descending order of power ratings.

The highest power dissipation is in the primary, equal to the sum of power ratings of all the secondaries plus transformer losses, a few percentages of the total, followed by the high voltage winding that needs to supply power to the output and all preamp and driver stages. Heater secondaries are usually of the lowest power rating, although in this case, their power is slightly higher than that of the high voltage winding.

Another reason heater winding(s) are wound last is their large wire size. If 6.3V heaters are used, that secondary winding usually has so few turns that it consists of one layer only, and when wound on top of all other windings, that layer pushes down on them and gives the whole coil mechanical strength and minimizes coil buzz.

Going back to the first reason for the slightly higher TPV figure, the clue lies, as always, in the fundamental transformer equation $TPV=N/V=10^4/(4.44BA_{EF}f)$.

As windings are wound on top of one another, the mean length of a turn increases, so their DC resistance increases. The turns also gradually encompass a larger and larger area, which includes the cross-section of the core's center leg and all the layers previously wound underneath. The magnetic flux (the number of magnetic lines) that couples all the windings is constant. Thus, the increase in the cross-sectional area encompassed by the upper windings means that B, the magnetic flux density, is dropping slightly, and since B is in the denominator of the TPV formula, TPV is increased to compensate for that small reduction in B!

The higher TPV is also used as compensation for copper losses under load, so the voltages are higher than nominal with no load and drop to their required (nominal) values under full design load.

Estimating the power rating and current capacities of the two secondary windings

Based on the given wire diameters, we can estimate the nominal current as $\mathbf{d=\sqrt{(1.27*I/J)}}$, where d is wire diameter in mm, I is current in Amperes, and J is current density in A/mm^2. We can calculate current as $I=Jd^2/1.27$ [A].

We don't know the current density the designer chose all those years ago. For small M cores J is usually between 2.7 and 3.5 A/mm^2, so if we assume 3.175 A/mm^2 for the heater winding with $1mm^2$ wire, we get I=2.5 A! The power rating of this winding is thus $P_{S2}=6.3*2.5 = 15.75$ W.

Likewise, the HV winding can supply $I=Jd^2/1.27 = 3.2*0.15^2/1.27 = 57mA$, so its power rating is $P_{S1}=250*0.057 = 14.25$ W, and the total power capability of secondary windings is $P_S= P_{S1}+P_{S2}=14.25+15.75 = 30W$!

German literature and DIN 41302 standard specify that this core (3.4cm stack is one of the standard bobbin sizes) is approximately 84% efficient, which means the primary power rating must be $P_P=P_S/0.84 = 30/0.84 = 35.7$ Watts.

The same literature lists this core as having a maximum power capacity of 50 Watts, which means the Loewe Opta transformer designers were very conservative in their approach, using only 36 Watts out of the possible 50!

DESIGN EXAMPLE: HEATER TRANSFORMER FOR A HIGH-POWER AMPLIFIER

The heater supply is usually obtained from one or more secondaries of a mains transformer that provides a high voltage (anode supply). However, as an educational exercise, let's start with the simplest possible transformer, supplying $6.3V_{AC}$ heating voltage to all tubes in an amplifier. One primary, one secondary, and the current is continuous and sinusoidal (unlike short current pulses with rectifiers and capacitive filters).

We want to construct a push-pull amplifier that could use any of the most common octal tubes, namely 6L6, EL34, 6550, KT88, KT90, KT100, and even KT120. The stereo amplifier will use two duo-triodes of 12A*7 variety in the input stage (12AU7, 12AT7, 12AX7) and two larger duo-triodes such as 5687 or 6CG7 in the driver stage.

Step 1: Load calculation

PREAMP TUBES: The 12A*7 tubes at 6.3 V draw 0.3 A each, 6CG7's heater draws 0.6A, while 5687 needs 0.9A. The lowest current draw is 2*0.3 + 2*0.6A=1.8A, while the highest current draw would be 2*0.3 + 2*0.9A=2.4A

POWER TUBES: The inclusion of 6L6 and KT120 tubes widens the current range considerably, from 6.0-6.4A to 3.6-7.8A (more than a 2:1 ratio)! If all transformers were ideal, we would get $6.3V_{AC}$ regardless of the load, so there would be no need to worry. Alas, the output voltage will depend on the current flowing through it. Most transformer manufacturers specify the heater voltage as 6.3V +/- 0.5V, meaning the allowable range is 1.0 Volt (5.8-6.8 V).

So, the lowest heater load will be 1.8+3.6=5.4A, while the highest current draw would be 2.4+7.8=10.2A.

Let's design our transformer for the highest load, 6.3V@ at 10.2 A We will get a higher voltage at 5.4A (which is almost half of the load used in the design), but hopefully, it will still be within prescribed tolerances.

There is no point in using ordinary silicon steel, especially if you make your own output transformers. Use the same GOSS material, buy the same size laminations in larger quantities, and then adjust the stack thickness to suit your purpose. You can even use it for chokes as well. Let's choose the most common GOSS material, M6.

OUTPUT TUBES	6L6	EL34	KT88/6550/KT90/KT100	KT120
Heater current [A]	0.9	1.5	1.6	1.7-1.95
Current for 4 tubes [A]	3.6	6.0	6.4	7.8

Copper losses, core losses and efficiency

Secondary power: $P_2=V_2I_2=6.3*10.2=64$ VA. The problem is that we don't know the primary power needed, so we have to assume the efficiency of the finished transformer. We have to consider power losses in the primary and secondary windings. Primary current I_1 flows through the primary winding with resistance R_1, so the power loss in the primary is $P_{L1}=I_1^2R_1$. Likewise, the secondary current I_2, which is also the load current I_L, creates a power loss in the secondary winding of $P_{L2}=I_2^2R_2$. The total copper loss is $P_C=I_1^2R_1+I_2^2R_2$.

The efficiency also takes into the account the core losses P_M (sometimes called iron or magnetic losses), which are also thermal-type losses (heat). Just as copper losses heat the windings, core losses heat the core. In lamination catalogs the core loss is expressed as W/kg at a certain magnetic induction level (1.11 W/kg for M6 @ 1.5T in our case). Once we know the lamination size and the stack thickness, we will be able to calculate the number of EI pieces used, the total core weight, and, therefore, the total core loss.

The total losses are copper plus magnetic losses: $P_L=P_C+P_M$

The ratio between the useful (output) and total (input) power is efficiency, whose symbol is the Greek letter η (eta): $\eta = P_{OUT}/P_{IN} = P_{OUT}/(P_{OUT}+P_L) = P_{OUT}/(P_{OUT}+P_C+P_M)$.

Power transformers are most efficient and economical when the primary copper loss equals the secondary copper loss. The core loss equals the total copper loss: $P_{L1}= P_{L2}$ and $P_M=P_C$. In this case, it can be shown that $V_2/V_1=\sqrt{(\eta)}N_2/N_1$. The secondary no-load-voltage is then $V_2 = V_1/\sqrt{(\eta)}$.

The efficiency of mains transformers greatly depends on their size. Simply, the smaller the core, the lower the efficiency. Transformers with a cross-sectional area of less than 5cm^2 have efficiency in the order of 70-75%, and for those with A of 5-10 cm^2, we can assume an efficiency of 75-85 %.

For GOSS laminations, efficiency is higher since core losses are lower, and with higher levels of B, fewer turns per volt are needed, so copper losses are lower.

Let's assume the efficiency of 85%: $P_1= P_2/\eta = 64/0.85 = 75.3$ VA.

Step 2: Core sizing

Now $A_{EF}=\sqrt{P_1} = \sqrt{75.3} = 8.67$ cm^2 and the gross area is A= 8.67/0.96 = 9.13 cm^2. Again, we have assumed a stacking factor of 96%, which is reasonable for currently produced laminations. Use 90% for older stampings.

We have two variables to play with, the center leg width a and the stack thickness S, so there is an infinite number of possibilities. However, pre-made bobbins come only in a few standard S-sizes, narrowing our choice. To get the lowest winding resistance and the best regulation, aim for a square stack or close to it.

Looking around EI tables and bobbin catalogs, we could use EI84 laminations (a=28mm) with 35 mm bobbins, which would give us A= 2.8*3.5=9.8 cm^2 or A_{EF}= 0.96*9.8 = 9.4 cm^2.

Step 3: Number of turns

For a GOSS EI core working on 50Hz mains, TPV=40/A_{EF}= 40/9.24= 4.33.

For 230 V_{AC} primary N_1=230*4.33= 996 and N_2=6.3*4.33= 27.3 = 27 .

If you live in USA or another 115V_{AC} country, you can repeat this practice calculation for a 115_{AC}/60Hz mains transformer. Likewise for Australian nominal 240V_{AC} mains or the actual 250V_{AC} in many areas.

Step 4: Wire sizing and window fitting

PRIMARY CURRENT & WIRE: $I_1=P_1/V_1$= 75.3/230 = 327 mA

Wire diameter d=$\sqrt{(1.27*I/J)}$, and if we choose current density J=2.5 A/mm^2 we get d =$\sqrt{(1.27*0.327/2.5)}$ = 0.41 mm, so we will use d=0.45 mm wire.

SECONDARY WIRE: For 10.2 A wire d=2.28 mm is needed. This is the net diameter, without insulation. For thicker wires, we can assume the insulation thickness of around 0.06 mm, so the actual wire size is 2.34 mm.

Coil length will be CL=WL-2*B = 42-2*2= 38mm (B is bobbin thickness), so TPL_{MAX}= 38/2.34 = 16.2 If we reduce 27 turns down to 26 and split them into two layers, we can fit 13 turns easily in one layer.

PRIMARY FIT: 0.9*38/0.45= 76 turns per layer (TPL), so 996/76 = 13 layers

PRIMARY HEIGHT: PH= 13*0.45mm = 5.85 mm

SECONDARY HEIGHT SH= 2*2.34 = 4.68 mm

TOTAL COIL HEIGHT WITHOUT INSULATION: CH = PH + SH = 5.85+4.68= 10.53 mm

The maximum coil height CH_{MAX}= WH-B = 14-2 = 12mm, CH/CH_{MAX}= 10.53/12 = 0.8775 = 88 %

This is too high. If a bulge factor of 15% is included, the actual coil height is 10.53*1.15= 12.1 mm so even before the thickness of the insulation is added, the winding is already too thick to fit!

What if the winding does not fit into the window?

We have several options to consider:

1. Increase the cross-sectional area by increasing S, the stack thickness.
2. Let the transformer operate at a higher level of flux density B. Both of these options would reduce TPV and the total number of primary and secondary turns.
3. Reduce the wire sizes, but this would increase the current density and wire resistance and increase copper loss, reducing efficiency.
4. Finally, we could choose a larger lamination size to increase the window dimensions.

Since our k=10,000/(4.44*50*1.126) = 40, we based our design on B=1.126 T, which is quite low for M6 GOSS material. To cut the primary layers from 13 down to, say, 10, we would need to reduce N_1 down to 10/13= 77%, so B would be increased from 1.126 to 1.125/0.77 = 1.46T, which is still way below the saturation level for M6 material.

Recalculating turns with higher magnetic flux density B

New k=10,000/(4.44*50*1.46)= 30.85 so new TPV = 30.85/9.24 = 3.3

For 230 V_{AC} primary N_1=230*3.3= 760 and N_2=6.3*3.3= 21

The secondary will have two layers of 11 turns each, equally spread across the whole bobbin width.

SECONDARY HEIGHT: SH = 2 layers * 2.34 = 4.68 mm

Primary fit: 0.9*38/0.45= 76 turns per layer (TPL), so 760/76 = 10 layers

PRIMARY HEIGHT: PH = 10 layers *0.45mm = 4.5 mm

TOTAL COIL HEIGHT WITHOUT INSULATION: CH= PH+ SH= 4.5+4.68= 9.18 mm

Allowing for 15% bulging factor CH=1.15*9.18 = 10.56 mm

Since the maximum coil height CH_{MAX}= WH-B = 14-2 = 12mm, 10.56/12 = 0.88 or 88% With tight winding (to reduce the bulge factor) and Mylar® insulation used instead of thick paper one, this can be done.

Step 5: Calculating winding resistances

PRIMARY: The Mean Length of Turn is MLT_1= 2*(a+S+4*BT)+2*π*W_1/2 = 2*(28+35+4*2)+π*4.5 = 85.14 mm.
SECONDARY: MLT_2=2*(a+S+4*BT)+2*π*(W_1+W_2/2)= 2*(28+35+4*2) + 2*π*(4.5+4.68/2) =114 mm

Primary's total length = 760*85.14= 64,706 mm = 64.7m

Secondary's total length = 21*114= 2,394 mm = 2.4m

$R_1=4\rho/\pi l/d^2= 2.126*64.7/0.4342*10^{-2} =7.3\ \Omega$

$R_2= 2.126*2.4/2.282*10^{-2} = 9.815\ m\Omega$

These are "cold" resistances at 20°C (room temperature). In normal operation, due to heat, the resistances will increase. The rule of thumb says that "hot" resistances are about 15% higher: R_{1H}=1.15*R_1= 8.4Ω and R_{2H}=1.15*R_2= 11.3 mΩ.

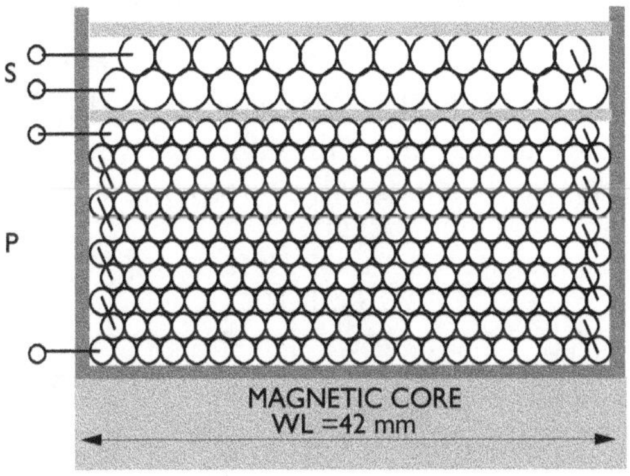

Step 6: Voltage drops, copper and core losses

Voltage drop on the primary resistance R_1 is $V_{R1}=I_1*R_{1H}$= 3.12V, and on the secondary $V_{R2}=I_2*R_{2H}$= 0.115V.

Copper losses are $P_{C1}= I_1^2*R_1$=1.16W, and $P_{C1}=I_2^2*R_2$= 1.18W, so the total copper loss is 2.34W.

To estimate the core loss we need the total weight of the laminations. From catalogs, we get the weight of 12.5 kg/1000 EI pieces. Since the stack is 35mm wide and each piece is 0.35mm thick, we need 35/0.35= 100 EI pairs, which weigh 100/1000*12.5 kg= 1.25 kg.

Depending on the induction level, the losses are between 1.11 and 1.65 Watts/kg. Our B of 1.46T is almost in the middle of the two given values, so let's say that losses are the mean of the two (1.11+1.65)/2 = 1.38 Watts/kg. So, the core losses are estimated at 1.25*1.38= 1.725W, bringing total losses up to 4.1 Watts.

Step 7: Efficiency and regulation

Let's check the efficiency: $\eta = P_{OUT}/P_{IN} = P_{OUT}/(P_{OUT} + P_L) = 64/(64+4.1)*100\% = 93.98\% \approx 94\%$.

Our 85% efficiency assumption was too pessimistic. Had ordinary 3% silicon steel been used, the efficiency would drop below 90% and close to the assumed 85%, but the GOSS laminations minimized losses.

Regulation can be calculated from efficiency $REG = (1/\sqrt{(\eta)} - 1)*100 = (1/\sqrt{(0.94)} - 1)*100 = 3.14\%$

Since $REG=(V_O-V_L)/V_L*100$ [%], we can calculate the no-load output voltage V_O as $V_O = (REG/100+1)*V_L = (1+0.0314)*6.3 = 6.5V$.

Obviously, for other tubes such as 6L6, EL34, KT88, etc., the voltage under load will be between 6.3 and 6.5V, which is within 3%, so our design criteria have been met.

DIY DESIGN: EI POWER TRANSFORMER FOR A 300B SE STEREO AMPLIFIER

In this power transformer, suitable for a stereo 300B SET amplifier, we have over-specified most current draws. The actual bias current is only a few mA, not 100 mA as specified. With up to 100mA for each 300B anode current and a few mA for each of the preamp stages, we need up to 250mA at the output of the voltage doubler, which means we need 500mA on the secondary winding. The assumed power dissipation (the sum total of all secondaries) is 144VA, while the real dissipation is probably just over 100VA.

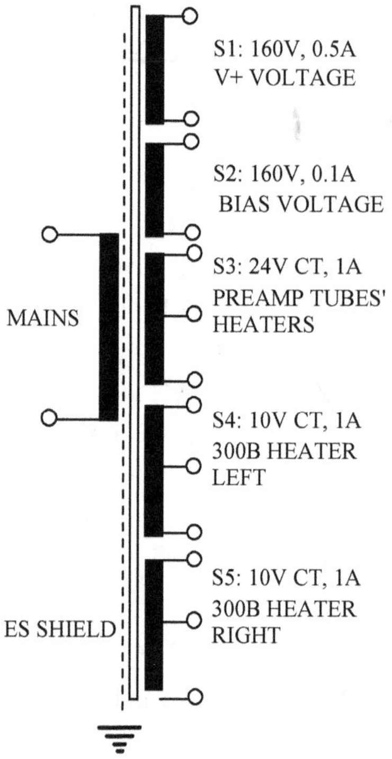

The power rating of the EI114 laminations (a=3.8cm) with stack thickness S=4.5cm is $P = A^2 = (aS)^2 = (3.8*4.5)^2 = 17.1^2 = 292$ VA. Our margin is larger than 2:1! This power transformer will not buzz or get hot; it will have an easy life and last a very, very long time.

As an exercise, take a vintage tube amp such as Eico, Fisher, or Dynaco, and estimate the power rating of the power TX core used (invariably, that would be EI imperial lamination size). Then look at the amp's circuit diagram and estimate its total power draw. How much margin did these amp makers allow?

Determining TPV (Turns-Per-Volt) and wire sizing

The formula for turns-per-volt for EI laminations is **$TPV=k/A_{EF}$** (effective cross section in cm^2). For 50Hz mains k=45, so we get TPV = 45/17 = 2.64. For 60Hz mains k=37.5, so we get TPV = 37.5/17 = 2.2.

Our design will be calculated for 240V, 50Hz mains in Australia. As an exercise, once you understand the methodology, repeat these calculations for your local mains voltage and frequency.

Primary for 240V/50Hz: $N_P=240*2.64= 634$ turns of 0.5 mm dia. wire

S4 & S5: 300B filament requires 5V@ 1.2A Since we are using a CT secondary, each half of the winding provides half of the load current, so the minimum would be 0.6A. Let's dimension it for 0.8A each half, or 1.6 A in total. That means these two windings will be operating at only 1.2/1.6 = 75% of their nominal current.

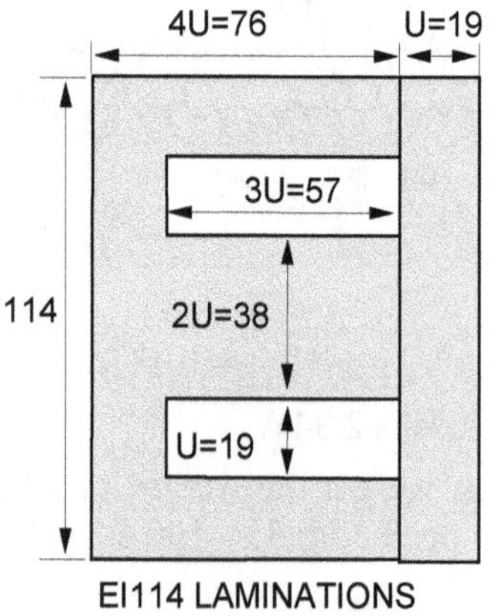

EI114 LAMINATIONS

TURNS-PER-VOLT FROM CROSS-SECTIONAL AREA (EI laminations)

$TPV \approx 45/A_{EF}$ (for 50Hz) or $TPV \approx 37.5/A_{EF}$ (for 60Hz)

$TPV = N_1/V_1 = 10^4/(4.44fBA_{EF})$

If we consider all factors except A_{EF} constant to get $TPV=k/A_{EF}$

For f=50 Hz and B=1T k=10,000/(4.44*50*1)=10,000/222 = 45

For f=60 Hz and B=1T k=10,000/(4.44*60*1)=10,000/266.4 = 37.5

For GOSS k=40 (50Hz), and for GOSS C-cores and toroid cores k = 30

Wire diameter d=√(1.27*I/J) = √(1.27*1.6/2.5) = 0.64mm, so we will use 0.6mm wire. We need 10V*2.64T/V = 26.4 turns .This assumes there is no loading effect; however, this voltage will drop up to 5-10% under load.

Also, we should allow another 10% as a margin. It is critical to have exactly 5V for 300B heaters! You can always drop some of the voltage off on a resistor, but you cannot raise the voltage if it's too low.

Add 20% on: 26.4*1.2 = 31.8, so we will use 32 turns.

S3: Using 12.6V heaters, the duo triodes we are using (12AU7) draw 0.15A of heater current each, and since there are three of them, we need 0.45 A current capacity. Let's keep things simple and allow for up to 1 A current draw. Now we can use the same wire as for the 300B heater, secondaries S4 and S5.

N_3 = 24V*2.64TPV = 64 turns. Since the winding is lightly loaded, don't add the 20% margin. It is better to use slightly lower heater voltages for preamp tubes anyway, since that prolongs their life and reduces noise.

S2: d=√(1.27*I/J) = √(1.27*0.1/2.5) = 0.225mm. However, the current load on the bias winding is not nearly 100mA as allowed, so we will use 0.16mm wire simply because we have this size in our workshop.

N_2 = 160V*2.64TPV = 422 turns

S1: The same voltage as S2, so we need 422 turns as well, but at a higher current, so the wire diameter needed is d=√(1.27*0.5/2.5) = 0.50mm, and that is what we will use.

The window length is WL = 57mm, and with 2mm bobbin thickness on each side, we get the coil length (CL) = WL - (2 x BT) = 57 - (2 x 2) = 57-4 = 53 mm, so we will work with 52 mm effective winding width. BT is the thickness of the wall of the bobbin.

PRIMARY TURNS-PER-LAYER:

TPL_{MAX}= 52/0.5 = 104

This is the maximum number of turns we can fit in one layer. 634 turns / 104 TPL = 6.1 layers

We will use 6 layers.

S1 TURNS-PER-LAYER:

TPL_{MAX} = 52/0.5 = 104

422 turns / 104 TPL = 4.06 layers

We will use 5 layers. If we get a higher voltage, that is OK, we can always reduce it after rectification and smoothing.

Actual TPL: 422/5 = 85

S2 TURNS-PER-LAYER: TPL_{MAX}= 52/0.16= 325

422 turns / 325 TPL = 1.3 layers

We will use 2 layers, so actual TPL= 422/2 = 211

S3 TURNS-PER-LAYER: TPL_{MAX} = 52/0.6 = 87 and since 64 turns/87TPL=0.74 layers, we will use a single layer.

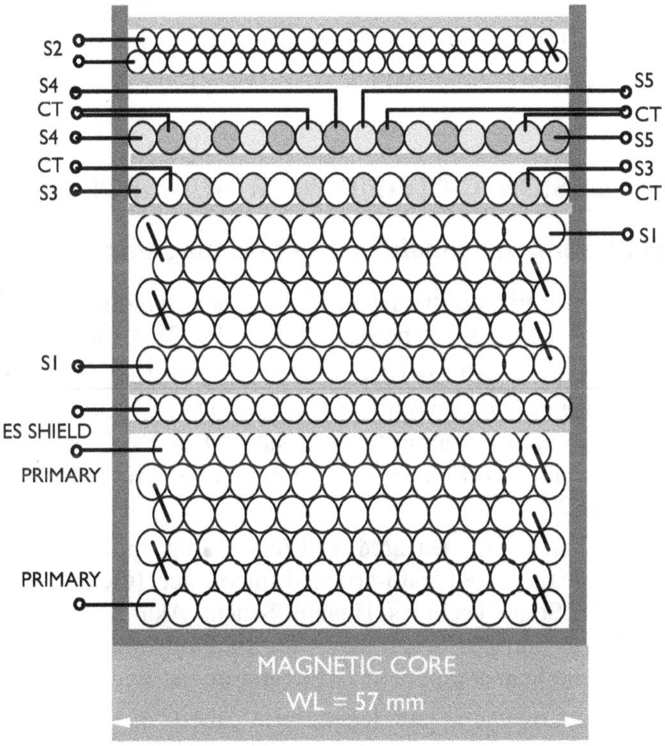

ABOVE: Detailed winding diagram indicating the order of winding, number of layers and insulation placement.

NOTE: The wire diameters are not to scale and the indicated number of turns per layer are not accurate, they are qualitative depictions only. The numbers of layers are accurate.

S4 and S5 TURNS-PER-LAYER: TPL_{MAX} = 52/0.6 = 87. Since 32 turns/87TPL=0.37 layers, we can fit both S4 and S5 onto the same layer.

ES is the electrostatic shield, a single full winding of any wire size but usually the same as the primary winding. Instead of a single layer of wire, a conductive nonferrous foil (aluminium or copper) can also be used. Its ends must not be joined (soldered); otherwise, it would constitute a shorted transformer turn!

ES shield reduces electromagnetic noise by diverting primary noise and interference to the ground. It also reduces the capacitive coupling between the primary and secondary windings. The electrostatic shield is earthed (grounded) at one end only.

BIFILAR WINDING: S3, S4 & S5 require a center tap. The two haves of the winding should be well balanced. The best way to make sure they are is to use bifilar winding.

Two identical wires are wound concurrently (in parallel), and the two opposite ends (not adjacent ends on the same side!) of each wire are then connected to form the CT! The two wires that need to be externally connected to the center tap are marked CT on the winding diagram.

DESIGN PROBLEM

A 200mA power transformer operating on 230V/50Hz mains needs to provide 700V on the center-tapped secondary (2x350V). Non-oriented silicon steel laminations (EI114 size) are used with a 4.5 cm stack. Calculate the number of primary and secondary turns and TPV (Turns-per-Volt).

What diameter wire should be used for primary and secondary winding? Assume power loss of 10% and a flux density B=0.9 Tesla.

The center-leg cross sectional area for EI114 laminations with a 45 mm stack is A = 3.8*4.5 = 17.1 cm^2. The effective area is A_{EF}=0.96A = 16.4 cm^2 so the power capability of this core is P = A^2 = 270 VA. Since the power delivered to the secondary is only P_2 = 700*0.2 = 140 VA, the chosen lamination stack is suitable.

$N_1=V_1 10^4/(4.44fBA)$ = 230*10^4/(4.44*50*0.9*16.4) = 702 turns, TPV = 702/230 = 3.05

The turns ratio is TR= V_1/V_2 = 230/700 = 0.3286, so $N_2=N_1/TR$ = 702/0.3286 = 702 * 3.05 = 2,141 turns. Thus, each half of the center-tapped secondary will have 1,070 turns (half of N_2).

With 200mA secondary current, the wire diameter needed is d=√(1.27I/J) = √(1.27*0.2/2.5) = 0.32mm

With 10% loss, $P_2=0.9P_1$ so primary power dissipation is $P_1=P_2/0.9$ = 140/0.9 = 156 VA

The primary current will be $I_1=P_1/V_1$ = 156/230 = 0.68 A, so its wire needs to be 0.6 mm in diameter!

DIY DESIGN: C-CORE POWER TRANSFORMERS FOR MONOBLOCK PP AMPLIFIERS

Evaluating the suitability of an existing transformer core or laminations

Two power transformers were needed for two monoblocks, each with a pair of PL519 power tubes producing up to 100 Watts in push-pull mode.

We had a few salvaged Japanese mains transformers from old color TVs. They used C-cores with "strip width" D=1.5" (38.1mm) and "build up" E=1" (25.4mm). The other dimensions were F=22.2mm, C=2E+F=73mm, G=63.5mm, and B=2E+G=114.3mm. Please refer to the C-core dimensions on page 22.

The transformers measured well (a very high primary inductance), so we were sure they used grain-oriented silicon steel (GOSS). Anyway, what was good for Sony, Technics, Hitachi, Sansui, Akai and other Japanese brands in the 70s and 80s, when they were running "total quality circles" around their American competitors, will certainly be fine in your DIY amplifier.

The mean magnetic path length, also called "the length of flux", was 25.1 cm, gross cross-sectional area A=ED = 9.68 cm^2, effective area A_{EF}=9.29 cm^2. A stacking factor SF=A_{EF}/A=9.29/9.68 = 0.96 was assumed).

A double-coil arrangement was used. The core's power rating was $P=(2A_{EF})^2=4A_{EF}^2$ = 345 VA.

As an interesting comparison, an EI-core with the same A_{EF}=9.29 cm^2 would have the power rating of only 86 VA, or four times lower.

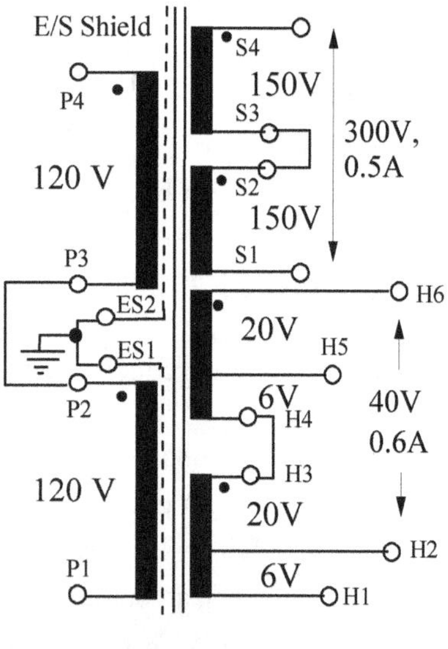

ABOVE: Electrical diagram including the required voltages

Determining TPV (Turns-Per-Volt) used

When you have a salvaged transformer in your hands, it is easy to do a bit of forensic work or "reverse engineering." Those old TV sets also used vacuum tubes, so we unwound the top 6.3 V_{AC} heater winding and counted 19 turns, meaning that the original designer used 19 Turns/6.3 Volts or 3 Turns-Per-Volt.

From the basic transformer equation **TPV=N/V=10^4/(4.44BAf)**, so substituting 3 for TPV, 50 for f, and 9.68 for A, we get the flux density that the original Japanese designer chose all those years ago: B=10^4/(3*2,065) =1.6T! While it seems high, this flux density is quite conservative for C-cores which can go up to 1.8T!

In our formula for TPV, factor 30 should be used for GOSS C-cores. Remember, a much higher factor of 45 is used for EI cores. For 60 Hz mains frequency, these factors are approx. 20% lower, 37.5 for EI cores and 25 for C-cores.

We carefully unwound the original coils and even reused most of the original winding wire and insulation. How is that for an eco-friendly approach?

Number of turns

We split all voltages in half and wound two identical coils. Each winding has a 120V primary, giving us the flexibility to connect them in series for 240V use, as illustrated, or in parallel to use in 120V mains countries. The HV secondary of 300V is made up of two 150V coils in series, as is the 40V for the heating of power tubes (2 x 20V in series).

PL519 or PL509 heater is rated at 40V @ 300 mA so we needed a minimum of 0.6A current capacity. A 6.3V heater voltage for the preamp tubes was also needed.

PRIMARY: 120 V x 3 TPV = 360 Turns, wire diameter 0.6mm; Electrostatic shield: one layer of d = 0.6 mm wire; S1: 150V x 3 TPV = 450 Turns, d = 0.5mm; S2: 20V x 3 TPV = 60 Turns, d = 0.6mm; S3: 6.5V x 3 TPV = 20 Turns, d = 0.6mm.

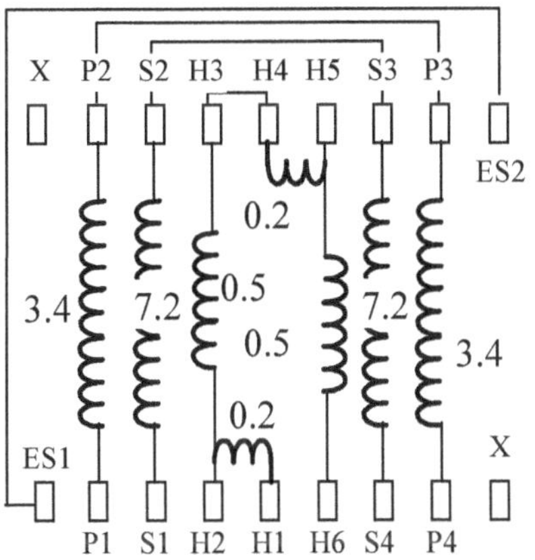

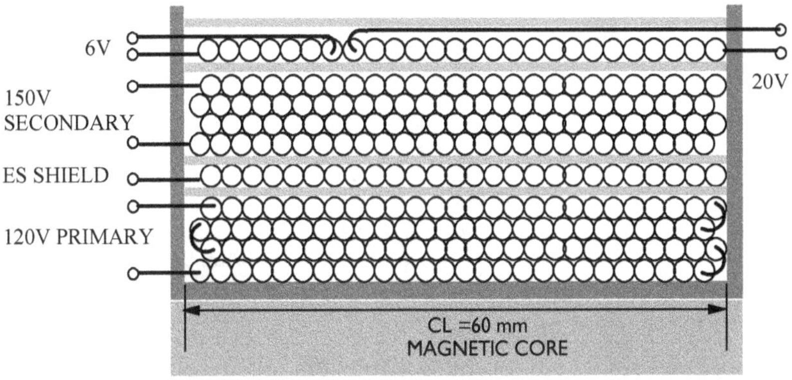

ABOVE LEFT: Connection diagram and measured DC winding resistances in Ω

ABOVE RIGHT: The cross-section of the wound coil. The other coil is identical.

Layering

The coil or winding length is CL=60mm. Maximum primary TPL= 60mm/0.6mm = 100. Since 360 turns / 100 TPL = 3.6 layers, we will use 4 layers with an actual TPL = 360/4 = 90!

SECONDARIES: S1: Maximum TPL (Turns Per Layer): 60mm/0.5mm = 120 We need 450 turns, so 450/120 = 3.75 layers. We will work with 4 layers. Actual TPL = 450/4 = 112

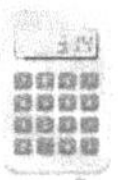

WIRE SIZING

$d=\sqrt{(1.27*I/J)}$

d=wire diameter [mm], I=current [A]

J=current density in A/mm^2

for $LI^2<0.2$: J=2.0 - 2.5

for $LI^2>0.2$: J=1.5 - 2.0

S2 and S3: TPL_{MAX}=60/0.6 = 100

60/100 = 0.6 and 20/100 = 0.2, so S2 and S3 together occupy 0.8 or 80% of the coil length CL, meaning we can wind both in one layer (top layer).

NOTE: For 115-120 V_{AC} mains operation, the two primary halves would be connected in parallel (the ends with a dot joined and two ends without the dot joined together).

TURNS-PER-VOLT FROM CROSS-SECTIONAL AREA OF C-CORES

TPV≈30/A (for 50Hz)

TPV≈25/A (for 60Hz)

The efficiency of power transformers used in tube amplifiers

One of the most important and most often used shortcuts during the design of a small power transformer is the initial assumption about the efficiency of the finished product. Since efficiency depends on losses and their relative ratio to the output power, it is intuitive to assume that the efficiency will increase with higher power ratings. The graph on the right will save you lots of time and iteration steps in your design efforts.

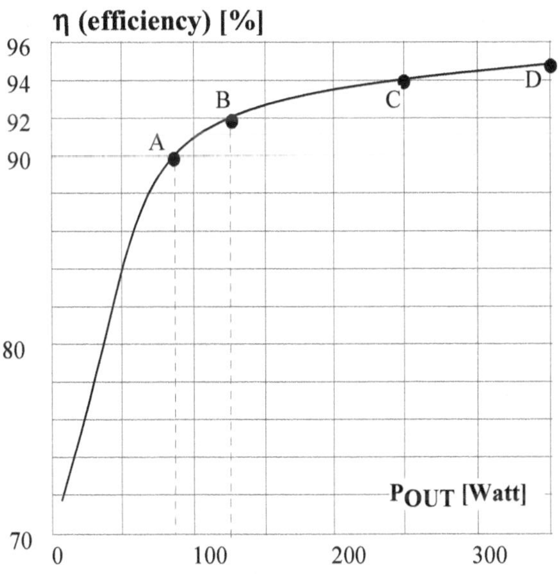

We see that 90% efficiency is reached at about 75 Watts (A), while 125 W transformers are approx. 92% efficient (B). Since most power transformers in hi-fi amps are way above 100 Watts, of most interest are points C (94% efficiency is reached once output power reaches 250 Watts) and D, indicating a 95% efficiency of 350 W transformers.

MAINS TRANSFORMER DESIGN SEQUENCE

Experienced transformer designers can usually accurately forecast the required size of the laminations and the stack thickness for a mains transformer of a particular rating. Beginners may have to go through two or three iteration steps outlined in this design sequence.

Unless you are extremely gifted or talented, the intuitive feel for things does not happen overnight. As with most other things in life, the more you practice, the better you'll get.

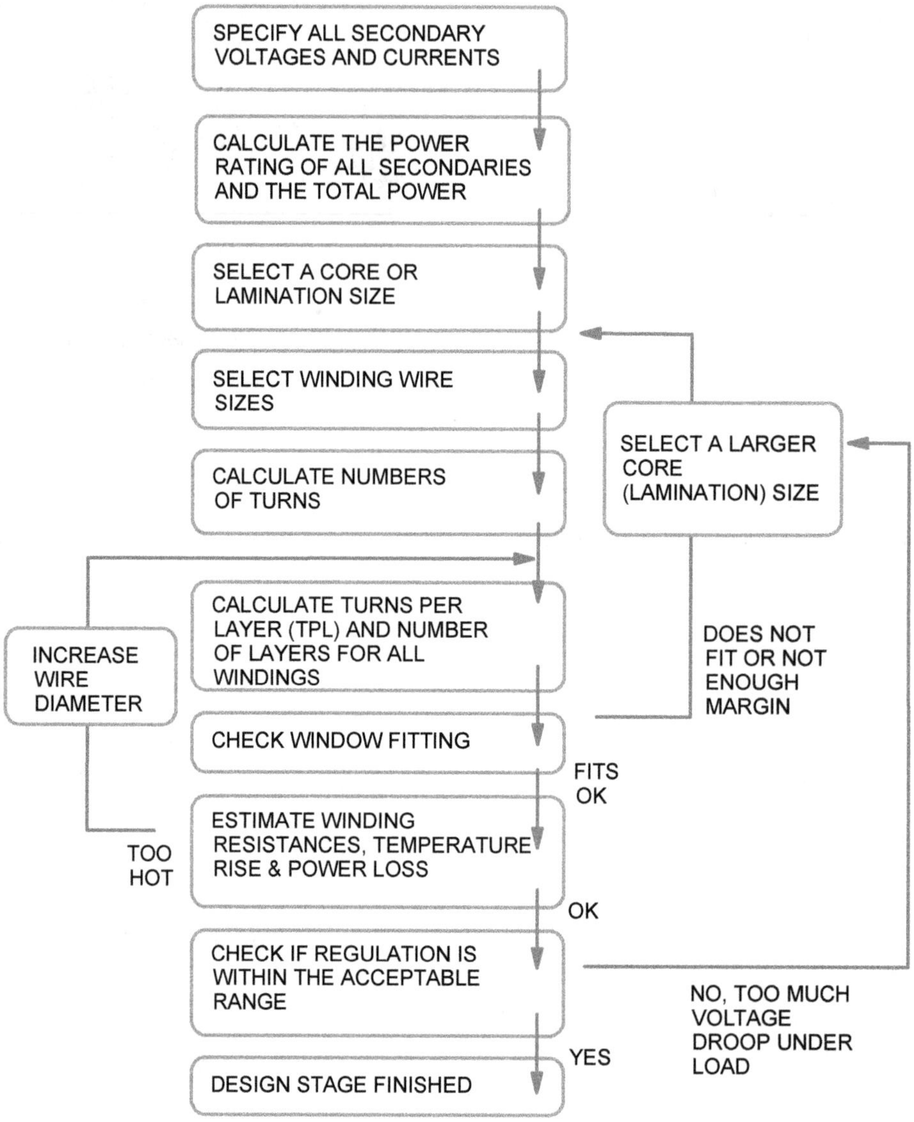
SPECIFY ALL SECONDARY VOLTAGES AND CURRENTS
CALCULATE THE POWER RATING OF ALL SECONDARIES AND THE TOTAL POWER
SELECT A CORE OR LAMINATION SIZE
SELECT WINDING WIRE SIZES
CALCULATE NUMBERS OF TURNS
SELECT A LARGER CORE (LAMINATION) SIZE
CALCULATE TURNS PER LAYER (TPL) AND NUMBER OF LAYERS FOR ALL WINDINGS
INCREASE WIRE DIAMETER
DOES NOT FIT OR NOT ENOUGH MARGIN
CHECK WINDOW FITTING
FITS OK
ESTIMATE WINDING RESISTANCES, TEMPERATURE RISE & POWER LOSS
TOO HOT
OK
CHECK IF REGULATION IS WITHIN THE ACCEPTABLE RANGE
NO, TOO MUCH VOLTAGE DROOP UNDER LOAD
YES
DESIGN STAGE FINISHED

PHYSICAL FUNDAMENTALS OF AUDIO TRANSFORMERS

- THE REAL TRANSFORMER MODEL & VECTOR DIAGRAMS
- FREQUENCY RESPONSE OF TRANSFORMER-COUPLED OUTPUT STAGES
- THE SHUNT-FED OUTPUT STAGE
- LEAKAGE INDUCTANCE
- THE INSERTION LOSS (ATTENUATION) OF AN AUDIO TRANSFORMER

6

THE REAL TRANSFORMER MODEL & VECTOR DIAGRAMS

Now that we understand transformer principles and the physical properties of magnetic cores, we can delve into the behavior of transformers operating across a wide band of audio frequencies, namely input, output, and interstage transformers, collectively called audio transformers. Everything we said so far still applies to audio transformers.

For instance, single-ended audio transformers have an air gap just as chokes do, and power (mains) transformers can be considered a special case of an audio transformer operating at a single fixed (mains) frequency, 50 or 60Hz, which are indeed audio frequencies!

Real transformer windings use copper wire, and such wire has resistance. R_P in the model represents the DC resistance of the primary winding and R_S of the secondary winding.

Not all flux produced by the primary current goes through the secondary coil. The same applies to the secondary coil. This "leakage" of flux into the surrounding air is modeled by a leakage inductance (symbol without the magnetic core). L_{LP} is the leakage inductance of the primary winding, and L_{LS} is the leakage inductance of the secondary.The excitation or primary current i_1 has two components. Current i_M is needed to magnetize the core. L_P is the primary inductance. The other component of the primary current, i_C, is the core-loss current.

Together with resistor R_C it symbolizes core power losses due to eddy currents and hysteresis.

"Copper losses" is the power dissipated in R_P and R_S. While R_P and R_S can easily be intuitively understood and measured, the R_C and L_P in the vertical branch are not real (actual) components. They are a symbolic representation of the physical phenomena in the magnetic core.

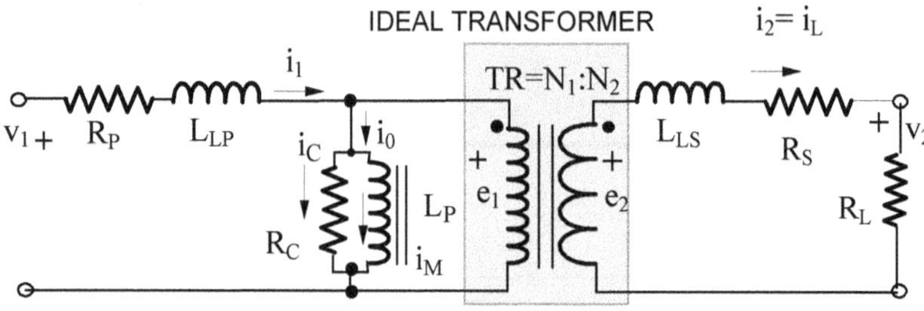

ABOVE: The equivalent diagram of an audio transformer at low and midrange frequencies.

No-load vector diagram

No-load conditions: $i_1=i_0$ and $i_L=0$. Without the load on secondary ($i_L=0$), the idle current i_0 is a vector sum of the magnetizing current i_M and the core loss current i_C. The two primary voltage drops, i_0R_P and i_0L_{LP} are so small (shown here much larger for clarity purposes) that the difference between $-e_1$ and V_1 is less than 0.1%, so they can be neglected.

i_C is in phase with $-e_1$, while i_M lags the induced voltage $-e_1$ by 90 degrees ($\pi/2$) and is in phase with the flux Φ.

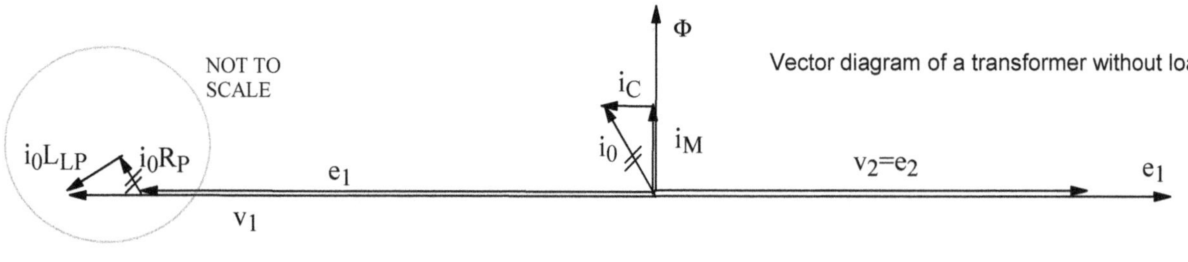

Vector diagram of a transformer without load.

Vector diagram of a transformer with an inductive load (a dynamic loudspeaker)

With the secondary current flowing through the load and the secondary winding, instead of i_0, we now have i_1 on the primary side, so all voltage drops are higher. A lagging power factor is assumed, meaning that the secondary current i_2 lags the secondary voltage v_2; this angle is named θ_2.

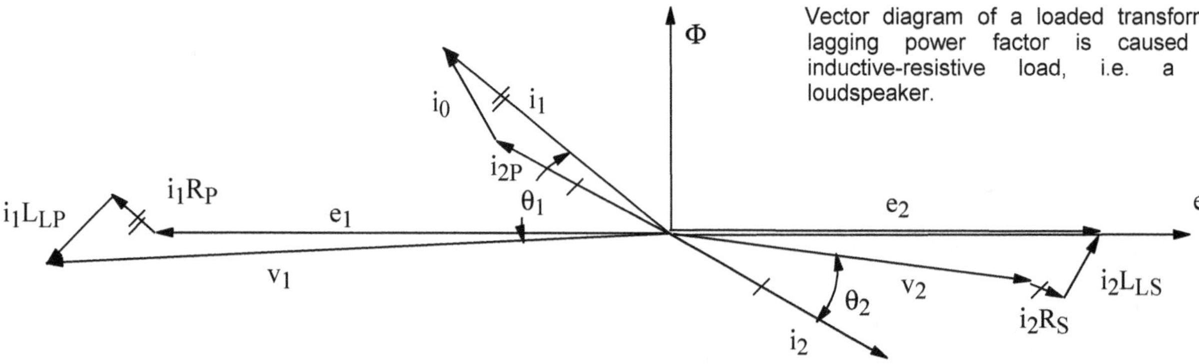

Vector diagram of a loaded transformer. The lagging power factor is caused by an inductive-resistive load, i.e. a dynamic loudspeaker.

The load $X_L = R+\omega L$ is partially inductive, for instance, a typical dynamic loudspeaker. With a capacitive or partially capacitive load (such as electrostatic speakers), the current would be leading the voltage, and the power factor would be leading, so the vector diagram would look different (draw it as a practice exercise).

This secondary current is reflected onto the primary side as i_{2P}, and together with i_0 makes the primary current i_1. The angle between v_1 and i_1 is θ_1.

FREQUENCY RESPONSE OF TRANSFORMER-COUPLED OUTPUT STAGES

Parasitic capacitances

The model is complicated by another type of transformer imperfection, the capacitances between each winding and the core and distributed capacitance between adjacent turns, between layers within each winding, and between windings. Each layer and the whole winding acts as a capacitor, and the insulation between layers or windings is its dielectric. By winding your transformer, you have just wound a few unwanted capacitors!

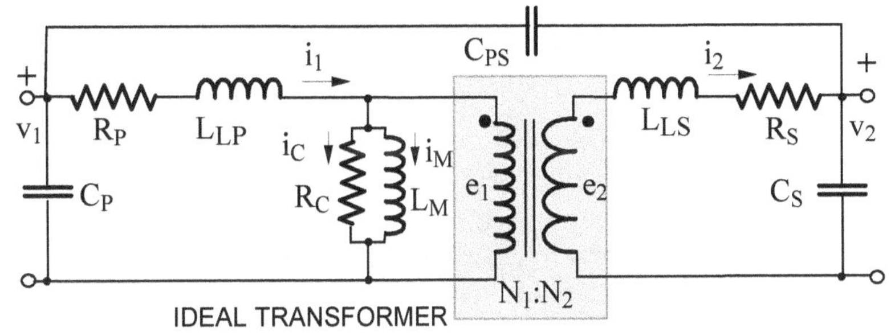

The complete equivalent diagram of an audio transformer

The first simplification is representing all these distributed capacitances with only three lumped parameters: C_1 (primary winding's capacitance), C_2 (capacitance of the secondary), and C_{PS} (between the primary & secondary windings). From the circuit analysis point-of-view, C_{PS} complicates things a lot; the whole circuit becomes a messy network requiring complex mathematical modeling. Luckily, in the next approximation, it can be omitted. Likewise, the C_P and C_S are irrelevant for power transformers and have little impact on the behavior of output transformers. They are mostly of interest in the design of input and interstage transformers.

All values reflected onto the primary side

It is convenient to reflect all values onto either the primary or the secondary side. We will reflect all secondary parameters to the primary side, meaning that our revised model will eliminate the shaded ideal transformer featured in the original model.

Both RS and LS will be multiplied by the IR (impedance ratio) or the square of the turns ratio $TR^2 = (V_1/V_2)^2$. This secondary shunt capacitance reflects over to the primary as divided by TR^2. The secondary voltage is reflected onto the primary side as $TR*V_2$ while the secondary current i_2 reflects as i_2/TR!

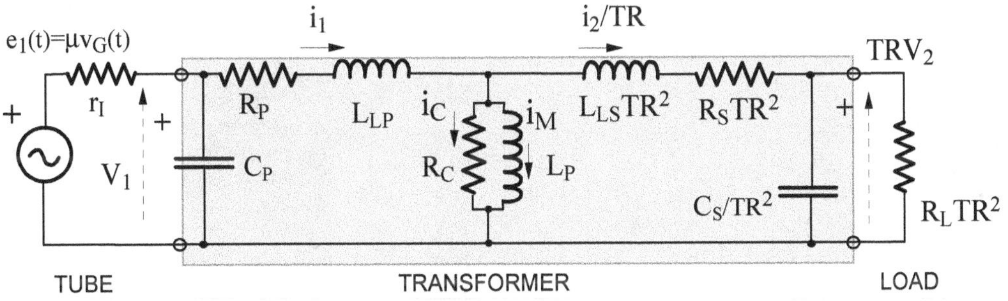

The complete equivalent diagram of an audio transformer with secondary parameters reflected onto the primary side, including the source (driver tube) and load.

The midrange model of an audio transformer

At midrange frequencies (approx. 200 Hz to 5kHz), the impedances of all reactive components (inductances and capacitances) can be neglected. Those in series (leakage inductances) present a minuscule impedance at such low frequencies and can be replaced by a short circuit. The impedances of reactive components in parallel, such as the capacitances and the primary inductance, are very high and can be replaced by an open circuit.

The transformer is simply two resistances in series (R_P and R_STR^2), so the whole output stage of an amplifier acts as a voltage divider.

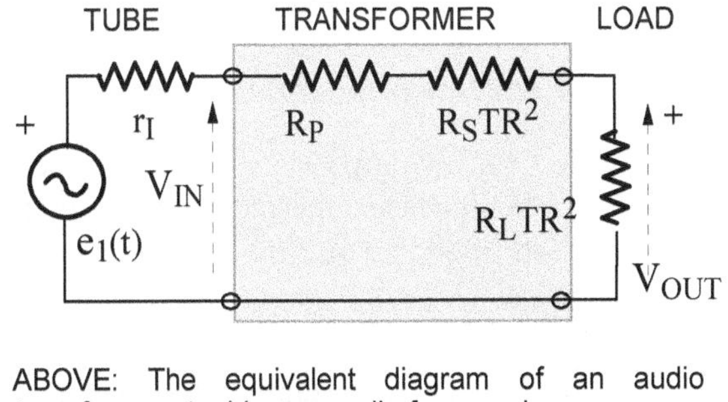

ABOVE: The equivalent diagram of an audio transformer at midrange audio frequencies.

The lower the internal resistance of the tube r_I and the lower the resistance of the transformer's windings, the better, meaning the less of the audio signal will be wasted on those resistances, and the higher percentage of it will be transferred to the loudspeaker or the grid of the following tube (for interstage transformers).

The LF model of an audio transformer

At low frequencies, the capacitive effects and leakage inductances have no bearing on the response since their impedances (reactances) at such low frequencies are negligible. However, the primary inductance L_P is now the most important parameter. Together with R_1 and R_2, it attenuates low frequencies. $R_1=r_I+R_P$ and $R_2=(R_S+R_L)TR^2$, and their parallel combination is $R_{PAR}=(R_1R_2)/(R_1+R_2)$. R_P can be neglected for high impedance valves, and for good quality output transformers, R_S is much lower than the loudspeaker impedance, so $R_1 \approx r_I$ and $R_2 \approx R_L$.

Of most interest is the frequency at which the inductive reactance or impedance Z_{LP} becomes equal to R_{PAR}. The output voltage will drop to $1/\sqrt{2} = 0.71$ (71%) of its midrange level at that frequency, called lower -3dB frequency, lower half-power frequency, or f_L. Together with an upper -3dB frequency f_U, it defines the bandpass of a transformer or a whole amplifier.

$Z_{LP}= \omega L_{PMIN}= 2\pi f L_{PMIN} = (R_1+R_2TR^2)/(R_1R_2TR^2)$

$f_L=(R_1+R_2TR^2)/(R_1R_2TR^2)/(2\pi L)$ or

$L_{PMIN} = (R_1+R_2TR^2)/(R_1R_2TR^2)/(2\pi f_L)$

If we use R_{PAR}, then we simply have $f_L= R_{PAR}/(2\pi L_{PMIN})$ and finally, $L_{PMIN} = R_{PAR}/(2\pi f_L)$.

L_{PMIN} is the minimum primary inductance value needed to get the lower -3dB frequency down to the desired level. This is the first choice we need to make when designing audio transformers! The table gives you approximate values of the primary inductance L_{PMIN} for a dozen or so commonly used power tubes for three f_L frequencies (5, 10, and 20 Hz).

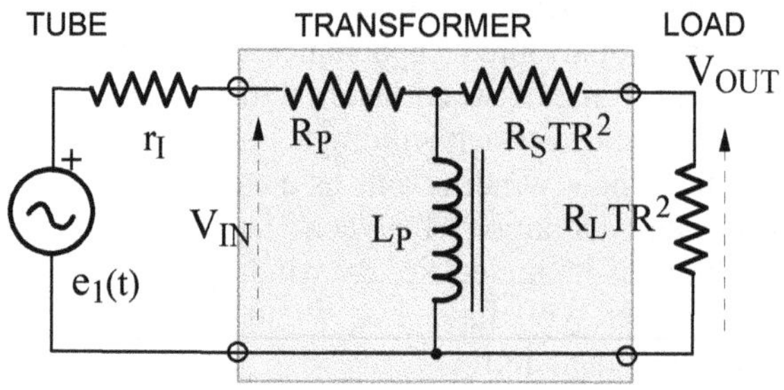

ABOVE: The equivalent diagram of an audio transformer at low frequencies.

BELOW: The simplified LF model

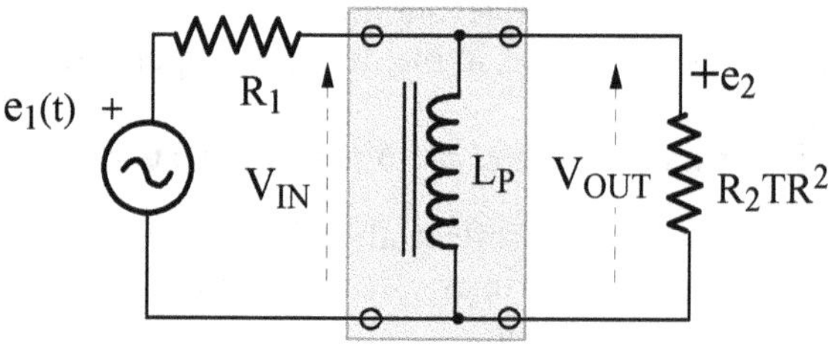

* Both sections in parallel.

\# The listed values for primary impedance Z_P are not necessarily optimal, but commonly used in practice.

				f_L (Hz)		
				20	10	5
TUBE	r_I (Ω)	Z_P (Ω)#	R_{PAR} (Ω)	L_P (H)	L_P (H)	L_P (H)
6C33C-B	100	1,000	91	0.7	1.4	2.9
6080 (6AS7)*	150	1,000	130	1.0	2.1	4.2
PL519 (triode)	275	1,500	232	1.9	3.8	7.6
2A3	700	2,500	547	4.4	8.7	17.4
300B	800	3,500	651	5.2	10.4	20.7
KT88 (6550)	1,300	4,500	1,009	8.0	16.1	32.1
EL34 (triode)	1,500	4,500	1,125	9.0	17.9	35.8
6L6 (triode)	1,700	4,500	1,234	9.8	19.6	39.3
845	1,700	8,000	1,402	11.2	22.3	44.7
6BQ5 (triode)	2,000	5,000	1,429	11.4	22.7	45.5
SV572-10	2,100	6,500	1,587	12.6	25.3	50.5
211	3,300	10,000	2,481	19.8	39.5	79.0

LEFT: The required output transformer's primary inductance as a function of desired lower -3dB frequency for common triodes and triode-connected pentodes. The values are for 5, 10 and 20 Hz, but inductances for any other frequency can be calculated.

$L_{PMIN} = R_{PAR}/(2\pi f_L)$

$R_{PAR}=(r_IZ_P)/(r_I+Z_P)$

The beauty of low impedance power tubes

Transformers for low impedance triodes are the easiest and fastest to make since the number of primary turns can be low. Thus, a beginner should start with designing and winding a transformer of the low impedance kind. Tubes with low r_I, such as 6080 and 6C33C-B triodes, require low values of L_P to achieve low f_L compared to high impedance tubes such as 6BQ5 pentode or directly-heated transmitting triodes (or pentodes connected as triodes) such as SV572 series, 211, 811, 813, GM70 and 845.

The primary impedance of output transformers for these tubes is usually 10 kΩ or more, and such transformers are hard to wind because they require a high number of turns. The achieved primary inductance is usually low, resulting in a relatively weak bass.

The HF model of an audio transformer for triodes

If C_S is low, or if the internal resistance of the power tube is low, as is the case with triodes, we can simplify the model further by neglecting the impact of the shunt capacitances.

The circuit is now a simple 1st order low-pass LR-type filter with the upper -3dB frequency f_U, at which the reactance of the total leakage inductance L_L becomes equal to the resistance of the two resistances in series: $R_{SER}=r_I+R_1+R_2TR^2$ and $L_L=L_{LP}+L_{LS}TR^2$

For -3dB attenuation of the midrange level $Z_L=R_{SER}$ or $2\pi f_U L_L=R_{SER}$ so $f_U=R_{SER}/(2\pi L_L)$ or $L_{LMAX}=R_{SER}/(2\pi f_U)$.

This inductance is the maximum value of the permissible leakage inductance to achieve the desired f_U. Any larger value of L_L will lower the upper -3dB frequency below the desired level.

The output transformer behaves as a first-order system (an LR filter) at high frequencies. $R_2=R_S+R_L$ (the secondary winding's DC resistance plus load resistance), r_I is the internal impedance of the output tube.

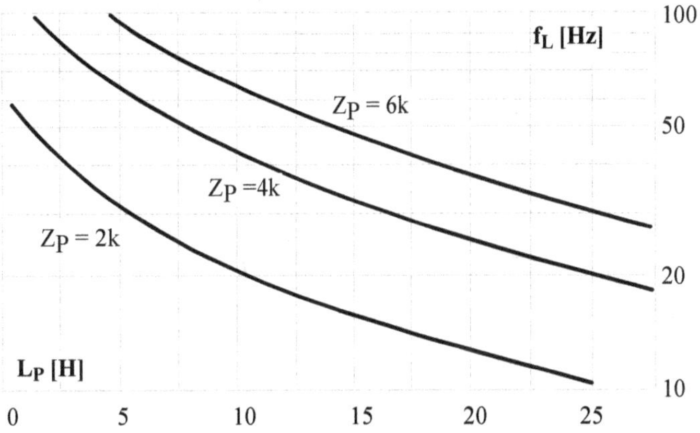

ABOVE: The lower -3 dB frequency for various primary impedances as a function of the output transformer's primary inductance L_P

BELOW: The simplified HF model of a transformer-coupled triode output stage, neglecting the effect of stray/shunt capacitances.

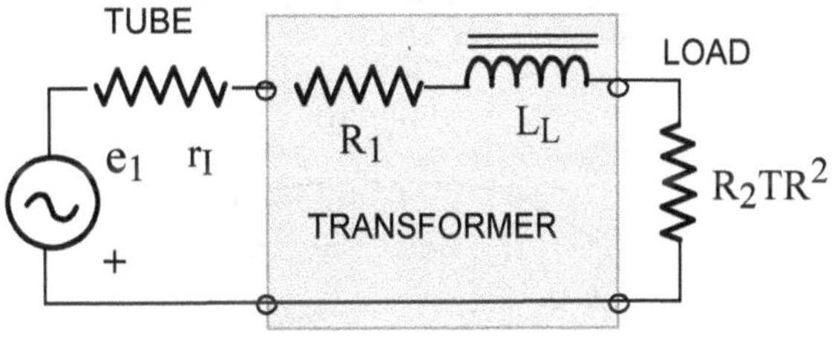

The HF model of an audio transformer for pentodes and beam power tubes

At high frequencies, the reactance of L_P is very high, so it represents practically an open circuit. Again, we can lump the internal tube resistance and primary winding resistance together ($R_1=r_I+R_P$) and, likewise, the load resistance and secondary winding resistance: $R_2=R_S+R_L$. We can also reflect the secondary leakage inductance onto the primary side and get the total leakage inductance $L_L= L_{LP}+L_{LS}TR^2$.

For step-down transformers, it is more appropriate to place the stray capacitance C_{EQ} on the high impedance side of the leakage inductance (the primary side) since it has a greater effect in high impedance circuits. Once the secondary shunting capacitance is reflected onto the primary side $C_{EQ}=C_P+C_S/TR^2$.

Because of the relatively large internal resistance of pentodes and beam tubes, this shunt capacitance cannot be neglected as it can with triodes.

The leakage inductance and secondary shunt capacitance form a series LC circuit with its resonant frequency f_R. Depending on its quality factor ($Q = \omega L/R$), this circuit will produce a more or less pronounced peak in the transformer's and amplifier's frequency response.

In our case $Q = \omega L_L/R_1$ where $R_1=r_I+R_P$, or $Q = 2\pi f_R L_L/R_1$! This effective resistance R often differs markedly from the DC resistance because of associated AC effects such as eddy currents and the skin effect. Therefore, R varies in a highly-complex manner with frequency.

The two models are illustrated on the next page.

The graph shows that higher primary impedances can tolerate lower-quality output transformers (higher L_L). To achieve f_U of 20 kHz with 6k primary impedance, leakage inductance can be relatively high 50 mH (A). To get the same f_U with Z_P=2k, L_L must be below 17 mH.

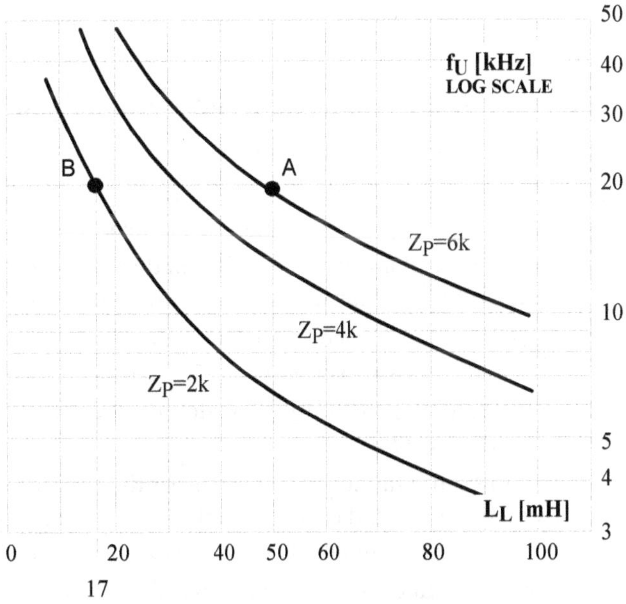

ABOVE: The upper -3 dB frequency for various primary impedances as a function of the output transformer's leakage inductance L_L

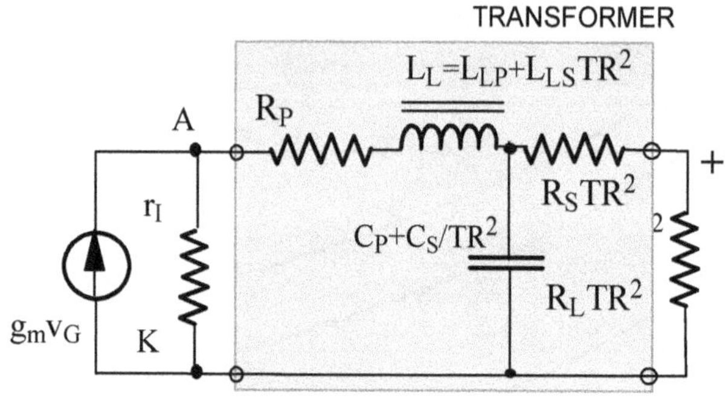

The high frequency model of a pentode output stage

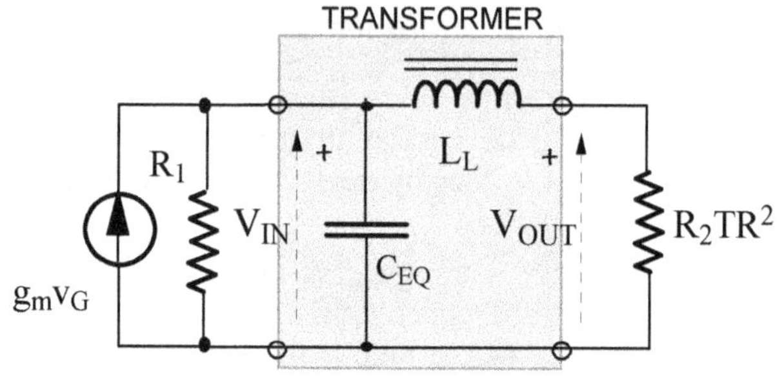

The high frequency model of with stray capacitance C_{EQ} moved to the high impedance side of L_L (the primary side).

The universal frequency and phase response curves of an output transformer

If we look at the values of R_{PAR} and R_{SER}, R_{SER} is usually about 5 times larger than R_{PAR}. The equations $f_L = R_{PAR}/(2\pi L_P)$ and $f_U = R_{SER}/(2\pi L_L)$ then become $f_L \approx R/(2\pi L_P)$ and $f_U \approx 5R/(2\pi L_L)$. If we divide the two equations, we get $f_U/f_L \approx 5L_P/L_L$. This is an extremely important conclusion. It says that an audio transformer's frequency range or bandwidth depends on the ratio of the primary inductance and the leakage inductance between the primary and secondary windings! The higher the L_P and the lower the L_L, the wider the transformer's frequency range!

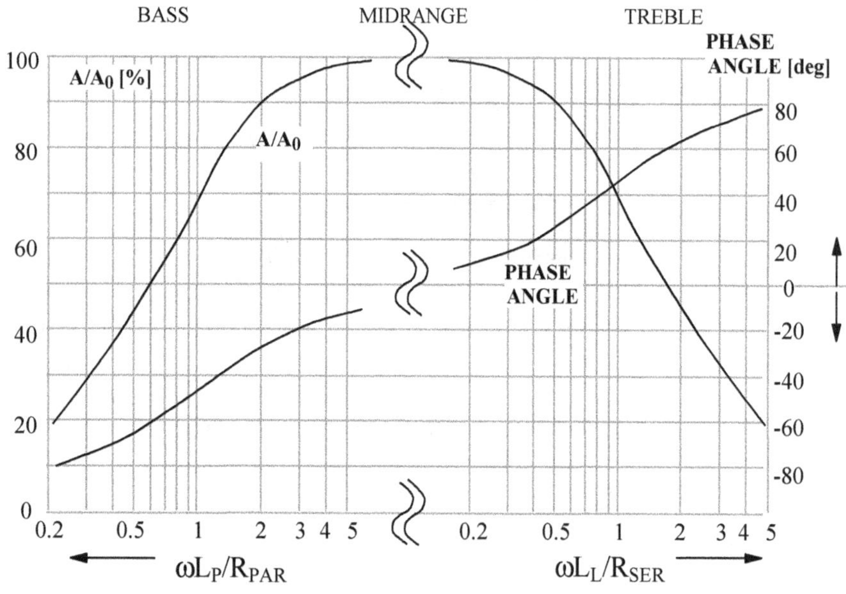

V_{OUT} lags V_{IN}

V_{OUT} leads V_{IN}

The universal attenuation and phase curves for an audio transformer

A = voltage gain at any frequency ω, A_0 = midrange gain

$R_{PAR} = (R_i + R_P)||[(R_S + R_L)TR^2]$

$R_{SER} = R_i + R_P + (R_S + R_L)TR^2$

THE FREQUENCY BANDWIDTH OF AN AUDIO TRANSFORMER

$f_U / f_L \approx 5L_P / L_L$

$f_U / f_L \approx 5\mu_R A^2/(\ell_{MP} * V_W)$

Since $\mathbf{L_P = N_P^2 \mu_R \mu_0 A/\ell_{MP}}$, and $\mathbf{L_L = \mu_0 N_P^2 V_W/A}$, by substituting those formulas into $f_U/f_L \approx 5L_P/L_L$ and dividing the two equations we get $f_U/f_L \approx 5\mu_R A^2/(\ell_{MP} * V_W)$.

Note that V_W here is not voltage but the volume of the transformer's winding, A is the cross-section of the core's center leg, and ℓ_{MP} is the mean length of the magnetic path, all three determined by the chosen lamination size. So, the frequency range of an audio transformer depends on its relative permeability (effective permeability if there is an air gap!), the size of the transformer core, and the geometry of the windings.

Using larger laminations and thicker stacks increases cross-sectional area A. The bandwidth increases with A^2, which is great news. However, ℓ_{MP} also increases, as does V_W, those two all but canceling any impact of larger A^2.

So, the only realistic possibility of widening the bandwidth is to use magnetic materials with as high effective permeability as possible!

The graph shows attenuation at both ends of the spectrum in terms of the parameters discussed and how the phase shift between the input and output signal becomes larger and larger at frequency extremes. Clearly undesirable, the phase shift is one of the main causes of distortion and instability, especially if negative feedback is used.

The primary inductance should be as high for superior bass and the tube's internal impedance as low as possible. That is easier to achieve using low impedance tubes such as triodes and transformers with low impedance ratios.

For wide frequency range on the treble side, the leakage inductances should be as low and the internal source resistance and reflected load impedance should be as high as possible.

This favors beam tetrodes and pentodes compared to low impedance triodes. Another contradiction of the audio field raises its ugly head. You cannot have your cake and eat it too! Remember, tetrodes and pentodes were invented precisely for that reason, to overcome the poor high-frequency extension of triodes.

THE SHUNT-FED OUTPUT STAGE

The main problem with single-ended output transformers is the high primary DC current which magnetizes the core and lowers primary inductance L_P, which should be as high as possible to get a good bass response.

The primary DC current is the main reason single-ended transformers need to be large, so lots of primary windings can fit in their windows to get a decent primary inductance and a decent bass response. However, this increases leakage inductance and parasitic capacitances, worsening the high-frequency response.

The parafeed arrangement (from the *para*llel *feed*) eliminates DC current from the output transformer. The coupling capacitor C prevents the DC plate current from flowing through the transformer's primary; there is no DC magnetization, and LP is increased significantly. The transformer can be made smaller and with no air gap.

Leakage inductance and parasitic capacitances are reduced, and bandwidth is widened. However, we need an anode choke now, which must be as big as the output transformer since all the plate current flows through it. It has to be designed and dimensioned properly, including the air gap.

Also, we introduce that unfortunate coupling capacitor that will color the sound. This capacitor forms a resonant circuit with the inductance of the anode choke, clearly an undesirable situation. So, as always in electronics and life, we solve one problem but create two or three new ones!

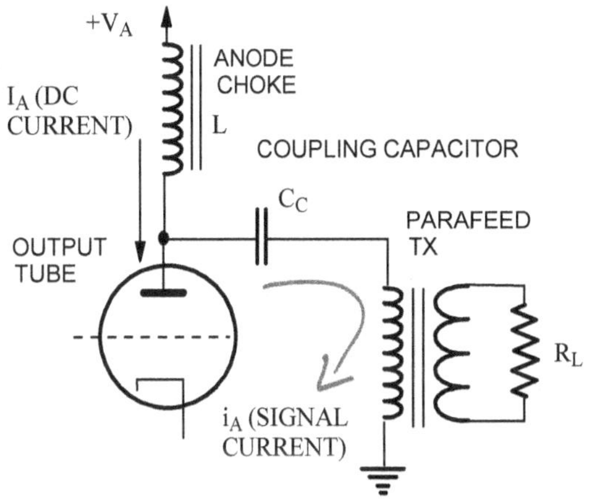

ABOVE: The shunt-fed stage or "parafeed" circuit: No air gap in the output transformer, but anode choke and coupling capacitor are now required.

The LF model of a shunt-fed output stage

The model at low frequencies includes the resistances of the primary and the resistance of the secondary winding reflected onto the primary side and the reflected load resistance. In reality, the load is not a pure resistance but a complex impedance. The model can be simplified by neglecting the reactance of the coupling capacitor C_C at such low frequencies and by combining R_P with it with r_I into one resistance R_1.

The equivalent inductance of L in parallel with L_P is $L_{EQ} = L_P||L = L_PL/(L_P+L)$.

This is bad news since the inductance of this parallel combination is smaller than either inductance by itself. If the driver or output tube is loaded with impedance Z_L, the minimum equivalent inductance L_{EQ} of the choke and the transformer's primary in parallel is $\mathbf{L_{EQ}= Z_L/(2\pi f_L)}$, where f_L is the lower -3 dB frequency.

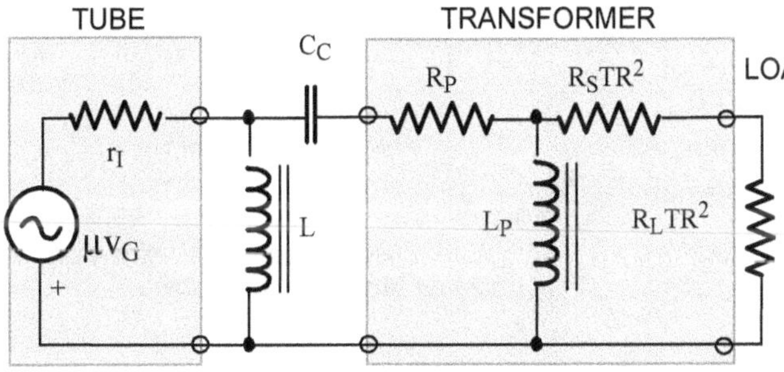

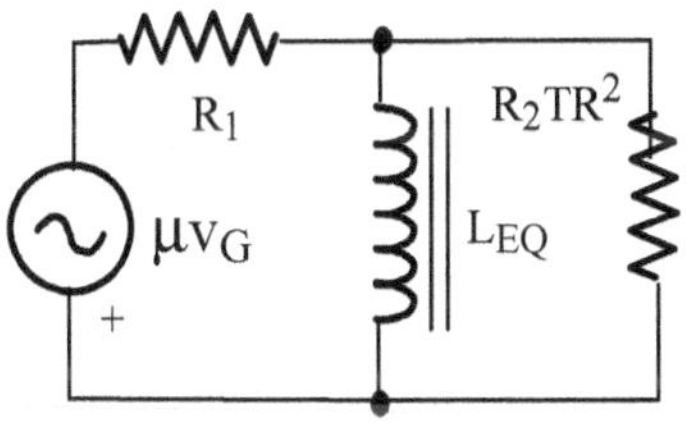

The low frequency model of shunt-fed stage (LEFT) and the simplified or combined model (ABOVE)

HIGH DEMANDS PLACED ON THE ANODE CHOKE AND OUTPUT TRANSFORMER CALCULATION

Determine the required value of anode choke's and output transformer's primary inductance for a parafeed 300B SET amplifier with 3,500Ω load to achieve a lower -3dB frequency of 20 Hz.

$L_{EQ}= Z_L/(2\pi f_L) = 3{,}500/2*3.14*20 = 27.9H$

To simplify things, assuming equal inductances of the two magnetic components, $L=L_P$, so $L_{EQ}= L/2 = L_P/2$, thus $L=L_P=2L_{EQ}= 55.7H$! This is a very high inductance, difficult to achieve in the output transformer without the gap, let alone in a choke *with* an air gap.

LEAKAGE INDUCTANCE

The physical character of leakage inductance

Leakage inductance is a parameter that indicates how much magnetic flux between the primary & secondary winding is not coupled through their shared magnetic core but rather leaked into the surrounding air. Ideally, L_L would be zero, meaning there is no leakage, and the windings share 100% of the flux.

As the audio signal's frequency and thus the magnetic flux it creates in the audio transformer increases, L_L also increases; that is why it should be measured using the highest test frequency.

To understand leakage inductance and the empirical formulas for its estimation (since the exact calculations are impossible), consider the simplest transformer structure with one primary and one secondary winding. CL is the coil length, d is the distance between the windings, or the thickness of the insulation between them.

For simplicity's sake, let's consider the windings as two flat conductive sheets, with currents $N_P I_P$ and $N_S I_S$ flowing through them. If CL is the length of the sheets, then the strength of the magnetic field between them is $\mathbf{H=N_P I_P/CL}$, and the energy accumulated in the magnetic field between the two conductive sheets is $\mathbf{E=\mu_0 \mu_R H^2 V/2}$.

Since the leakage flux circulates mostly through the air around the coils and not through the core, $\mu_R=1$ for air, so we have $E=\mu_0 H^2 V/2$, where V is the volume between the two windings. When the secondary is short-circuited, the same energy is accumulated in the leakage inductance and can also be expressed as $E=L_L I_P^2/2$. From these two equations, we can express L_L as $\mathbf{L_L=\mu_0 H^2 V/I_P^2}$.

If we substitute $N_P I_P/CL$ for H, $L_L=\underline{\mu_0 V/CL^2} N_P^2 = \underline{k} N_P^2$. Factor k is constant for a certain transformer, and L_L increases with a square of primary turns (meaning fast!), which is bad news.

From this important equation, we see that leakage flux does not depend at all on the permeability or the induction levels, but it does depend on the size of the laminations since it varies inversely with the coil length CL. It also increases as the volume of the windings increases.

Finally, the thicker the insulation, the larger the L_L. That is why we prefer bulk winding to layered arrangements, where insulation between layers quickly adds up to a significant thickness, resulting in much higher L_L!

As we increase N_P to increase the primary inductance L_P and get better bass, we also increase the leakage inductance L_L and affect the transformer's high-frequency response and the whole amplifier.

So, the two requirements are clearly contradictory. As we lower f_L we are also lowering the f_U, keeping the bandwidth of a transformer more or less constant.

The smaller the transformer, the lower these leakage inductances and parasitic capacitances usually are.

So, you can now see the frustrating compromise needed in the design and construction of audio transformers. The transformer should be as large as possible to get good bass (superior low-frequency reproduction). However, it should not be physically too large for an extended high-frequency range (sparkling and detailed treble).

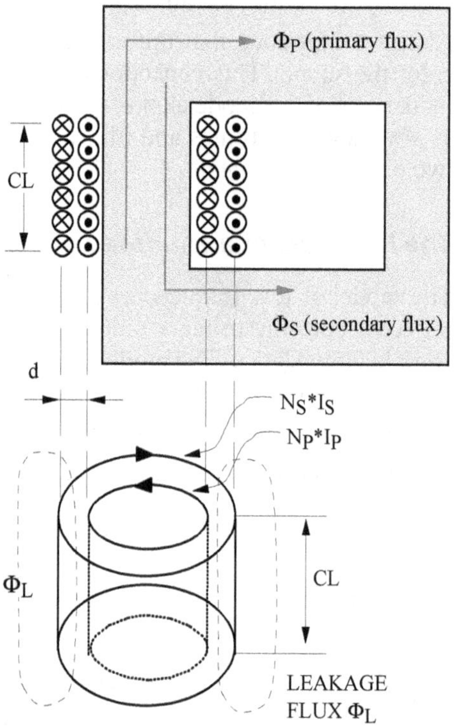

ABOVE: The simplified audio transformer structure used to derive an approximate formula for leakage inductance

Vertical and horizontal sectionalizing

Depending on the transformer's purpose and usage, the designer has to decide on the optimal winding arrangement to minimize leakage inductance and shunt capacitance. The leakage inductance can be reduced by sectionalizing, or interleaving the primary and secondary windings in horizontal layers, stacked one on top of the other.

However, by solving one problem (or at least by keeping it under control), we create another.

The more layers we have, the more paralleled capacitors we create, and the higher the shunt capacitance! To reduce the stray capacitance, we can sectionalize the secondary windings (or even the whole bobbin, including the primary) vertically.

Alas, such division increases the leakage inductance, requires more complex bobbins, and slows down the winding process, increasing cost. Also, the total coil length is reduced due to the added multiple thicknesses of bobbin walls so that fewer turns can be wound. These two methods are illustrated on page 40.

THE INSERTION LOSS (ATTENUATION) OF AN AUDIO TRANSFORMER

Below left is the ideal case of a single-ended output transformer coupling a tube to its load R_L without any losses in the transformer, while the diagram on the right takes into account winding resistances. Both simplified models only apply at the mid-band of frequencies, with primary parameters reflected to the secondary side.

For an ideal transformer $V_{OUT}=V_{IN}$, but for a lossy (real transformer) $V_{OUT} = V_{IN}R_L/(R_L+R_S+R_P/IR)$. The insertion loss or attenuation in dB is $A=20\log(V_{OUT}/V_{IN}) = 20\log[R_L/(R_L+ R_S+R_P/IR)]$. Thus, the whole circuit can be reduced to a simple resistive voltage divider.

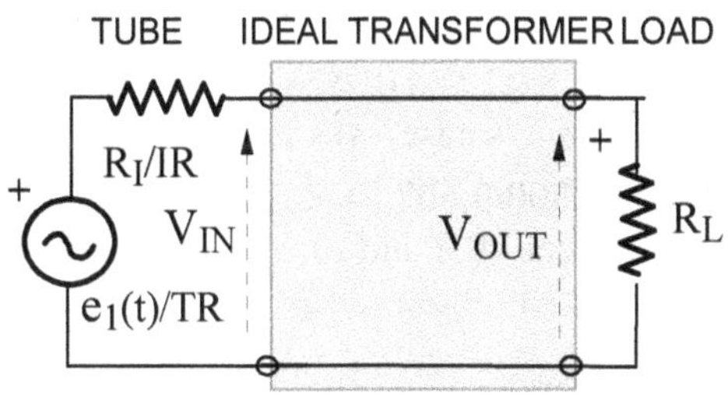

LEFT: An ideal transformer coupling a single-ended tube with its internal resistance RI to the load R_L.

RIGHT: A real SE transformer at midrange frequencies with primary parameters reflected onto the secondary side.

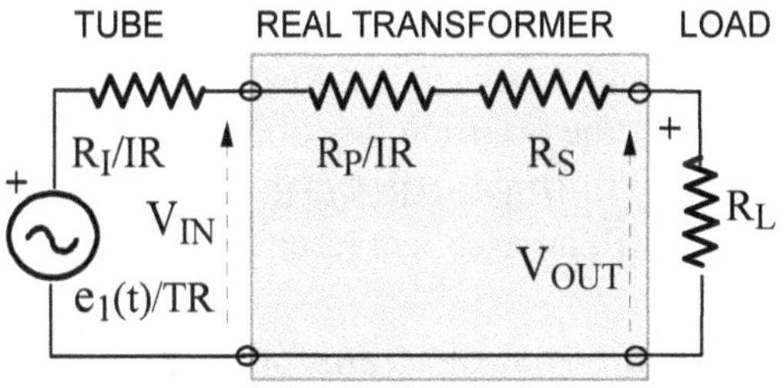

COMMERCIAL EXAMPLE: Tango SE output transformers

Output transformers by the Japanese transformer maker Hirata Tango sell on eBay for US$2-3,000 a pair. Can you make a transformer with equivalent or better performance parameters? You can! Will your transformers sound good, even great? Yes! Will they sound the same as these Tango transformers? No. Every transformer sounds different, their sonic imprint is determined not only by design (the number of turns, the size of the laminations and the wires, etc.) but also by the materials used (type of insulation, magnet wire used, magnetic lamination material, bobbin material).

Tango X-2.7S is one of their standard models with Z_P= 2.7k and three secondaries (4, 8, 16 Ω). Its power output is specified as 40 W @ 30Hz and frequency response as 15Hz - 80kHz (-2dB, 4V input, $R_I=Z_P$, DC 120mA).

Interestingly, instead of a single primary inductance figure as most transformer makers do, Tango specified the inductance of its primary winding as a range, 25 - 35 H, measured using a 5V signal of 50Hz frequency, with 100mA of DC current flowing through it.

Perhaps that resulted from complaints of pedantic buyers who measured lower primary inductance than the single specified figure, so Tango decided to indicate that a wide range and variation of L_P values is unavoidable. Remember, in SE transformers, a small variation in gap thickness has a profound impact on L_P, as does the stacking of the laminations and even the degree of tightness of the bolts that hold the transformer together!

The dimensions were given as 125 x 125 x 138 mm, but these are of their steel covers and not the actual lamination stack and are thus meaningless in evaluating the design.

The specified power or insertion loss was 0.2 dB, with DCR values of R_P=54 Ω and R_S=0.4 Ω.

CALCULATION: The insertion loss of Tango X-2.7S SE output transformer

For Tango X-2.7S the impedance ratio is IR=2,700/16=169, while R_P=54 Ω, R_S=0.4 Ω were specified. A 16Ω load must be used in these calculations, since R_S=0.4 Ω is DCR for the whole secondary winding. $A= 20\log[R_L/(R_L+R_S+R_P/IR)] = 20\log[16/(16+ 0.4+0.3195] = 20\log(16/16.7195) = 20\log 0.95696 = - 0.38$ dB! The negative sign in front of dB figure means attenuation or loss. However, Tango claimed a -0.2dB power loss. How could that be?

It seems that Tango did not consider the loss of the output transformer in isolation but expressed a loss of the whole test circuit, using 2,700 Ω for R_I since that was the value of the series resistance used in their measurements, simulating the internal resistance of the power tube.

Most power triodes (such as 300B and 2A3) will have a much lower internal resistance (under 1kΩ, typ. around 800Ω). Second, such a definition of insertion loss yields a lower figure, so their transformers look better on paper!

$V_{OIDEAL}=e_1TRR_L/(R_L+ R_I/TR^2)$ and $V_{OREAL}=e_1/TRR_L/(R_L+ R_S + R_I/TR^2 + R_P/TR^2)$. Since TR^2=IR, and when we divide V_{OREAL} by V_{OIDEAL} the common term e_1/TR gets canceled and we get the **insertion loss or attenuation** $\mathbf{A=(R_L+ R_I/IR)/[R_L+ R_S + (R_I+R_P)/IR]}$, or in decibels A[dB]=20*logA.

Now, substituting Tango's parameters we get $A=20\log\{(R_L+ R_I/IR)/[R_L+ R_S+(R_I+R_P)/IR]\}= 20\log[(16+2,700/169)/(16+0.4+2,754/169)] = 20\log[(16+15.8)/(16+0.4+16.3)] = 31.8/32.7 = 20\log 0.9725 = -0.24$dB, very close to the claimed value of 0.2dB power loss.

CALCULATION: The insertion loss of our design & built SE output transformer

Instead of reflecting primary values onto the secondary side, the secondary and load resistances can be referred to the primary side. The formula is slightly different, but the results obtained must be the same both ways:

$V_{OUT}/V_{IN} = R_L TR^2/(R_P+R_S TR^2+R_L TR^2) = R_L IR/[R_P+IR(R_S+R_L)]$.

The insertion loss is defined as: $A=20\log(V_{OUT}/V_{IN})$, so $A=20\log\{R_L TR^2/[R_P+TR^2(R_S+R_L)]\}$

Let's calculate the insertion loss of one of our designs, a single-ended output transformer intended for a 300B output stage. The primary impedance is 3,500Ω with a nominal 8Ω load, and the measured DCR parameters are R_P= 20 Ω and R_S = 0.2Ω.

Before proceeding, compare the DCR values to Tango X-2.7S transformers in the previous example. Both the primary and secondary DCRs of our transformer are significantly lower, 20Ω versus 54Ω and 0.2Ω vs. 0.4Ω. Does that mean the insertion loss of our transformer will also be lower than -0.38dB in Tango's case? Yes it does!

$A=20\log\{R_L IR/[R_P+IR(R_S+R_L)]\}= 20\log\{[8*437.5/(20+437.5(0.2+8)]\}= 20\log(3{,}500/3{,}607.5) = 20\log0.9702=$ -0.2628 dB, an excellent result. Quality output transformers have insertion loss between -0.2 and -0.4 dB.

Distortion in audio transformers

This graph for three main types of magnetic materials illustrates the steep increase in voltage distortion due to increased levels of magnetic flux density B.

Up to 0.3T, the distortion is relatively low for any type of material. For M6 GOSS laminations, B should ideally be kept under 0.7T (for 0.1% distortion) and definitely under 1.0T (for 0.25% distortion). For 80% nickel materials, B is limited to 0.7T, and for 50% nickel cores, B can only go up to 0.9T.

RIGHT: Total harmonic distortion levels for three main types of magnetic materials as a function of magnetic flux density B

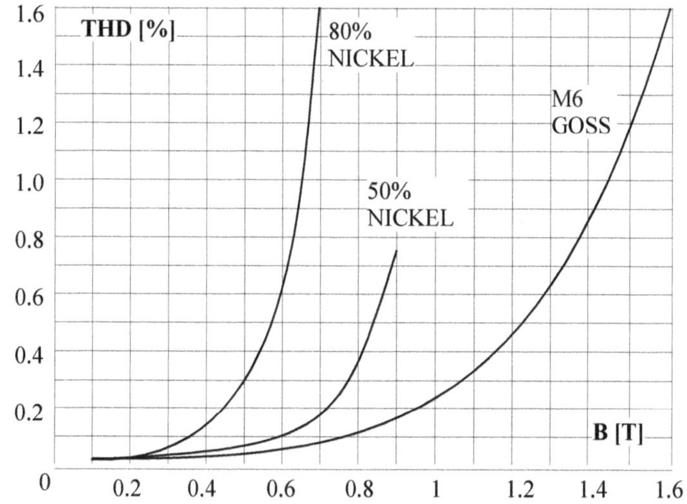

Optimizing or compromising? The rules of thumb

Optimists would call transformer design the art of optimizing; pessimists would rather talk about compromising. There is truth in both views. Many choices a designer faces are clearly contradictory, meaning that each choice regarding one parameter (such as f_L or L_P) has consequences that negatively affect other parameters (such as f_U).

There are so many parameters in play that it is almost impossible to provide a checklist of general rules. Nevertheless, here are the top three optimization rules:

1. The center-leg width (a) should be close to the stack thickness (S), ideally a 1:1 ratio. That ratio should never exceed 2:1 for audio transformers.
2. The copper losses in the windings should be roughly equal to the core losses (hysteresis and eddy current losses).
3. The mean length of the magnetic path (ℓ_{MP}) should be equal to the Mean Length of a Turn (MLT).

OPTIMIZATION RULES
1. Square cross-section (a=S)
2. Core loss = copper loss
3. ℓ_{MP} (mean length of the magnetic path) = MLT (the mean length of a turn)

MEAN MAGNETIC PATH LENGTH
EI LAMINATIONS: $\ell_{MP}=2(b+e-d)+\pi c$ or $\ell_{MP}\approx 6*CL$
FOR C-CORES: $\ell_{MP}=2(b+c)+pa$

SINGLE-ENDED OUTPUT TRANSFORMERS

- DIY DESIGN: SET OUTPUT TRANSFORMER FOR 6C33C-B, 6080, 5528, 6336
- DIY DESIGN: SE OUTPUT TRANSFORMER FOR TRIODE-STRAPPED PL519 AND E130L
- DIY DESIGN: 2.6kΩ SE OUTPUT TRANSFORMER FOR 2A3 and 300B
- DIY DESIGN: 300B SE OUTPUT TRANSFORMER - VERSION #2 (3-2 SECTIONALIZING)
- DIY DESIGN: 10kΩ SET TRANSFORMER FOR GM70, 211, 845 AND 813
- DIY DESIGN: 6kΩ SE OUTPUT TRANSFORMER FOR GM70, 211, 845 AND 813
- COMMERCIAL CASE STUDY: MUSIC ANGEL XD-SE 300B TRANSFORMERS
- DIY DESIGN: 15 WATTS 2k7 UNIVERSAL SE OUTPUT TRANSFORMER
- ANALYSIS: 2k3 TO 6Ω SE OUTPUT TRANSFORMER FROM SHINDO CORTESE
- DIY DESIGN: 1k8 SE OUTPUT TRANSFORMER FOR TWO 300B TRIODES IN PARALLEL
- COMMERCIAL SE TRANSFORMERS

7

DIY DESIGN: SET OUTPUT TRANSFORMER FOR 6C33C-B, 6080, 5528, 6336 TUBES

Now that we understand the basic parameters of audio transformers and related issues, let's see how we would go about designing a few different single-ended and push-pull output transformers. Let's start with a single-ended transformer design for low impedance triodes and duo-triodes such as 6C33C-B, 6080, 5528, 6336, and pentodes such as PL519 (or PL509) and E130L when connected as triodes. We need a 1,000Ω primary impedance transformer to produce an output power of P=15 Watts in an 8Ω speaker. EI GOSS laminations should be used.

Determining the required lamination size and stack thickness

First, we need to decide on the lamination size, determined from P=15W and Z_P=1,000Ω. The empirical formula for effective cross-section needed is $\mathbf{A_{EF}=k\surd(P/f_L)}$.

Since this formula contains the desired lower -3dB frequency f_L, we also need to decide how low do we want our transformer to go, the frequency at which it will be able to transfer half of the full midrange power to the load (f_L). The design brief does not mention the lowest frequency at which the output power must be achieved. Let's say we are ambitious and want to get 15 Watts down to 5Hz.

Now we have to choose "k." More than 60 years ago, Telefunken suggested the range for k between 20 and 30. Higher k results in a larger cross-section, larger window area (more turns can be fitted), and better bass (lower f_L). Let's be modest and select the lowest k=20. In that case we would need $A_{EF}=k\surd(P/f_L) = 20*\surd(15/5) = 34.6\ cm^2$.

The effective cross-sectional area is always smaller than the actual gross area of the lamination stack; how much smaller would depend on the quality of the lamination stamping and the thickness of the insulating layer on one or both sides of each lamination. These factors are collectively lumped into the so-called "stacking factor" SF, which we can only estimate at this stage.

Typically, the effective area AEF would be 3-6% smaller than the gross physical cross-section size, meaning SF is usually around 0.97 for newly produced laminations and 0.94 for older laminations from the 1950s and 1960s. We have $A_{EF}=A/0.96 = 34.6/0.96 = 36.1cm^2$.

We are talking about a 6x6 cm square. A 6cm thick stack is reasonable; bobbins around that size are available. What size lamination would have a central leg around 6 cm wide? From EI lamination tables, we find that EI181.2 has a center leg 60.4mm wide (1/3 of the overall length, which is, you've guessed it, 181.2 mm). That beast of a transformer would be 18cm tall and 15cm wide!

Remember, we've used factor "k" at the bottom of the range to calculate the lamination size required. What if we used k at 30, the top of the range? How large would that monster of a transformer be?

The answer for the slackers amongst you: $A_{EF}=k\surd(P/f_L) = 30\surd(15/5) = 52\ cm^2$! That giant would weigh more than 10 kg!

Anyway, back to reality, now we have to scale down our expectations vis-a-vis f_L of 5 Hz (which also contributed to such unrealistic size requirements) and choose a higher f_L which would result in a smaller lamination size. Ordering small quantities of laminations is expensive, and most suppliers have a minimum order quantity, so you may have to order 100 kg of laminations or more. A prudent approach is to see what size laminations you have lying around or can get easily, such as buying a ready-made couple of transformers or salvaging a pair that you can take apart and reuse their laminations.

We had EI114 GOSS laminations (since we use them for power transformers as well) and 50mm bobbins, meaning our core's cross-sectional area was $A=aS=3.8*5.0=19cm^2$. "a" is the width of the center leg (2U for scrapless laminations, U=19mm for EI114), and S is stack thickness, 50 mm (the width of the bobbin used).

Let's see what f_L can we get with it: $f_L=P(K/A_{EF})^2$, and since $A_{EF}=0.96A = 18.24cm^2$ we get $f_L=P(K/A_{EF})^2 = 15(20/18.24)^2 = 1.2*15 = 18$ Hz.

To get an f_L of 18 Hz without any negative feedback is a great result, so we will proceed with the EI114 laminations and a 50 mm stack.

SIZING OUTPUT TRANSFORMERS

$\mathbf{A_{EF}=k\surd(P/f_L)}$
k is chosen between 20 and 30
P is the output power in Watts
f_L is the desired lower -3dB frequency

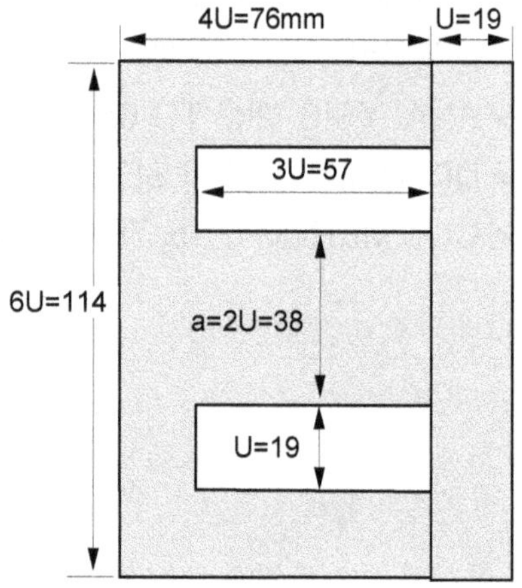

ABOVE: EI114 laminations

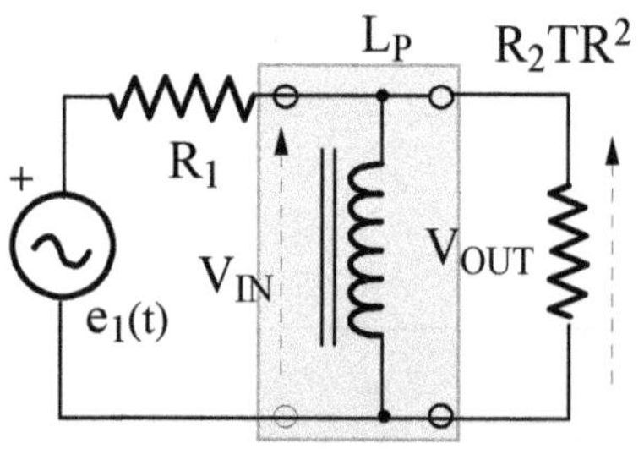

ABOVE: The low frequency model of a single-ended output stage

We know that $\mathbf{f_L = R_{PAR}/(2\pi L_P)}$ or $\mathbf{L_{MIN} = R_{PAR}/(2\pi f_L)}$.

We don't know what will be the primary or secondary resistances of the finished transformer, but this is an estimate anyway, so let's assume the worst case (the highest resistance), the tube's internal impedance r_I by itself, without the reflected secondary resistance R_2TR^2 in parallel. For R_{PAR}=100Ω (for 6C33C-B tube) we get $L_{MIN} = R_{PAR}/(2\pi f_L) = 100/(2*3.14*5) = 3.18H$! Anything under 5H is a modest requirement, meaning even the lowest quality magnetic laminations could be used.

Determining the primary and secondary number of turns

Based on the required or anticipated output power of 15 Watts, the primary voltage will be $V_1=\sqrt{(PZ_P)} = \sqrt{(15*1{,}000)} = 122V$, again, a low figure due to low primary impedance. Now we need to choose B_{MAX}. The lower the B_{MAX} we choose, the higher N_1 and, consequently, N_2, and the larger the winding area needed. Low B_{MAX} means low distortion, but you risk that your finished coil will not fit into the window.

If we select 0.85 Tesla as AC magnetic flux density, once the DC flux density is added, we should still be under the saturation level of around 1.8T for grain-oriented stainless steel, so there is plenty of headroom.

$N_1=V_1 10^4/(4.44 f_L B_{MAX} A_{EF}) = 122*10^4/(4.44*18*0.85*18.24) = 985$ turns! Again, a very low number of primary turns due to low Z_P. You can see why we selected this design as the first practical project; it is the simplest and easiest to design and wind.

Let's round 985 turns up to 1,000 turns and see what we get. With Z_P=1,000 Ω and Z_S=8 Ω load we get ZR= 1,000/8 = 125 and TR=$\sqrt{(IR)} = \sqrt{(125)}$=11.18. Finally, N_2=1,000/11.18 = 89.4 turns. Let's choose 1,000 and 90 turns.

Wire sizing

PRIMARY: The primary must be able to take I_0 = 200mA, and up to I_{MAX} = 280mA of current at full power. The needed primary wire diameter is $d_1=0.71\sqrt{(I_P)} = 0.71\sqrt{(0.28)} = 0.38$ mm.

We had some 0.45 mm wire, so lets use that. The calculated figures are not Holy Scriptures; they are just a starting point. You can always use a larger diameter wire and often even a smaller diameter wire.

In the first case, the number of layers will increase, and if you take things too far, there is a danger that the coil height will exceed the window height and that your finished coil will not fit into the window. In that case, you'd have to start the whole winding job again, using a smaller diameter wire, so don't get carried away, especially not on your first projects. Later, as you gain more experience, your estimation skills will also improve, and your "gut feeling" (a highly technical term used by psychologists and transformer makers) will become more and more accurate.

In the latter case, using a smaller diameter wire, the winding resistance will increase, as will the copper losses and the insertion loss of the transformer. The temperature rise will also increase, but output transformers never work at their maximum power level anyway, so that is not an important issue, as it would be with mains or power transformers which would eventually overheat and fail if the primary wire was too thin.

SECONDARY: With an 8 Ω load and 15 Watts of power, the secondary AC current will be $I_2=\sqrt{(P/R)} = \sqrt{(15/8)}$ = 1.4A, so the smallest wire diameter we should use is $d_2=0.71\sqrt{(I_2)}$ =0.71*0.615= 0.84 mm Again, to minimize the secondary resistance and because we had 1.0mm wire in the workshop, let's up that to 1.0mm!

Sectionalizing and layering

The window length of EI114 laminations is 57 mm. The winding length or Coil Length (CL) is CL= 57 - (2 x 2) mm = 57-4 = 53 mm.

The walls of a typical bobbin are 2mm thick, plus we could allow 1 mm of clearance for the insulation (cut slightly wider than the CL, to ensure the last turn of the next section does not drop down onto the previous layer). So, CL=52mm.

PRIMARY: Maximal number of turns per layer TPL_{MAX}=CL/d = 52mm/0.45mm = 115

Allowing a horizontal fill factor of 0.9, we can fit 115*0.9 = 103 turns, so 100 turns can fit in one layer nicely, so let's split the primary into 300-400-300 turns and have 10 primary layers (3+4+3 =10).

The three primary sections, P1, P2, and P3, will be connected in series, as will the two secondary sections, S1 and S2.

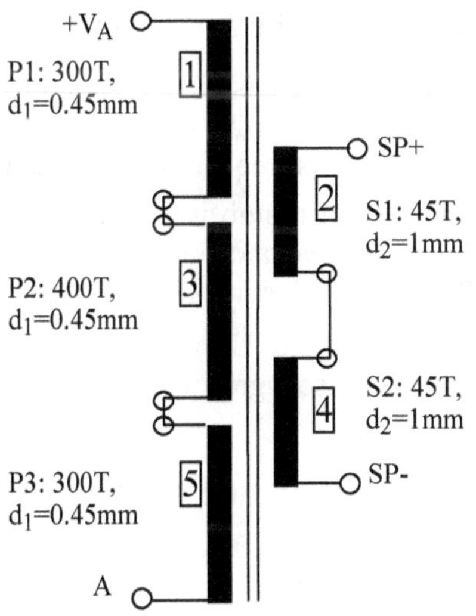

RIGHT: The winding diagram showing the winding order of sections (framed numbers), number of their turns and wire diameters.

SECONDARY: TPL_{MAX}= 52mm/1.0 mm = 52 turns. We need only 45 turns in each of the two sections (connected in series), so each secondary section will easily fit in one layer! Check: HFF = 45/52 = 86.5% , perfect!

For inexperienced coil winders, the HFF (Horizontal Fill Factor) should be under 90% and, if possible, even under 80%. As your winding skills improve, you'll be able to achieve HFF above 90%. Some professional winders using the layered winding method (where there is insulation between all layers, not just between sections as in "bulk" winding) achieve HFF close to 98%.

Insulation

Two layers of 0.15mm Mylar® insulation film were used between sections (a total of five insulation layers). The total insulation thickness (height) is IH =0.3 mm*5 = 1.5 mm, which is about 10% of the net window height (16 mm).

Vertical fit

The window is 19mm tall, but we lose 2mm on bobbin thickness and another 1mm on top for clearance, so our total winding height must not exceed 16mm!

Primary height: 10 layers*0.45 mm = 4.5 mm

Secondary height: 2 layers*1 mm = 2 mm

Insulation thickness: 1.5mm

Total winding or coil height is CH = PH+SH+IH = 8 mm. We need to allow for the "bulging factor" or "bowing." The winding is always taller in the middle of the window; it bows or bulges out. This undesired effect depends on the tensioning skills of the person winding the transformer but is usually 10-15%, so 8mm*1.15 = 9.2mm.

We have filled only 9.2/16 = 58% of the allowable height. The ideal is 80-90%; anything under 70% is considered too low. We could increase the number of primary and secondary turns and increase the primary and secondary wire diameter. Do that as a practice exercise.

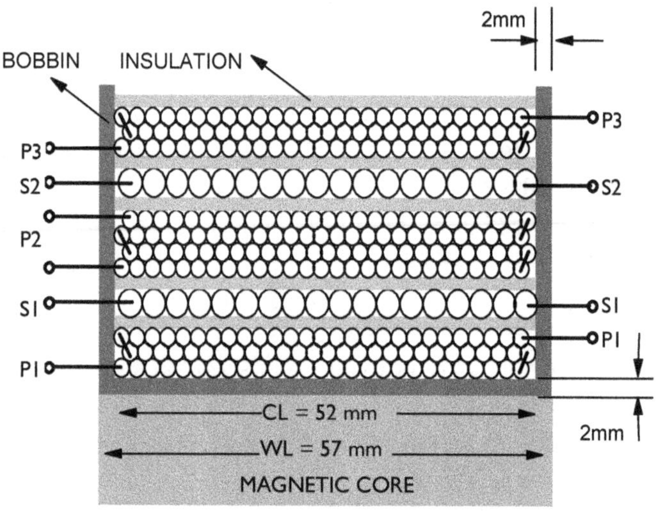

ABOVE: The cross-section of the coil

Measured results

If you recall, using the worst-case scenario, the required primary inductance was 3.18H, and the achieved L_P was 3.8H (at 120Hz test frequency), which is fine.

DESIGN PARAMETERS:	MEASURED RESULTS:
• EI114, 50mm stack	• R_P = 30 Ω
• Z_P=1,000 Ω	• L_{P1000}= 3.62 H, L_{P120}=3.8H, L_L=2.4 mH
• P_{OUT}=15 Watts	• C_{ps1000} = 2.63 nF
	• BW @1W: 5 Hz - 112 kHz (-3dB)
	• BW @15W: 10 Hz - 67 kHz (-3dB)

Estimating the frequency range and resonant frequency

The ratio L_P/L_L is the indicator or measure of the bandwidth of an audio transformer at its rated power output. We had L_P/L_L = 3.62/0.0024 = 1,508, which is an excellent result. The higher the number, the wider the audio transformer's frequency range (bandwidth).

The rule of thumb is that the ratio of upper and lower -3dB frequency limit is about five times the primary and leakage inductance ratio, or $f_U/f_L \approx 5L_P/L_L$! In this case f_U/f_L=5*1,508 =7,541 and f_L= f_U/7,541 = 67,000/7,541 = 8.9 Hz, which comes very close to our measured result of 10 Hz.

Using values of L_L and C_{PS} at 1kHz, we can estimate the resonant frequency as f_R= $1/[2\pi\sqrt{(L_L C_{PS})}]$ = 63,348 Hz, which is also a great result, way above the audible range and also very close to the measured f_U of 67 kHz.

SE TRANSFORMER POWER BANDWIDTH

$f_U/f_L \approx 5L_P/L_L$

RESONANT FREQUENCY OF AN OUTPUT TRANSFORMER

$f_R \approx 1/[2\pi\sqrt{(L_L C_{PS})}]$

DIY DESIGN: SE OUTPUT TRANSFORMER FOR TRIODE-STRAPPED PL519 AND E130L TUBES

PL519 and E130L tubes, when triode strapped, have a low internal impedance, so a transformer for 6C33C-B or 6080 tubes (with Z_P=1kΩ) could be used, although we've found that 1.4-1.5kΩ primary sounds better and results in lower distortion. This is for the ordinary control grid driven output stage, not for the screen-driven triode mode (so called "enhanced triode"), which requires an output transformer with about three times higher primary impedance.

Allowing for the crossover into class A_2 operation and the increase of output power from around 15W to 20W, the primary voltage will be $V_1=\sqrt{(PZ_P)} = \sqrt{(20*1{,}400)} = 167$ V.

If we select 0.9 Tesla as AC (maximum signal amplitude) flux density, once the DC flux density is added, we should still be under the saturation level of 1.8T for grain-oriented stainless steel. $N_1=V_1 10^4/(4.44 f_L B_{MAX} A_{EF})=$ $173*10^4/(4.44*18*0.9*16.5)$ =1,407 turns. If we use 3 primary sections, we need 1,407/3 = 469 turns in each section.

For our favorite EI114 laminations with a 45mm stack, A_{EF}=0.96*A = 0.96*3.8*4.5= 16.5 cm^2.

For an 8 Ω speaker to be reflected to the primary as Z_P =IR*8=1,400 Ω, the impedance ratio IR must be 1,400/8=175. The voltage/turns ratio is thus $TR=\sqrt{(IR)}= \sqrt{(175)}=13.2$ and $N_2=N_1/TR$= 1,404/13.2 = 106 turns.

The four secondary sections are bifilar-wound. Two 0.65mm diameter wires are wound together and joined in parallel, effectively doubling the cross-section and halving the secondary's DC resistance.

The window length is 57 mm. The winding length or Coil Length (CL) is CL= 57 - (2 x 2) mm = 57-4 = 53 mm.

PRIMARY: Maximal number of turns per layer: 53mm/0.42 mm = 126 TPL

469/126=3.72 so we will use 4 layers: 469/4=117 turns/layer, or 117*4 = 468 turns/section or N_1=1,404 turns.

SECONDARY: Theoretical maximum TPL_{MAX} (turns per layer): 52mm/2/0.65 = 40 turns. We need 53 turns, so each secondary section will easily fit in two layers. Each layer will have 54/2 = 27 turns of bifilar wire.

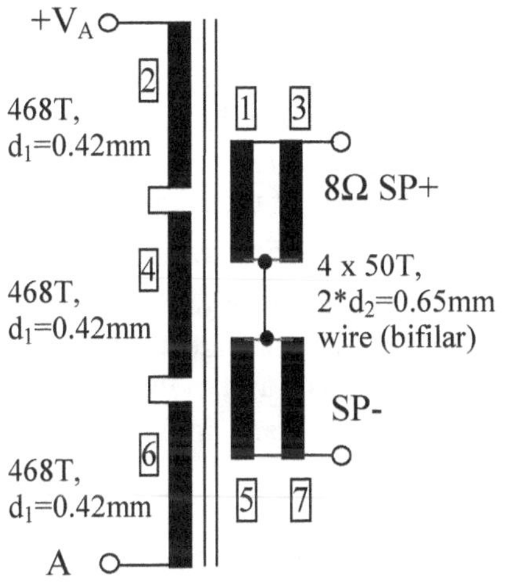

The winding diagram

DIY DESIGN: 2.6kΩ SE OUTPUT TRANSFORMER FOR 2A3 AND 300B

Our goal is to design a SE output transformer that can be used for both 2A3 and 300B tubes, capable of delivering up to 12 Watts into a 2,600 Ω reflected load (8 ohm speaker). 3k5 primary impedance is normally used for 300B tubes, but most speakers' impedance is way above the nominal 8 ohms over most of the frequency band, so for most frequencies the reflected impedance will be closer to 4 or 5kΩ rather than 2k6 or even 3k5!

Lamination size and stack thickness

The empirical formula for effective cross-section needed is $A_{EF}=k\sqrt{(P/f_L)}$. We want to get 12 Watts down to 15Hz, so now we have to choose "k ." Let's be prudent and select k in the middle of the range (between 20 and 30), k=25. $A_{EF}=25\sqrt{(12/15)}$ = 22.4 cm^2.

Luckily, 22 cm^2 was the cross-section of our most-commonly used lamination choice, EI114, with S=58mm stack thickness (for which we had bobbins)! A=3.8*5.8= 22 cm^2.

Of course, that is the gross area; depending on the quality of the lamination stamping and the stacking factor, the effective area A_{EF} would be 3-6% smaller than the physical cross-section size (gross), A_{EF}=0.96A = 21.1cm^2. That is close to the calculated 22.4 cm^2, so we will proceed with the EI114 lamination and 58 mm stack.

Primary and secondary number of turns

From maximum power $P_1=V_1^2/Z_P$ we get the voltage on the transformer's primary: $V_1=\sqrt{(PZ_P)} = \sqrt{(12*2{,}600)} = 176$ V.

Since $N_1=V_1 10^4/(4.44f_L B_{MAX} A_{EF})$, now we need to choose B_{MAX}, the level of induction when the amp is working at its maximum power, which hopefully won't be very often. Again, don't go overboard and choose B_{MAX} too low. If we choose 0.75T, we get $N_1=V_1 10^4/(4.44f_L B_{MAX}*A_{EF}) = 176*10^4/(4.44*15*0.75*21.1) = 1{,}670$ turns.

With Z_P=2k6 and 8Ω load, the impedance ratio is $IR=Z_P/Z_S = 2{,}600/8 = 325$ and the turns ratio $TR=\sqrt{IR}=\sqrt{325}\approx 18$, so with 1,670 primary turns, we need $N_2=N_1/TR = 1670/18 = 92.78 \approx 93$ turns. 1,670 and 93 are awkward numbers; we need "easy" figures to remember and, more importantly, easy to divide into sections and fit into layers.

Sectionalizing

Now we have to choose how to sectionalize the windings. Again, don't run wild and choose 8 primary and 9 secondary sections. The more sections, the more insulation layers, and the fatter your total winding height, which may quickly exceed the 16 mm net window height!

The 3/4 design is usually a good starting point, 3 primary sections and 4 secondary sections. Since output transformers are always of a step-down kind, with much more primary than secondary windings, we could have opted for a 4-3 design, but it is better to have more sections in the secondary winding to parallel them and lower the secondary's resistance, thus improving the damping factor of the amplifier.

1670 turns/3 = 556 turns, so let's choose 550 turns for each section.

The nominal connection is for 8 Ω secondary and 2k6 primary impedance. Since there are four identical windings, you have the luxury of connecting them in various ways, as illustrated below. Depending on the nominal speaker impedance, primary impedances of between 0.66 kΩ and 10.5 kΩ are obtainable!

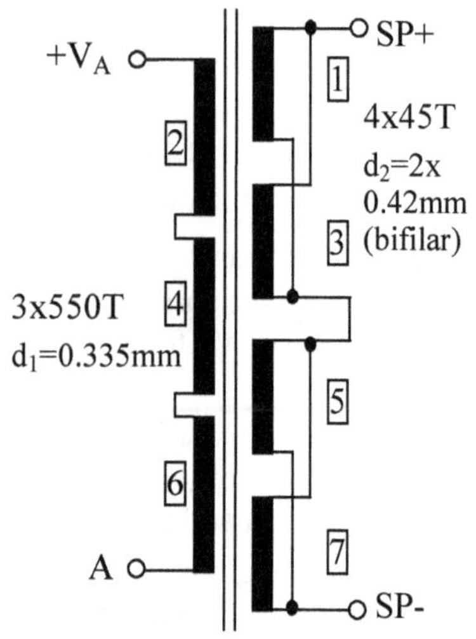

The winding diagram

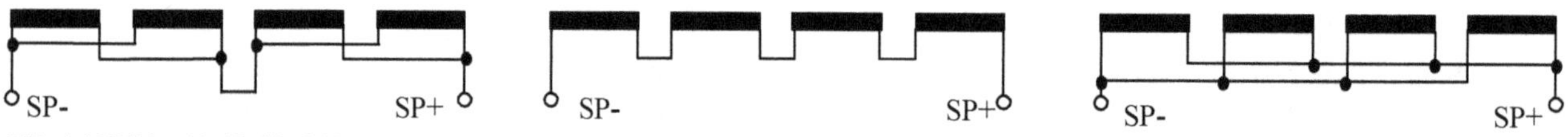

TR=1,632/90= 18.13, IR=336
8 Ω load reflected as 2,630 Ω.

TR=1,632/180= 18.13, IR=82.2
8 Ω load reflected as 658 Ω.
16 Ω load reflected as 1,315 Ω.

TR=1,632/45=36.27, IR=1,315
4 Ω load reflected as 5,260 Ω.
8 Ω load reflected as 10,520 Ω.

ABOVE: Optional secondary connections for other speaker loads

Primary and secondary currents and wire sizing

About 65-70 mA of DC current will flow through the primary at idle and 100mA at full power. Primary wire diameter: $d_1=0.71\sqrt{I_1} = 0.71*\sqrt{0.1} = 0.22$ mm To minimize primary's resistance, let's choose 0.335mm wire. Just kidding, that was the wire size we had a large roll of, but minimizing primary resistance sounds like a much more professional and highbrow design criterion.

SECONDARY: $P=V^2/R$ or $P=I_2{}^2R$, so $I_2=\sqrt{(P/R)} = \sqrt{(12/8)} = 1.23$ A

Since two and two sections will be paralleled, the current through each section will be halved, i.e. 1.23/2 = 0.615 A, and the minimum size wire we need is $d_2=0.71\sqrt{I_2} = 0.71*\sqrt{0.615} = 0.56$ mm.

CALCULATING THE REQUIRED WIRE DIAMETER FROM CURRENT

$d=0.71\sqrt{I}$ for $J=2.5\ A/mm^2$ OR $d=1.128\sqrt{(I/J)}$ for any J

Current density = current / cross-sectional area of the wire or J= I/S, since $S=\pi d^2/4$, $d=2\sqrt{(I/\pi/J)}$

For $J=2.5A/mm^2$ $d= 2S\sqrt{(I/3.14/2.5)} = 2\sqrt{(I/7.85)} = 2*0.357*\sqrt{(I)}$, so $d=0.71\sqrt{I}$

For any current density $d=2\sqrt{(I/\pi/J)}$ or $d=1.128\sqrt{(I/J)}$

Primary layering

Coil Length (CL) is CL= 57 - (2 x 2) mm = 57- 4 = 53 mm We will work with 52 mm effective winding width.

Maximum number of turns per layer: 52mm/0.335mm = 155 TPL, 1,650 turns / 155 TPL = 10.6 layers

Since we have 3 identical primary sections, our total number of layers has to be divisible by 3, so we will use 12 layers in total, 4 layers in each section! Actual TPL: 1,650/12 = 137.5 We will use TPL=136 !

The new N_1=12*136 =1,632 turns, or 1,632/3 = 544 turns per section

With N_1=1,632 and TR=18, we get N_2=1,632/18= 90.67 so we will use N_2 = 90. The final turns ratio is TR=1632/90= 18.13, which with an 8Ω load gives us a primary impedance of 2,630Ω.

Secondary layering

For mechanical and insulation reasons, we should aim to fill as much of the winding length as possible so the subsequent layers wound on top of each secondary layer sit nice & straight, without any valleys or peaks. The coil should be firm, all the layers should be of the same length and wound as uniformly as possible.

Maximum wire diameter d_{MAX}=50mm/45 turns = 1.11 mm

We could fit 45 turns of 1 mm wire, instead of the 0.56mm required. This would reduce the secondary resistance and amplifier's output impedance. However, to reduce the winding height, we will use two 0.42 mm wires in parallel. Each secondary section would then have one layer of 2x0.42 mm wires wound in a bifilar fashion.

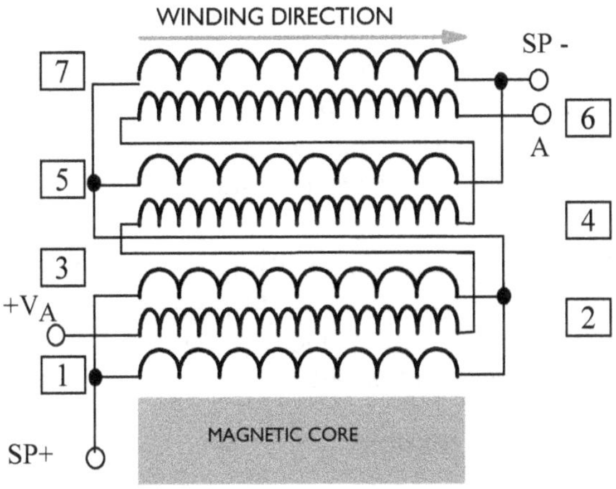

PRIMARY: 3x544T, d_1=0.335mm SECONDARY: 4x90T, d_2=2x0.42mm

Another way of drawing winding and interconnection diagrams. The framed numbers indicate the order of winding but aren't necessary in this kind of diagram, the order is obvious from the depiction of the coil.

Insulation

Two layers of 0.15mm Mylar® insulation were used between sections and one on top of the winding. There are seven sections and seven insulation layers. The total insulation thickness is d_I=0.3*7 = 2.1 mm.

Window fitting using the WW method (winding area vs. window area)

The alternative method of calculating the exact winding height is to estimate the winding area in relation to the window area. The gross window area F_W=19mm*57mm = 10.8 cm^2

The primary area F_P=1,632*0.335^2*π/4 = 144mm^2 = 1.44 cm^2 and the secondary area is F_S=4*90*2*0.42^2*π/4 = 99 mm^2 ≈ 1 cm^2.

Total copper area is F_C=F_P+F_S=2.44 cm^2 or F_C/F_W=2.44/10.8 = 22.5% of the window area, so we should have plenty of space to fit our windings. The empirical (from experience) benchmark is around 35%. If your copper area exceeds 35% of the window area, you should consider choosing a larger size lamination or core.

Estimating DC resistances of the windings

PRIMARY: R_P=0.045N_1(a+S+1.5b)/100/d_1

For scrapless lamination sizes (a=2U, b=U): R_P=0.045N_1(3.5U+S)/100/d_1 and R_S=0.045N_2(4.5U+S)/100/d_2

For EI114 laminations: U=1.9cm, S=5.8 cm, N_1=1,632 turns and d_1=0.335mm we get R_P=81.5 Ω

SECONDARY: R_S=0.045N_2(a+S+2.5b)/100/d_2

By using d_2= 1.05 (the equivalent to 2x0.42 mm wires in terms of the cross-sectional area) and N_2=45, we get R_S= 0.26 Ω . The equivalent resistance of two sections in parallel is half of the resistance of one section, but then we have another identical pair of paralleled sections added in series, so we end up with the equivalent resistance of one section.

Measured results

Before stacking the laminations, always test the finished coil with an LCR meter, just in case something went wrong during the winding process. At 120Hz test frequency we got L_{120}= 0.147 mH.

At 1kHz test frequency we got L_P = 0.142 mH. These are typical inductance figures without the magnetic core.

DESIGN PARAMETERS:	MEASURED RESULTS:
• EI114, 58mm stack	• R_P= 88Ω, L_P= 11.2 H, L_L=23mH
• Z_P=2,600 Ω	• C_{PS} @1,000 Hz = 2.88 nF
• P_{OUT} = 12 Watts	• BW @1W: 6 Hz - 40 kHz (-3dB)
	• BW @10W: 12 Hz -28 kHz (-3dB)

DIY DESIGN: 300B SE OUTPUT TRANSFORMER - VERSION #2 (3/2 SECTIONALIZING)

Let's see how changing a few principal factors changes the design and performance of the 300B SE transformer. Still using the same EI114 laminations as in the previous design, let's reduce the stack thickness from 58 mm to 44 mm (another standard bobbin size). That reduces the gross cross-sectional area from 22 to 17 cm^2.

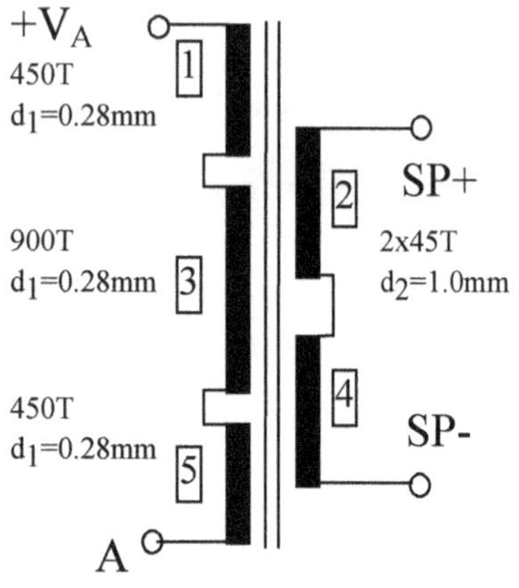

The winding diagram using simpler 3/2 sectionalizing

Let's also simplify the sectionalizing from 3-4 to 3-2 (5 sections in total instead of 7) and divide the primary differently. Instead of 3x550 = 1,650 turns, lets use 450+900+450 = 1,800 turns. That will give us TR=1,800/90 = 20, the impedance ratio of 20^2=400, so the 8Ω speaker's impedance, when reflected onto the primary side, will result in the primary impedance of Z_P=3,200Ω.

Why the higher number of primary turns N_1? Let's consult the fundamental TX equation $N_1=V_1 10^4/(4.44 f_L B_{MAX} A_{EF})$. Since V_1 stays the same and A_{EF} is reduced, to keep f_L and B_{MAX} at previous levels, N_1 needs to be increased to compensate for smaller A_{EF}.

Measured results

Because we used a smaller diameter wire and had a higher number of primary turns, the primary DC resistance increased from 88Ω to 120Ω. Once the transformer was assembled, the primary inductance was 10.3H, lower than 11.2H in the previous case.

The primary-secondary capacitance increased from 2.88 nF to 4.15 nF, and the -3dB frequency range at 1 Watt shifted higher, 9 Hz - 47 kHz, and at 10 Watts to 16Hz - 32 kHz!

MEASURED RESULTS:	DESIGN PARAMETERS:
• R_P= 120 Ω, L_P= 10.3 H	• EI114, 44mm stack
• C_{PS1000} = 4.15 nF	• Z_P=3,200 Ω
• BW @1W: 9 Hz-47 kHz	• P_{OUT}=12 Watts
• BW @10W: 16 Hz-32 kHz	

DIY DESIGN: 10kΩ SET TRANSFORMER FOR GM70, 211, 845 AND 813

We need a SE transformer with 10kΩ primary impedance, good for 25 Watts using GM70, 211, 845 or 813 tube (triode connected) and an 8Ω speaker.

Sizing the core

Our commonly-used EI114 laminations may be too small for these high-power impedance tubes, so let's go one size up. What f_L can we get with EI120 laminations with a 50mm stack (A =4.0*5.0= 20 cm^2)?

$f_L=P(K/A_{EF})^2$, and since A_{EF}=0.96A=19.2 cm^2 we get $f_L=P(K/A_{EF})^2$ = $25*(20/19.2)^2$ = 27 Hz. Alternatively, we could go one metric size up again and use EI133.2 laminations. With the same 50mm stack, A =4.44*5.0= 22.2 cm^2, A_{EF}=0.96A = 21.3 cm^2 and f_L= $25*(20/21.3)^2$ = 22 Hz.

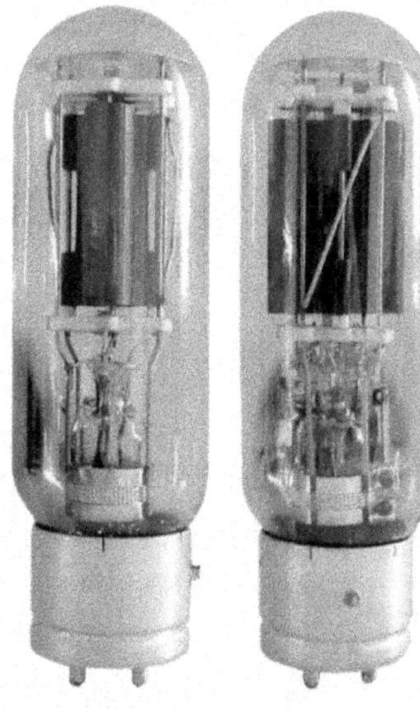

211 (VT4-c) triodes

Our listening tests show that mild negative feedback is usually better sounding than the same amplifier without any feedback (single-ended triode purists and horn junkies will disagree), and even such mild feedback will lower f_L from 27 Hz down to say, 12 Hz and all will be well again.

So, if you decide to use a mild NFB, go for the smaller lamination size, EI120, as we will do here. If you don't want to use any NFB in this design, use EI133.2 laminations.

Number of turns and sectionalizing

The primary AC voltage will be $V_1=\sqrt{(PZ_P)}=\sqrt{(25*10{,}000)}$=500V! Disregarding the previous f_L estimation, let's see if we can get f_L down to 15 Hz: $N_1=V_1 10^4/(4.44 f_L B_{MAX} A_{EF}) = 500*10^4/(4.44*15*1*19.2)$ = 3,910 turns.

Before deciding on the final number of primary turns, we should consider sectionalizing issues. With the winding length WL of 60mm, the useable window length or Coil Length is CL=55mm.

PRIMARY: From the wire tables, we can see that 0.3 mm wire (0.32 -0.34 mm with insulation) is good for at least 181 mA, so let's start with that size.

Maximum TPL (turns per layer): TPL_{MAX}=CL/d= 55mm/0.3mm = 183 T

3,910T / 183TPL = 21.36 layers We need 24 layers or 3,910/24 = 163 TPL

Primary horizontal fill factor check: 163/183 = 89%, perfect.

If we use 5/4 sectionalizing, 24 is not divisible by 5, so we can pull a trick out of our sleeve - have 4 sections and split the fourth one in half to make it into five sections (3 full sections and two sections with 1/2 number of turns).

24 layers = 3*24/4 + 2*24/4/2 = 3*6 layers + 2*3 layers = 3 * 6*163T + 2 * 3*163T = 3*978T + 2*489T

SECONDARY: $N_2 = N_1/VR$ = 3,910/35.36 = 110.6 turns

TPL_{MAX}= 55mm/0.73 mm = 75 turns

With insulation, the outside diameter of a 0.71mm wire is actually 0.73 mm. We will use 55 TPL, 2 layers for each 110 turns section; then, all four sections will be connected in parallel to reduce the resistance and increase the damping factor. Secondary horizontal fill factor check: HFF_S=55/75=0.74 = 74%, too easy.

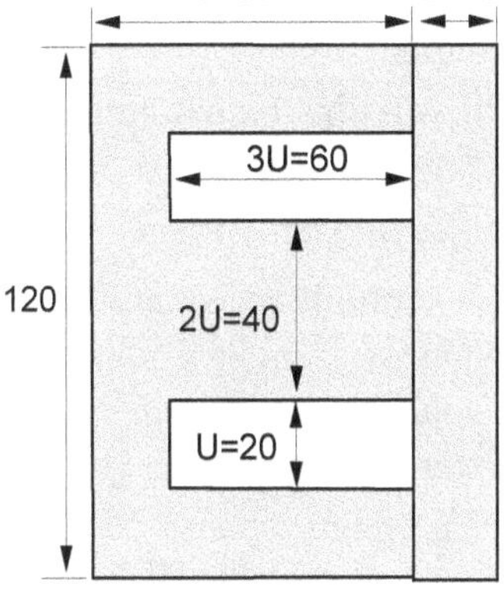

EI120 LAMINATIONS

Window fitting

Primary's height: PH = 24 layers *0.32mm = 7.7mm

Secondary's height: SH = 8layers*0.73mm = 5.8mm

The total winding height is 13.5 mm. Allowing for 15% bowing (bulging), we get coil height of CH = 1.15*13.5 = 15.5 mm.

Even with a bulk winding (no insulation between layers of each section) we need 5 layers of insulation between primaries and secondaries, at 0.3 mm each (due to high voltages involved), the total insulation will be 1.5 mm, bringing the total coil height to just over 17 mm.

The window is 20 mm high. The bobbin thickness and clearance will reduce the height to 17.5 mm, so a seasoned winder may be able to do it with tight and careful winding.

If you feel uneasy about such a small error margin, try increasing the lamination size or reducing the wire sizes. Instead of a 0.3mm primary wire, the 0.25mm wire would do the job; the amp will not work at the maximum power all the time anyway. As an exercise, repeat the calculations for 0.25mm primary instead of 0.3mm and see if we gain anything in terms of coil height clearance.

Sectionalizing, iteration #2 (2x0.43 mm secondary wire instead of 1x0.73 mm)

Before increasing the core's size, we can try reducing the diameter of the secondary wire. Instead of a single 0.73mm wire, let's use two thinner wires in parallel (a bifilar winding).

We can easily fit 110 turns of 2x0.43 mm wire, resulting in horizontal fill factor of HFF = (2*0.43*55)/55= 47.3/55 = 86%.

The secondary's height will now be 8 layers *0.43mm= 3.44 mm, and the total height of windings without insulation is now 9.11 mm, which is less than 50% of the window height, so we are now out of the danger zone for sure.

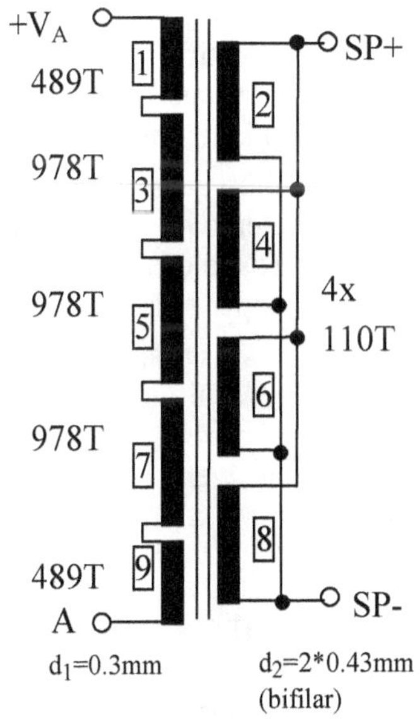

Winding diagram

Key design parameters

- EI120 laminations: a=4.0 cm, S=5.0 cm, so A=aS = 20 cm^2
- IR=10,000/8=1,250, TR=√(ZR)= √1,250 = 35.35
- I_0 = 110mA, I_{MAX}=180 mA
- 5/4 sectionalizing, primary winding with 3+6+6+6+3 layers of 163 turns in each layer, total primary turns N_1=3,912, d=0.25 mm wire
- Secondary: 4 parallel sections, 2 layers with 55 turns each, so 110 turns in each section, 2x0.43mm wires (bifilar winding)
- Air gap g = 0.18-0.20 mm

DIY DESIGN: 6kΩ SE OUTPUT TRANSFORMER FOR GM70, 211, 845 AND 813

The 6kΩ output transformer is a relatively simple, yet very well performing design with 4/3 sectionalizing on EI96 laminations: a= 32mm, S= 6.25cm, A= a*S= 20cm^2.

As a rule-of-thumb, the power rating of this transformer's core is $P \approx A^2 = 20^2 = 400W$, yet it works on audio power levels of only up to 40 Watts, so it is ten times oversized.

Remember this 10X criterion well; use it to evaluate commercial SE output transformers and estimate their oversize factors.

PRIMARY: 4 x 675 turns = 2,700 turns, wire d = 0.25 mm

SECONDARY: 3 x 99 turns in parallel, wire d = 1 mm, g = 0.15 mm (air gap)

TR=2,700/99 = 27.27, ZR=TR^2 = 744, Z_P (8Ω load) = 744 x 8 = 5,952 Ω

For multiple secondary taps, you need to change the sectionalizing to either 3/4 or 5/4, so you can have 4 secondary sections. Then you can connect them all in parallel, all in series or two and two in parallel, and then in series.

Should you want to increase the anode voltage to 1,000V to get higher output levels, increase the impedance of the output transformer to 8kΩ or even higher.

This design can also be used with high voltage transmitting power tubes of the GM70, 211, 845, and 813 kinds. The stack size remained the same (A=20cm^2), so the lower frequency limit will be in the same ballpark. However, a smaller laminations size, in this case, reduces leakage inductance, so the upper cutoff frequency is increased, resulting in a more detailed, shimmering treble, with overtones reproduced up to 70 or so kHz!

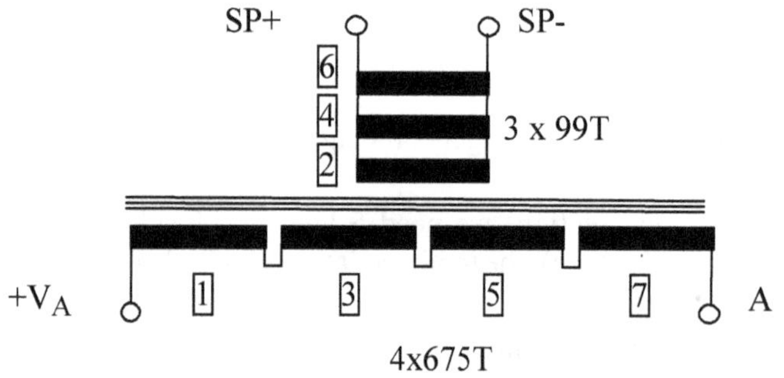

LEFT: The winding diagram of 6kΩ SET output transformer. Framed numbers indicate the winding order of the sections.

MEASURED RESULTS:

- BW: 10 Hz - 73 kHz (-3dB, 1W into 8Ω)
- BW: 13 Hz - 63 kHz (-3dB, 10W into 8Ω)
- V_{OUTMAX}=17V_{RMS}, P_{MAX} = 36 W

COMMERCIAL CASE STUDY: MUSIC ANGEL XD-SE 300B TRANSFORMERS

Although the sound of the Music Angel XD-SE 300B amplifier was warm and mellow, the more we listened, the less we liked it. It lacked upper frequencies; it wasn't transparent or detailed, nothing but a syrupy midrange.

The cause of its muffled sound was quickly identified on the test bench. The first amplification stage only had an upper -3dB frequency f_U of around 27 kHz, but that wasn't the main culprit. Overall, one channel had a measured f_U of only 9.6 kHz; the other went to only 14.4 kHz!

The suspicion quickly moved to its output transformers, so we opened them up. After removing the upper I-section from both output transformers (in situ), the cause of the problem became obvious: three very thin and flimsy adhesive paper strips were used as an air gap (the upper photo on the right).

Since there was practically no air gap, the magnetic core went into saturation very early, resulting in weak, almost nonexistent bass. Applying a new gap insulation paper fixed the saturation problem (right). However, the laminations were made of rather inferior material, no GOSS here, and rusty!

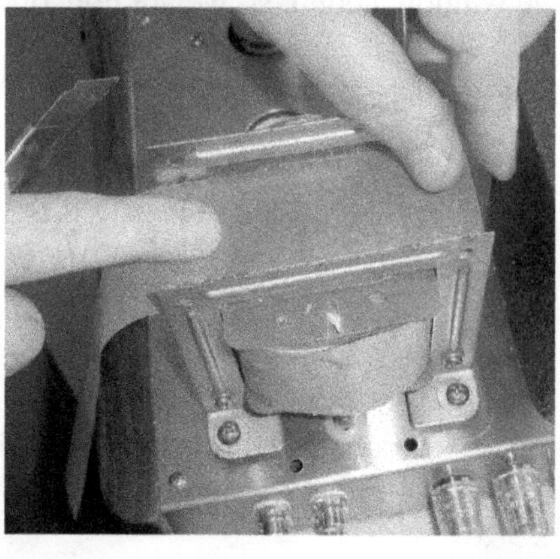

ABOVE RIGHT: The three flimsy strips of paper were too thin; there was practically no air gap at all!

RIGHT: Installing a proper air gap was an easy fix, but it only revealed other weaknesses in the transformers (low quality, high loss laminations) and the poor overall design!

DIY DESIGN: 15 WATTS 2k7 UNIVERSAL SE OUTPUT TRANSFORMER

A SE transformer of 2.5-3.5kΩ primary impedance and 15 Watts handling power can be used for many output triodes and triode-strapped pentodes, such as 2A3, 300B, EL34, KT88, F2a, EL153 and lots of others.

Core sizing

EI96 laminations are a reasonable starting point for most output transformers. EI114 size (assuming the same or similar stack thickness) gives you a much larger cross-sectional area, resulting in lower -3dB limit and improved bass performance.

However, the parasitic capacitances and leakage inductance also increase due to the larger dimensions, which negatively impacts the high-frequency performance, so a compromise is inevitable.

Now we need to determine the stack thickness using the empirical formula $A_{EF}=k\sqrt{(P/f_L)}$. Since k is usually around 20-25, let's be modest and choose k=20 and the lower -3dB frequency of 20Hz. Now the effective cross-sectional area is $A_{EF}=k\sqrt{(P/f_L)} = 20*\sqrt{(15/5)} = 17.3\ cm^2$.

The effective area A_{EF} is 3-6% smaller than the gross cross-sectional area A. We have $A= A_{EF}/0.96 = 17.3/0.96 = 18\ cm^2$.

Since the width of the center leg for EI96 lams is 3.2cm, the stack thickness we need is $S=A_{EF}/3.2 = 18/3.2 = 5.6$ cm. The standard bobbin widths are 50, 55, and 60 mm; any of them could be used here, but 55mm is the closest.

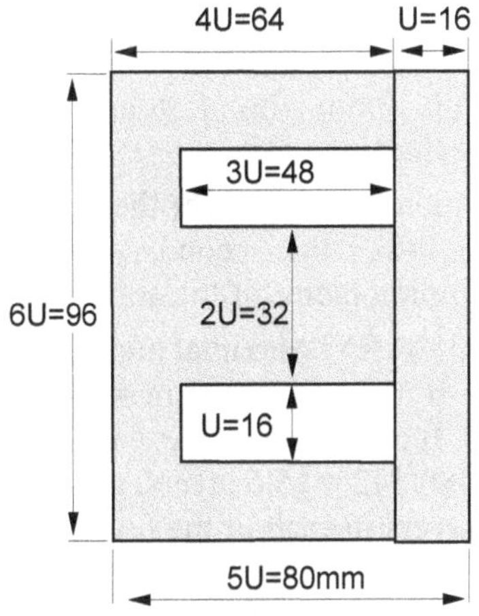

EI96 LAMINATIONS

Primary voltage, primary & secondary number of turns

To get 15 Watts of output power, the primary voltage needs to be $V_1=\sqrt{(PZ_P)} = \sqrt{(15*2,700)} = 201V$.

The choice of B_{MAX} is, as always with single-ended transformers with unbalanced DC current, based on the assumption that half of the total maximum B (saturation level of around 1.6T) will be the DC flux density and the other half (0.8T), the AC flux density produced by the signal voltage.

$N_1=V_1 10^4/(4.44 f_L B_{MAX} A_{EF}) = 201*10^4/(4.44*12*0.8*17.6) = 2,680$ turns. With Z_P=2,700 Ω and the load Z_S=8 Ω we get the impedance ratio of IR= 2,700/8 = 337.5 and turns or voltage ratio of TR=$\sqrt{(IR)} = \sqrt{(337.5)}$ =18.37.

Finally, we can calculate the required number of secondary turns as N_2=2,680/18.37 = 145.9 or 146 turns.

Sectionalizing

The leakage inductance of 1:2, 2:1, 3:2 and 2:3 designs is way too high, the magnetic coupling is inferior and the upper -3dB frequency would be under 10 kHz. That is fine for a guitar tube amp but not for an audiophile amp. The smallest number of sections you should ever use is 3:4, but in this case, we want a superior design, so let's choose a 4:5 arrangement, 4 primary and 5 secondary sections.

Wire sizing and window fitting

PRIMARY: The primary must be able to take I_0 = 80mA, and up to I_{MAX} = 160mA of current at full power. Primary wire diameter: $d_1=0.71\sqrt{(I_{MAX})} = 0.71\sqrt{(0.16)} = 0.28$ mm, but since no amplifier works at full power all the time, we can safely use 0.23mm diameter wire.

SECONDARY: With an 8 Ω load and 15 Watts of power, the secondary AC current will be $I_2=\sqrt{(P/R)} = \sqrt{(15/8)} = 1.4A$, so the smallest wire diameter we should use is $d_2=0.71\sqrt{(I_2)} =0.71*0.615= 0.84$ mm.

However, we will have 5 secondaries in parallel, so that the total current will be equally divided, 1.5A/5 = 0.3A through each secondary section, so $d_2=0.71\sqrt{(0.3)} = 0.39$ mm!

The window length of EI96 laminations is 48 mm. The walls of a typical plastic bobbin are at least 1.5 mm thick (on each side, so 3mm total).

We should also allow 1 mm of clearance for the insulation (cut slightly wider than the CL to ensure the last turn of the next section does not drop down onto the previous layer), a total of 4mm. So, the winding length or Coil Length (CL) is CL= 48-4 = 44 mm!

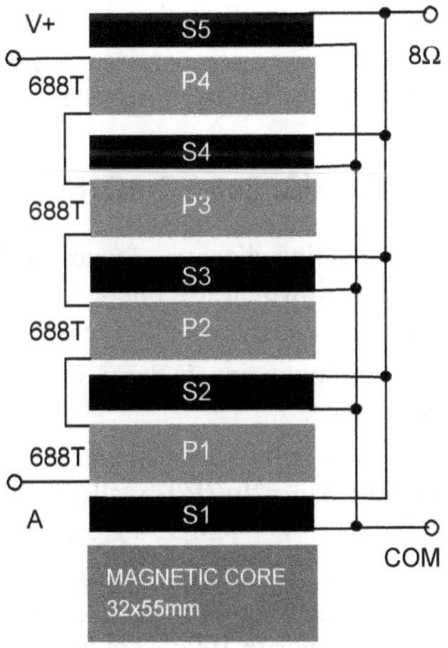

The winding diagram

We need 146 turns in the secondary and 44mm/146 turns = 0.30 mm diameter wire. So, if we use 0.3 mm wire with perfect winding (100% horizontal fill factor HFF), we could fit all 146 turns in a single row.

Since we are not professionals winders, let's assume 80% HFF and two layers per secondary section and see what size wire we could fit in: 44*0.8/146*2 = 0.48mm! So, let's use 0.45mm wire;. If you have 0.5mm wire you could also use that.

Remember, the larger the diameter of the secondary wire, the lower the secondary resistance and the higher the damping factor of the amplifier.

PRIMARY: Maximal number of turns per layer TPL_{MAX}= CL/d = 44mm/0.23mm = 191! Allowing for a horizontal fill factor of 0.9, we can fit 191*0.9 = 172 turns and 2,680/172 = 15.6 layers, meaning we need 16 layers. With 4 layers in each of the four sections, 4x4=16, perfect!

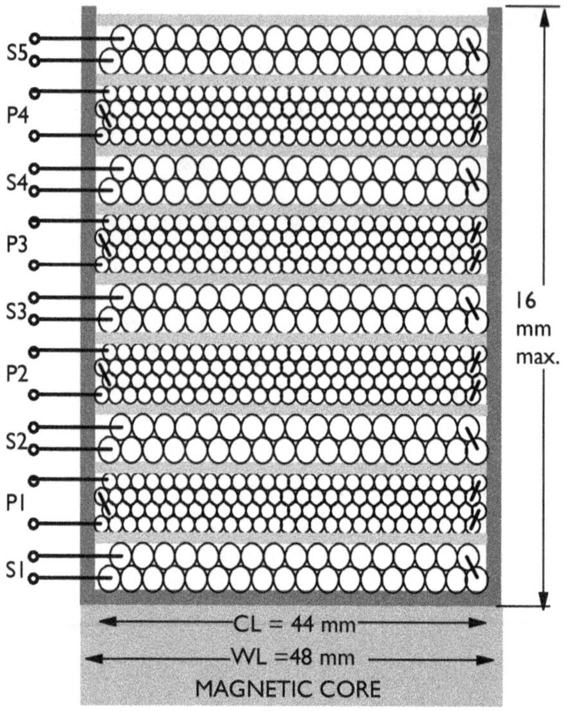

RIGHT: Cross-section of the coil (not to scale)

Checking the vertical fit

The window of EI96 laminations is 16mm tall, but we lose 2mm on bobbin thickness and another 1mm on top for clearance, so our total winding height must not exceed 13 mm! The height of the primary winding is 16 layers*0.23 mm = 3.7 mm, the secondary is 10 layers*0.45 mm = 4.5 mm high.

We have 8 insulation layers between primary and secondary sections, and with 0.35mm Mylar insulation, that is 2.8 mm, our total coil height is CH = PH+SH+IH = 3.7+4.5+2.8= 11 mm.

We need to allow for the "bulging factor" or "bowing." The winding is always taller in the middle of the window; it bows or bulges out. It depends on the tensioning skills of the person winding the transformer but is usually 5-15%, so assuming the worst case of 15%, 11mm*1.15 = 12.65mm, just under the 13mm maximum, perfect!

An alternative, formula-based way of sizing the airgap

Some designers determine the air gap's width based on the core's cross-sectional area (A=a*S), using the formula $g=0.2*\sqrt{A}$.

The main factor determining the required air gap size is the total DC magnetizing force, which may be expressed in AT (Ampereturns). The type of the magnetic material used (its permeability) and the size of the laminations (the length of the magnetic path ℓ_{MP}) also impact the size of the gap, but their effect is so small compared to that of the magnetizing force that it can be neglected, at least in the first approximation. Perplexingly, this rule-of-thumb formula does not involve AmpereTurns in any way!

Secondly, from our experience, the values obtained by this method are too high; if you are keen to use a formula, $\mathbf{g=0.05\sqrt{A}}$ would be more accurate.

ANALYSIS: 2k3 TO 6Ω SE OUTPUT TRANSFORMER FROM SHINDO CORTESE

This winding diagram, found on the internet, is claimed to be for output transformers used in Shindo Cortese amplifiers, which use one F2a power tube per channel, working as a pentode. Regardless of the accuracy of this claim, let's analyze it and see what kind of performance can be expected and if its design can be improved.

The sectionalizing is very simple, with three primary and two secondary sections (3/2) and a single output impedance.

The gross cross sectional area is A=aS = 3.8*6 = 22.8 cm^2. Assuming A_{EF}=0.96A = 21.9 cm^2, our rule of thumb formula $\mathbf{A_{EF}=20\sqrt{(P/f_L)}}$ tells us that at 10 Watt output level, the stack of that size can go down to $f_L=P(k/A_{EF})^2$ = $10*(20/21.9)^2$ = 8.3Hz, a great result.

Allegedly, a primary impedance of 9.5H was measured by one keen enthusiast, while the frequency response was specified as 8Hz-110 kHz. However, the referent power level was not specified. The bandwidth of our transformers never extends to 100kHz at full power, so it is safe to assume that this is a 1 Watt bandwidth.

Primary & secondary number of turns

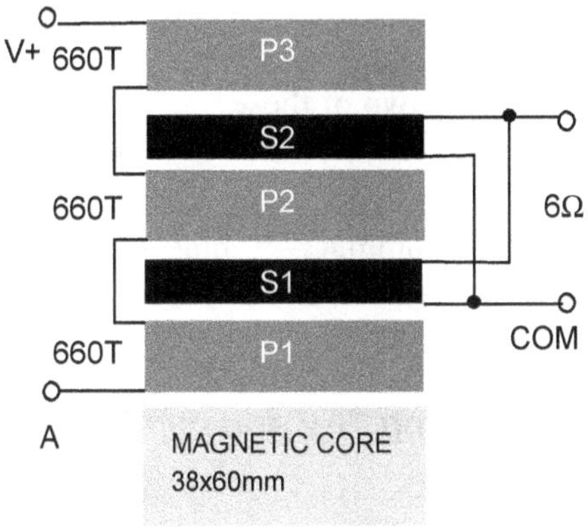

ABOVE: The winding diagram

BELOW: EI114 laminations

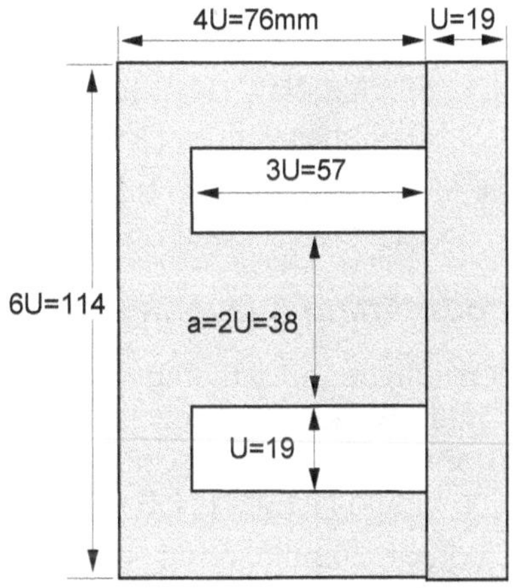

The primary signal voltage is $V_1=\sqrt{(PZ_P)} = \sqrt{(10*2,300)} = 151$ Volts. If we select 0.8 Tesla as AC magnetic flux density, once the DC flux is added, we should still be under the saturation level of around 1.6T for GOSS, so there is plenty of headroom.

Since $N_1=V_1 10^4/(4.44 f_L B_{MAX} A_{EF})$, we can express the product $f_L B_{MAX}$ as $f_L B_{MAX} = 10^4 V_1/(4.44 N_1 A_{EF}) = 6.75$ [HzT].

Again, assuming f_L of 10Hz, the induction of $B_{MAX}=0.675$ Tesla seems very conservative, keeping the distortion low.

Wire sizing and window fitting

There are 3x660 = 1,980 primary turns and 100 secondary turns. Thus, turns and voltage ratio is TR=1,980/100 = 19.8, while the impedance ratio is $TR^2 = 19.8^2 = 392$. A 6Ω speaker impedance would reflect onto the primary side as 6*392= 2,352 Ω.

The primary wire is specified as AWG30 size, meaning its diameter is 0.255mm, while the secondary wire is AWG 22 with 0.65mm diameter.

The window length of EI114 laminations is 57 mm. The winding length or Coil Length (CL) is CL= 57 - (2 x 2) mm = 57-4 = 53 mm.

The walls of a typical bobbin are 2mm thick, plus we should allow 1 mm of clearance for the insulation (cut slightly wider than the CL to ensure the last turn of the next section does not drop down onto the previous layer)! So, CL=52 mm.

PRIMARY: Maximal number of turns per layer TPL_{MAX}=CL/d = 52mm/0.26mm = 200 .

Since 660/200 = 3.3, let's assume 4 layers in each primary section. That would mean 660/4 = 165 turns in each layer and a Horizontal Fill Factor of HFF= 165/200 = 82.5%. That is reasonable.

SECONDARY: TPL_{MAX}= 52mm/0.65 mm = 80 turns, so to fit 100 turns 2 layers are needed, with 50 turns in each layer. HFF= 50/80 = 62.55, which is very low.

The secondary layers should always be fully filled since a large diameter wire gives the coil rigidity and mechanical strength. This means that the size of the secondary wire should be chosen that with the given number of turns per layer (50 in this case), the whole width of the bobbin is filled.

52/50 = 1.04, so 1mm diameter wire should be used.

The window is 19mm tall, but we lose 2mm on bobbin thickness and another 1mm on top for clearance, so our total height of the winding must not exceed 16mm.

Primary height: 12 layers*0.26 mm = 3.12 mm, secondary height: 4 layers*0.65 mm = 2.6 mm. Assuming 0.1mm insulation between layers and 0.2 mm between sections, the insulation height is 11*0.1 + 5*0.2 = 2.1mm.

The total winding height is 7.8 mm. Around 8 mm coil height in a window that is 19mm tall (16 mm useable height) means that Vertical Fill Factor or VFF is less than 50%, which is too low. Should you wish to use a similar design, let's see what will happen if we use larger diameter wires.

2nd iteration: Coil height with larger diameter primary and secondary wire

A larger diameter primary wire could also be used, say 0.35mm instead of 0.255mm.

For the primary, TPL_{MAX}=CL/d = 52mm/0.35mm = 148 and 660/148 = 4.6, so we can comfortably fit that into 5 layers in each primary section.

TPL=660/5 = 132, and HFF= 132/148 = 89%, perfect.

For the secondary, TPL_{MAX}=CL/d = 52mm/1mm =52 so with careful & tight winding we could fit 100 turns into 2 layers.

Primary height: 15 layers*0.35 mm = 5.25 mm, secondary height: 4 layers*1 mm = 4 mm. With an insulation height of 14*0.1 + 5*0.2 = 2.4 mm, the total winding height is 11.65 mm. Even with a 15% bulging factor, the coil height will not exceed 13.4mm, way under the 16 mm useable height.

DIY DESIGN: 1k8 SE OUTPUT TRANSFORMER FOR TWO 300B TRIODES IN PARALLEL

We bought two of these transformer sets (4 x C-cores in each) from an eBay seller in Bulgaria. The seller specified sheet thickness of 0.3mm, maximum flux density B_{MAX}=2.2T, and permeability μ=8,000 at B=1.1T.

The C-cores had "strip width" D=50mm, and "build up" E=12.5mm. The other dimensions were F=25mm, C=2E+F=50mm, G=50mm, and B=90mm. The mean magnetic path length, also called "the length of flux", was 25.1 cm, gross cross-sectional area was A = 2ED = 2*1.25*5=12.5 cm^2, so the core's power rating was approx. P=$(2A_{EF})^2$ =$4A_{EF}^2$ = 600VA. These would make a nice high-power mains transformer!

Determining TPV and number of turns

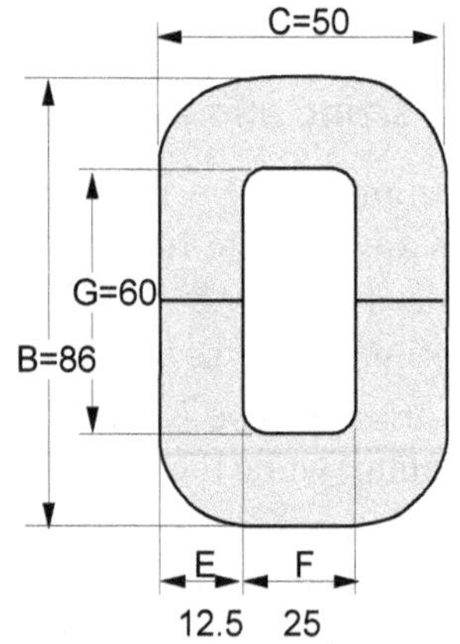

Assuming the stacking factor of 0.98, the effective cross-sectional area is A_{EF}=0.98A=12.25cm^2. From the basic transformer equation TPV=N/V= $10^4/(4.44BA_{EF}f)$, so now we need to decide on B and f. We want these transformers to transmit the full power down to 20Hz, so let's use f=20 and B=1.1 and see what we get.

TPV=N_1/V=10^4/(4.44*1.1*12*20) = 10,000/1,172 = 8.5.

It is possible to get 12W from a single 300B, so one would expect 24 Watts from two in parallel. The primary voltage will be $V_1=\sqrt{(PZ_P)}$ = $\sqrt{(24*1{,}800)}$ = 208V, which means N_1= TPV*V = 8.5*208 = 1,768 turns.

Since the impedance ratio is IR=1,768/8 = 221, the turns (voltage) ratio is TR= $\sqrt{IR}$ = $\sqrt{221}$ = 14.87, and finally, $N_2=N_1$/TR = 1,768/14.87 = 119 turns.

Window fitting & layering

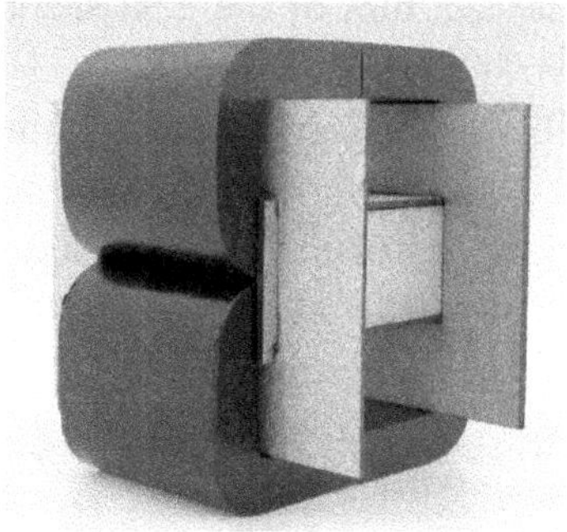

The width of the winding window is G=60mm, and the window height is F=25mm. We lose 4mm on the bobbin's walls thickness (horizontally) and 2 mm vertically, so the coil or winding length is CL=60-4=56 mm and the maximum coil or winding height is CH= 25-2=23 mm.

TPL_{1MAX}= 56mm/0.5mm =112. Since 1,768 turns/112 TPL=15.8 layers, we will use 16 layers with an actual primary TPL_1=1,768/16 = 110.

SECONDARIES:

TPL_{2MAX}= 56 mm/0.9 mm = 62 We need a total of 119 turns, 119/62 = 1.92, so we will work with 2 layers. Actual TPL_2 = 118/2 = 59.

Checking window height

With 4 primary & 3 secondary sections, there are seven insulation layers. Assuming 0.2 mm thick plastic sheets, the total thickness of insulation is 1.4mm.

The primary height is 16 layers*0.5 mm = 8 mm, secondary height is 6 layers*0.9 mm = 5.4 mm. We have 7 insulation layers between primary and secondary sections plus one final layer on top, and with 0.35mm Mylar insulation that is 2.8 mm overall.

Our total coil height is thus CH = PH+SH+IH = 8+ 5.4 + 2.8= 16.2 mm.

We need to allow for the "bulging factor" or "bowing effect of 15%, so 16.2 mm*1.15 = 18.6 mm. We have 23 mm to play with, so a very comfortable fit.

Measured results

Due to the large diameter of its wire, the primary's DC resistance is low, under 50 ohms, which is a great result. The leakage inductance is only 4 mH, another great achievement, as is the primary-to-secondary capacitance of 2.6 nF.

However, the primary inductance is also fairly low, less than 5H, which indicates that the magnetic material used in these cores has low permeability - the specified μ=8,000 is very low for C-cores.

The primary inductance L_P can be increased by increasing the number of primary turns N_1 (and N_2 as a consequence). We have an extra height of 4mm, so as an exercise, recalculate everything to fill the winding window vertically.

MEASURED RESULTS:

- R_P= 48 Ω
- R_S= 0.6 Ω
- L_P= 4.8 H
- L_L=4.1 mH
- C_{PS1000} = 2.6 nF
- BW @1W: 17Hz - 27kHz (-3dB)

COMMERCIAL SE TRANSFORMERS

James output transformers

Paradoxically, transformer winding is more difficult for most people than transformer design. If losing your hair and sanity over a winding machine does not appeal to you, these excellent budget transformers are just what the doctor ordered. James transformers are made in Taiwan and sold on eBay in black or silver finish. There are a dozen or so SE and as many PP designs. For low-powered SE amps, model 6123HS is of interest. Rated at 20 Watts, it has three primary taps (2.5, 3.5, and 5kΩ) and three secondary taps (4,8, and 16Ω). If you are unsure what load impedance, you should use with your design, this gives you lots of flexibility, more precisely, nine different possibilities.

The primary inductances at 120Hz and 1kHz are given in the table (below right), as are leakage inductances measured at 1 kHz. Notice a respectable result (7mH) for the 2k5 case and quite a high leakage for the 5k option, almost 33mH. This means the HF response of higher impedance taps will get progressively worse; the best obtainable is with 2k5 primary.

PRI.:	4Ω SEC.	8Ω SEC.	16Ω SEC.
2.5k	5k	2.5k	1.25k
3.5k	7k	3.5k	1.75k
5k	10k	5k	2.5k

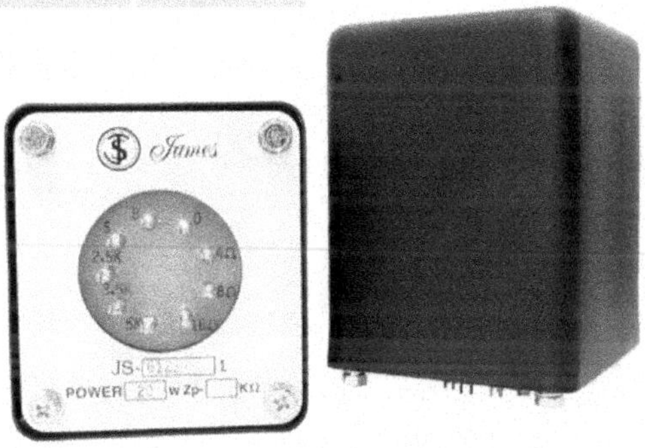

JAMES 6123HS: MEASURED RESULTS

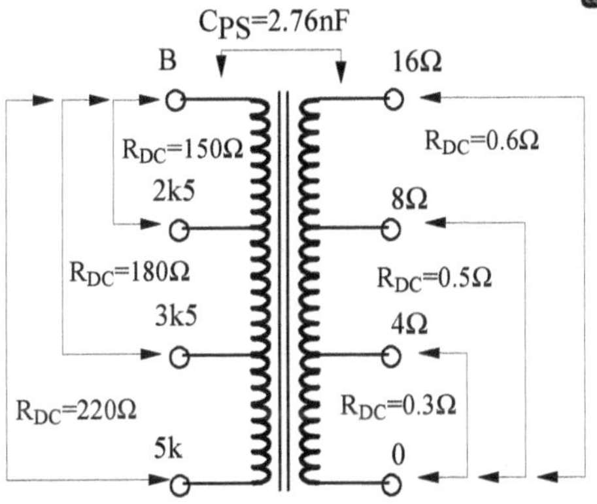

PRIMARY:	Lp [H] @ 120Hz	Lp [H] @ 1kHz	L_L [mH] @ 1kHz
2.5k	9.1	7.5	7.0
3.5k	10.6	10.3	14.7
5k	15.5	14.8	32.5

Amorphous cores: not recommended for output transformers

In 2018 an eBay seller from Hong Kong was selling three types of SE output transformers, 2.5, 3.5, and 5.0k primary impedance, using the same type and size of C-cores. They were made in China using amorphous C-cores allegedly sourced from the UK. It took the seller more than two weeks to ship our 3k5 pair, meaning he had none in stock and had to wait for them to be made.

The measured results show that these were not 3k5 but 4k5 transformers. The more problematic issue was the difference in primary inductance Lp between the two transformers, 18.3 versus 13.8 Henry. That indicates sloppiness during the core assembly, resulting in unequal air gaps between the two units and thus unequal Lp!

SPECIFICATIONS:

- Primary impedance: 3.5kΩ
- Primary DCR: 162Ω
- Primary inductance: 17H
- Secondary impedance: 0-4-8Ω
- Secondary DCR: 0.56Ω
- Maximum voltage: 2000V
- Maximum wattage: 32W
- Frequency range : 20 Hz - 35 kHz
- Maximum primary DC current: 80 mA

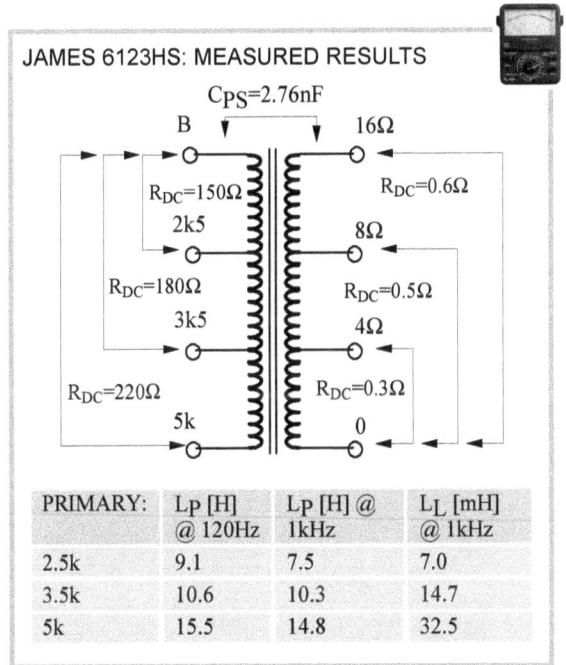

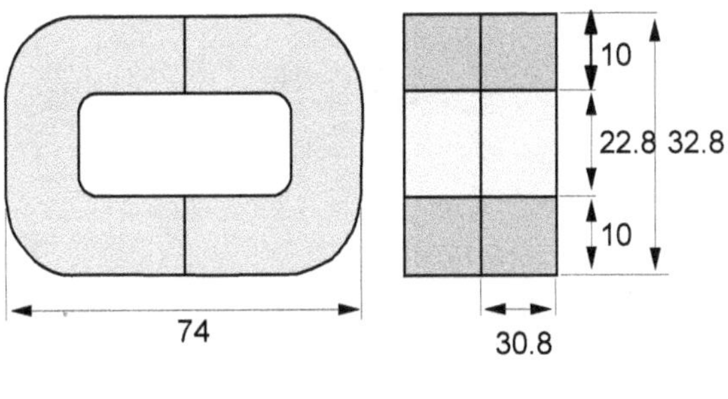

Indeed, there was no paper or plastic in the air gap, just two C-pieces hastily put together, not even aligned properly (1)!

Notice a large difference between primary inductance at 1kHz and 120Hz, 58.3H versus 18.3H, a factor of 3.2! That is bad news because it means the permeability is much higher at 1kHz than at 120Hz, and, as a result, the lower frequency limit of such a transformer will be compromised. The two inductance figures should be as close as possible, ideally equal.

Indeed, in both measurements and listening tests on a few 300B amplifiers, these transformers were inferior to even budget Chinese-made EI96 transformers, not to mention better SE transformers wound by us.

MEASURED RESULTS:

	TR. #1:	TR. #2:
Z_P [kΩ]	4.5	4.38
R_P [Ω]	173.8	174.0
R_S [W]	0.6	0.6
L_{P120} [H]	18.3	13.8
L_{P1000} [H]	58.3	42.0
L_{L1000} [mH]	11.4	13.2

For instance, one Chinese SE amplifier had a measured -3dB bandwidth of 14Hz -23 kHz with its original output transformers, as mentioned, but with these, its bandwidth shifted higher 31Hz-40kHz. These are relatively poor results (a jump from 14 to 31 Hz is a significant deterioration, a whole octave is lost), and the listening tests confirmed a weaker and softer bass, the last thing you need with SE amps whose bass is already inferior compared to push-pull tube amps, not to mention solid-state amps.

This was not the first Chinese-made output transformer with amorphous cores on our test bench. Such transformers cannot have a flat frequency response since their permeability changes not only with signal level (amplitude) but also with its frequency. For that reason, this type of (expensive) magnetic material is to be avoided in audio transformers.

PUSH-PULL OUTPUT TRANSFORMERS

- SPECIFIC ISSUES RELATING TO PUSH-PULL TRANSFORMERS
- COMMERCIAL CASE STUDY: RO-223 UNIVERSAL 10W HI-FI PP TRANSFORMERS
- DIY DESIGN: DUAL-COIL 4kΩ /1kΩ ULPP OUTPUT TRANSFORMER ON C-CORES
- THE WAY IT USED TO BE: WILLIAMSON OUTPUT TRANSFORMER (10kΩ PP)
- COMMERCIAL CASE STUDY: ERTOCO TOROIDAL OUTPUT TRANSFORMERS
- AUDIO CLASSICS: DYNACO ST70 ULPP OUTPUT TRANSFORMER (A-470)
- DIY DESIGN: HI-END 9kΩ ULPP TRANSFORMER (FOR EL84, 6V6)
- DIY DESIGN: LOW IMPEDANCE PP OUTPUT TRANSFORMER (6C33C-B and 6080)
- DIY DESIGN: 4 kΩ PP OUTPUT TRANSFORMER (2A3, 300B, EL34, 6L6)

8

SPECIFIC ISSUES RELATING TO PUSH-PULL TRANSFORMERS

Push-pull versus single-ended output transformers

In push-pull audio transformers, under ideal (balanced) conditions, there is no DC component through the primary. Thus, as the AC (signal) current reverses its direction, the magnetizing of the transformer core follows the major hysteresis loop and crosses the highly nonlinear region with low B values. As illustrated below, the magnetic flux (B) swings through all four quadrants.

The situation in single-ended transformers is different. The constant DC magnetization H_0 and its DC flux B_0 (magnetic bias) keep the core magnetized and never allow the flux to drop to zero. The AC signal current in the primary operates in the positive quadrant only and never reverses its direction as in push-pull conditions. The AC or signal flux "modulates" the steady-state or DC flux, as illustrated.

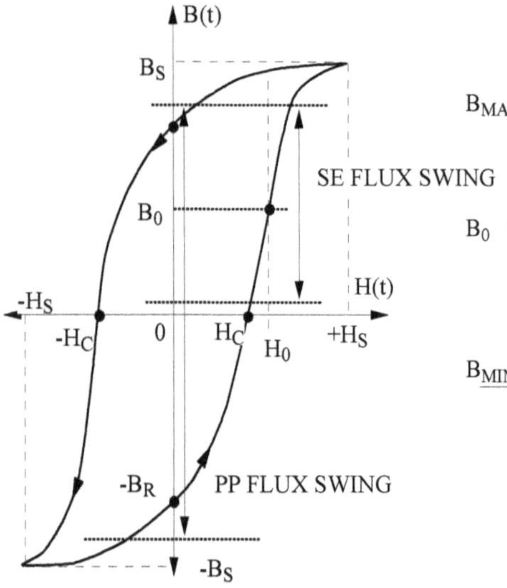

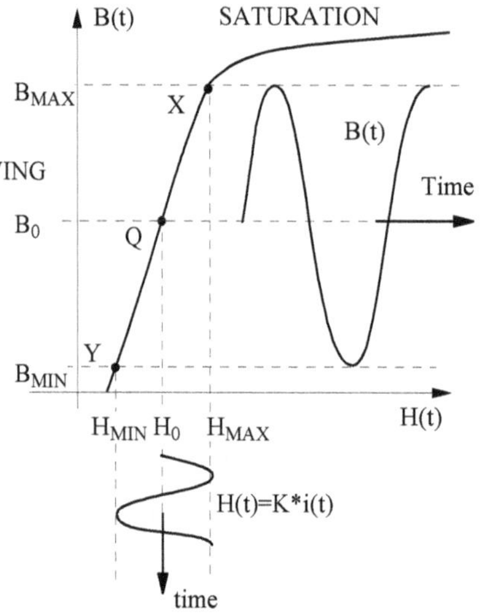

Providing that the DC flux level B_0 is properly chosen and that the maximum anode current never takes B_{MAX} into the saturation region, the transformer always operates in the most linear region of the B-H curve! Is that the secret behind the magic of single-ended sound?

FAR LEFT: With zero DC magnetization, the AC signal current in push-pull output transformers follows the major hysteresis loop.

LEFT: With a constant DC magnetization H_0 and its DC flux B_0 (magnetic bias!), the AC signal current in a single-ended transformers operates in the positive quadrant of the major hysteresis loop only and never reverses its direction as in push-pull conditions.

Push-pull winding arrangements

The first test you should perform on any push-pull transformer you have or are considering buying is to measure DC resistance of the two halves of the primary winding (anode #1 to CT and CT to anode #2). If there is a significant difference (say 180Ω compared to 130Ω) as illustrated below, the transformer is of "el cheapo" winding design and will always have a certain degree of unbalance.

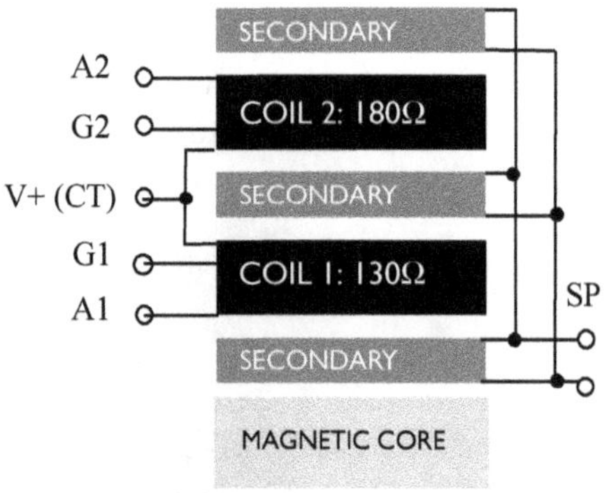

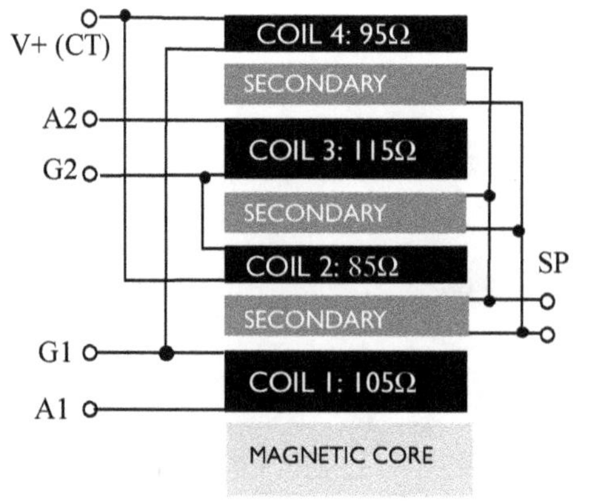

Apart from bifilar winding, which always ensures complete balance, both in DC and signal (AC) terms, connecting sections 1 and 4 together, as well as sections 2 and 3, ensures a much better DC and AC balance (left) compared to the simple winding used in budget transformers (far left)!

SIMPLE (UNBALANCED) WINDING

Its advantage is minimized primary capacitance due to fewer sections (five compared to seven for the balanced winding). However, the primary (coil 1 and coil 2) halves have different winding lengths. The upper coil 2 will have a higher DC resistance than the lower coil 1, which is closer to the core, so its mean length of a turn MLT will be much lower, and since the number of turns is the same, its DC resistance will be lower.

In one EL84 PP transformer, coil 2's DC resistance was 180 Ω, while coil 1's resistance was only 130 Ω! In such cases, the output stages will be dynamically unbalanced as well, since leakage inductances of the primary and secondary will not balanced, and that will affect upper-frequency behavior and result in higher distortion!

SUPERIOR (BALANCED) WINDING

The two halves are within a few ohms of one another; in another EL84 transformer, the primary sections had the following resistances (from the core upwards): 105, 85, 115, and 95 Ω. This means that the A1 half has a total resistance of 105+95=200 Ω and the A2 half's total resistance is 85+115=200 Ω, the same.

The price paid is increased complexity (seven sections instead of five) and time required for winding. Due to additional insulation layers between the sections and the higher number of sections, the overall coil height would increase, resulting in increased parasitic capacitances.

Now that you understand the basic difference between SE and PP transformers and the two basic winding topologies let's look at an illuminating commercial example from Japan.

COMMERCIAL CASE STUDY: RO-223 UNIVERSAL 10W HI-FI PP TRANSFORMERS

This vintage transformer by Tomoe Electro Eng. Works, Tokyo, is a good educational example; plus, if you wish to experiment with various primary and secondary impedances in your amplifier design, a pair of "universal" output transformers is a great tool!

These days most audiophiles would not go for a 10 Watt PP amplifier when you can get more power from most single-ended designs. However, the same design and winding principles apply to larger transformers, so some valuable lessons are to be learned here.

The largest scrapless EI lamination that would fit in the 73mm tall case would be EI66, with the other dimension of 55mm to fit the 63mm width of the case.

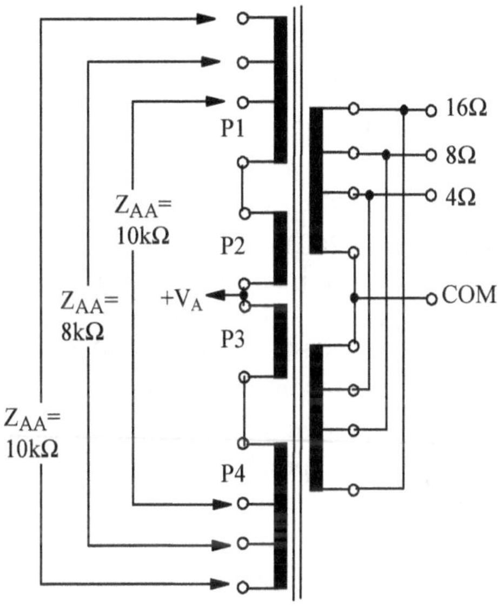

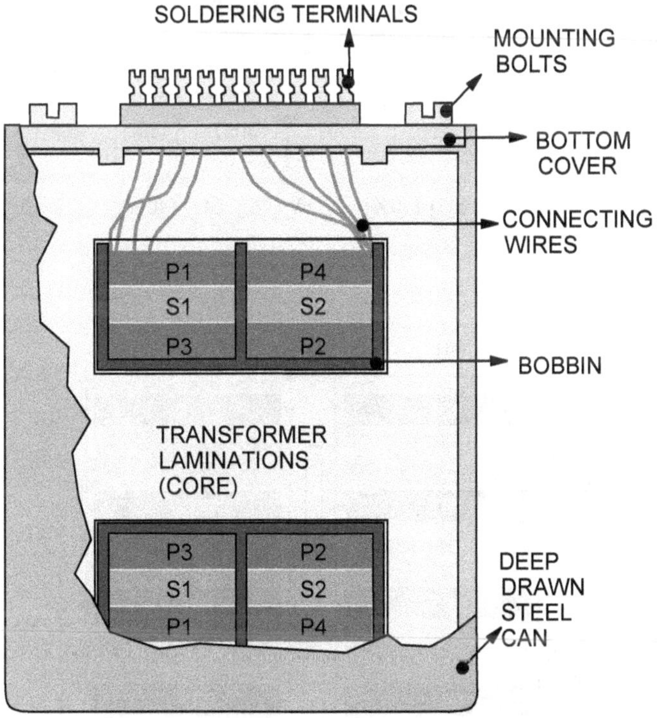

Vertically, one secondary section is sandwiched between two primary sections (2:1 sectionalizing), but 2:1 horizontal sectionalizing is also used, resulting in 4 primary sections and 2 secondary sections. Vertically splitting the winding in half reduces the shunt capacitance and increases the upper -3dB frequency f_U.

Notice the clever choice of the winding order. The two center sections of the primary (closest to the CT), P2 and P3, are wound first since they have no taps. Then, the two secondaries are wound side-by-side, followed by the topmost windings, primary sections P1 and P4, with 3 taps each.

The main benefit is the ability to connect all the taps on the two halves of the secondary windings in parallel (4-, 8- and 16Ω).

Since they were wound side-by-side, they have the same number of turns, equal DC resistances, and, most importantly, equal voltages. Only bifilar winding would achieve the same result.

The same benefit is obtained on the primary side; P2 and P3 windings are identical, as are P1 and P4. Without vertical split, the windings further from the core would have a higher MLT (mean length of turn) and higher DC resistance, so the two primary windings would be unbalanced!

MANUFACTURER'S SPECIFICATIONS:

- Primary impedances: 5, 8 & 10 kΩ
- Secondary impedances: 4, 8 & 16 Ω
- Power rating: 10 Watts @ 50 Hz
- Insertion loss: 1 dB (efficiency of 80%)
- Frequency response: 40 Hz - 50 kHz (+/- 1dB)
- Size: 80 x 63 x 58 mm, weight: 1.0 kg
- For use with: 45, 2A3, 6V6, 6F6, 6AR5, 6AQ5

DIY DESIGN: DUAL-COIL 4kΩ /1kΩ ULPP OUTPUT TRANSFORMER ON C-CORES

Since they were made from good quality GOSS material, the same Japanese C-cores from our power transformer example (D=38.1mm, E=25.4mm) can be used to make output or interstage transformers. The gross cross-sectional area is A=E*D= 2.54*3.81 = 9.68 cm^2 and A_{EF}= A*0.95 = 9.2 cm^2. We decided to make a universal ultralinear transformer, which can be used with 6L6, EL34 and is large enough even for KT88 and KT90 tubes.

For the maximum power of 45 Watts and a small cross-sectional area of under 10 cm^2, we need to choose a modest f_L of 30Hz; otherwise, the needed number of turns could be so high that it may not fit into the available window height of 22.2mm!

If 30Hz seems high, remember that negative feedback will lower that down to under 10 Hz in an amplifier, and, again, this is at peak power, and a hi-fi never runs at peak power level for long. Very few loudspeakers go down to 30 Hz anyway, so you need to keep requirements in perspective.

$TPV=10^4/(4.44BA_{EF}f_L)=$

$=10^4/(4.44*1.6*9.2*30) = 10{,}000/1{,}375 = 5.1$

$V_P=\sqrt{(PZ_P)}=\sqrt{(45*4{,}000)} = 424V$,

so $N_1=V_1TPV = 424*5.1 = 2{,}152$ turns!

This intuitively leads us to picture 1,000 turns on each bobbin, in 5x 200 turn sections. We could easily take the U/L taps after 400 turns, which would be at 40% of the primary winding (always counted from $+V_A$ start point!), or 400T/(400+200+400) =400/1,000 = 0.4 = 40% , which is very close to the optimal 43% for that tube.

Finally, instead of 5x200 sections, let's simplify things a bit and have only 3 primary sections per tube, of 400+200+ 400 turns.

To improve coupling and minimize leakage inductance, let's make things more interesting (after all, this is supposed to be an educational publication!) and intertwine the sections for tube 1 and tube 2 between the two coils, instead of all turns for tube 1 on coil A and all turns for tube 2 on coil B!

The primary has 2 x 1,000 turns of 0.42 mm wire; the secondaries (8 in parallel) have 90 turns of 0.48mm wire each.

The turns ratio is 2,000/90 = 22.22, and the impedance ratio is 493.8, which with an 8 Ω load gives us a primary plate-to-plate impedance of 3,950 Ω, very close to our starting point of 4,000 Ω.

We did try increasing the number of primary turns to get about 5,000 Ω Z_{PP}, but that skewed the frequency range too low; at 12 watts, fL dropped from 12Hz to 9 Hz, which was great, but the f_U dropped from 31 kHz to only 20 kHz. Also, the 4kΩ PP impedance option sounded better!

The two banks of 4 secondary windings can be connected in parallel (as in our case), or series.

In series, the turns ratio is 11.11 and ZR=123, which reflects an 8 ohm load as the primary impedance of just under 1 kohm plate-to-plate (Z_P=988 Ω), which would be fine for low impedance triodes such as 6080, E130L and 6C33C-B!

With 2 x 6L6GWT (USA) tubes, we got 30 Watts maximum power (1 kHz signal into 8W load) and 50 Watts with 2 x KT90 power tubes (Ei - Yugoslavia), very good results indeed.

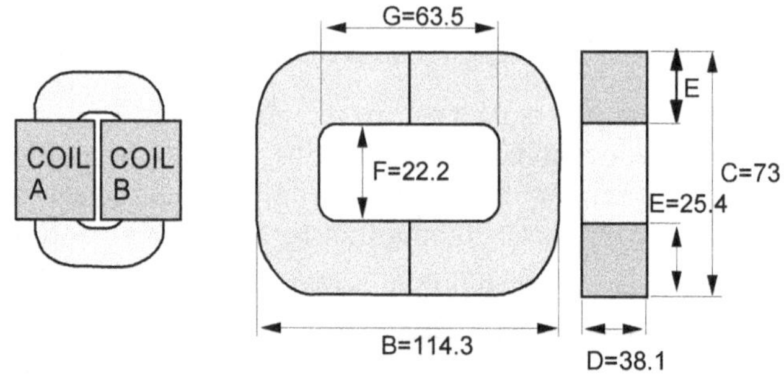

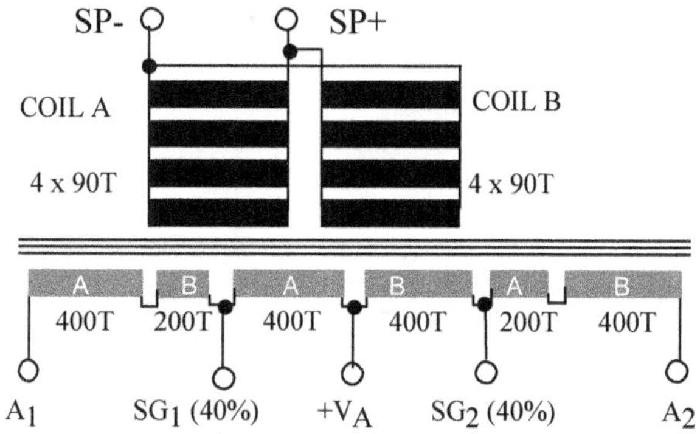

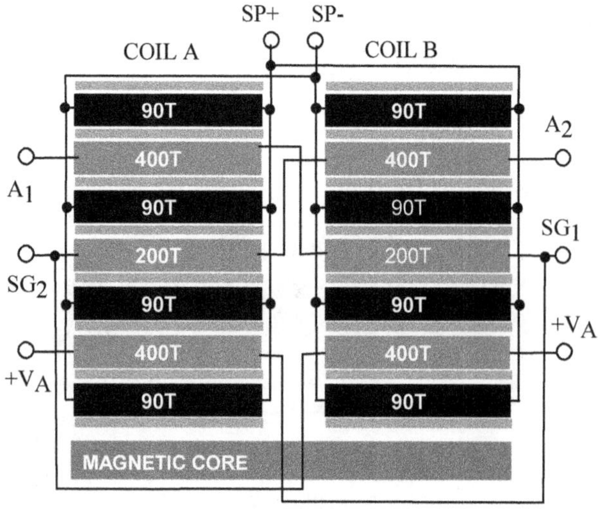

MEASURED RESULTS:

- L_P= 26 H, L_L=3.8 mH
- C_{PS1000} =5.6 nF
- BW @1W: 8 Hz-41 kHz (-3dB)
- BW@12W: 12Hz-33 kHz (-3dB)

The "phasing" procedure

To ensure you've connected the sections properly, check the phasing in the following manner.

Step 1: Wire all secondaries from coil A in parallel and coil B in parallel. Connect a small mains transformer's secondary (output) with a 3V-9V secondary (1) to the secondary coil A. Connect an AC voltmeter to monitor that test voltage (say $6V_{AC}$, for example).

Briefly connect the secondary from coil B in parallel to coil A (2). If there is a large voltage drop, the two windings oppose each other, disconnect immediately and connect them the other way around. If the voltage stays at the previous voltage ($6V_{AC}$ in our examples), the two secondary sections are connected in-phase. Don't touch the primary terminals during that test; high voltages will be present on the primaries.

Step 2: With the ancillary transformer still energizing the secondary windings (now properly phased and connected in parallel), connect the primary sections in order(+VA - SG - A, measuring voltages which should keep increasing if you connect them in phase.

Using a $6V_{AC}$ test voltage on the secondary, we got $120V_{AC}$ between +VA (center tap) and the SG tap, then $180V_{AC}$ between the +VA terminal and the end of the 200 Turns section, and then $300V_{AC}$ across the whole half primary (120V+60V+120V=300V), meaning all were connected properly, in phase.

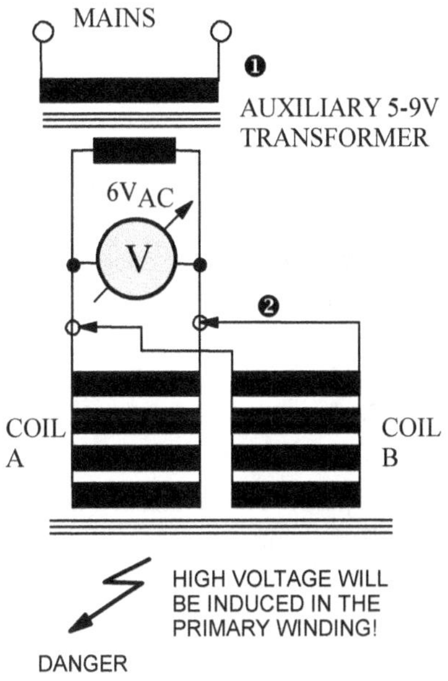

THE WAY IT USED TO BE: WILLIAMSON OUTPUT TRANSFORMER (10kΩ PP)

The Williamson amplifier was one of the seminal developments in hi-fi history. Let's look at how its output transformers were designed and wound, starting with their description from the original article published in The Wireless World:

"Core: 1.75 inch stack of 28A Super Silcor laminations. The winding consists of two identical interleaved coils each 1.5 inch wide on paxolin formers 1.25 x 1.75 inch inside dimensions. On each former is wound 5 primary sections, each consisting of 440 turns (5 layers, 88 turns per layer) of 30 SWG enameled copper wire interleaved with 2 mil. paper, alternating with 4 secondary sections, each consisting of 84 turns (2 layers, 42 turns per layer) of 22 SWG enameled copper wire interleaved with 2 mil. paper.

Each section is insulated from its neighbors by 3 layers of 5 mil. Empire tape. All connections are brought out on one side of the winding, but the primary sections may be connected in series when winding, two primary connections only per bobbin being brought out. Windings to be assembled on core with one bobbin reversed, and with insulating cheeks and a center spacer."

First, the conversion of dimensions and SWG (Standard Wire Gage) sizes. 1 mil is one thousandth (0.001) of an inch or 0.0254 millimeters, so 2 mil paper is 0.05mm. The thickness of 3 layers of 5 mil tape is 15 mil or 0.38mm. 1.75" stack thickness is 4.45cm, the coil width is 1.5" or 38mm. The primary is wound with wire diameter d=0.315mm, the secondary with wire d=0.71mm.

Secondly, we don't know what 28A size means, but we do know that the winding window is 3" or 7.62 cm wide. The current EI size table reveals that EI152.4 lamination has the same window width; we will base our analysis of this transformer based on that assumption.

The laminations in the original Williamson transformers were made of ordinary 4% silicon steel, so to get a high primary inductance of around 100H, Williamson had to use a large core. The cross section of the center leg had an area $A=aS = 5.08*4.45 = 22.6\ cm^2$.

With much better GOSS and amorphous materials nowadays, you should be able to achieve the same or even better specs with smaller laminations such as EI114. Remember, the larger the core, the higher the primary inductance, but also the higher the leakage inductance and the distributed capacitance, and the lower the upper -3dB frequency limit of the transformer. That is why the bobbin was split vertically in half - to reduce the distributed capacitance and improve the upper-frequency limit of the transformer.

> SWG is an abbreviation for Standard Wire Gauge, also known as Imperial Wire Gauge or British Standard Gauge. 30 SWG wire has a diameter of 0.315 mm , while 22 SWG is d=0.711mm. To add to the confusion, there is also AWG, which is an acronym for American Wire Gauge, which is, of course, different from SWG. For instance 30 AWG would be 0.255 mm, while 22 AWG is 0.644mm.

The winding arrangement

The primary impedance is specified as 10kΩ and the turns ratio as 76:1. That means the impedance ratio is $IR=76^2 = 5,776$! The article explains that the impedance of each secondary section is 10,000/IR = 10,000/5,776 = 1.73Ω.

Since there are five primary sections, all connected in series, each half has 5x440T=2,200 turns, but since there are two vertical halves, the total (plate-to-plate or anode-to-anode) number of primary turns is 4,400.

The number of turns in each secondary section should then be $N_2 = N_1/TR = 4,400/76 =$ 58, exactly as specified.

Once the bobbin walls are subtracted from half the window width of 38mm, each coil has a width of around 35mm. With 88 turns in each layer of d=0.315 mm wire, that is 88*0.315 = 28mm. With 42 turns in each secondary layer of d=0.711wire, that is 29*0.711 = 30 mm.

The physical diameters of magnet wires are always larger than their effective areas specified (copper area), so ϕ=0.315 wire may have a diameter between 0.33 and 0.36, depending on the number of insulation varnishing coats, one, two, or three!

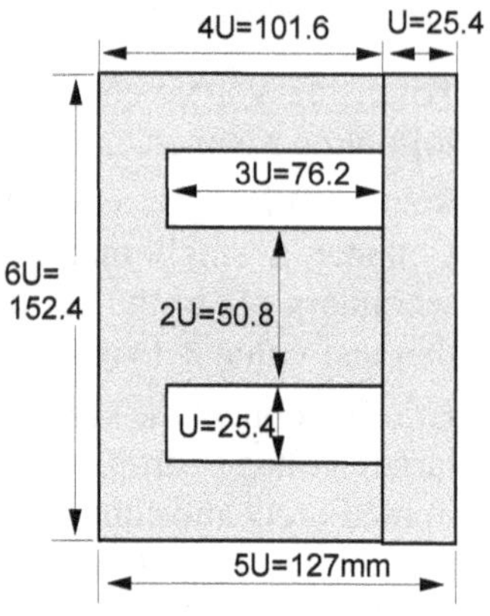

EI152.4 LAMINATIONS

RIGHT: The winding diagram of the original Williamson's output transformer

BELOW: Assuming the vertical secondary sections are connected in parallel (S1a and S1b, S2a and S2b, etc.), there are four possible ways of doing so, resulting in low output impedances of between 1.73 and 6.92Ω. Without paralleling vertical sections, higher output impedances are possible, but the coupling would be unbalanced and the performance diminished!

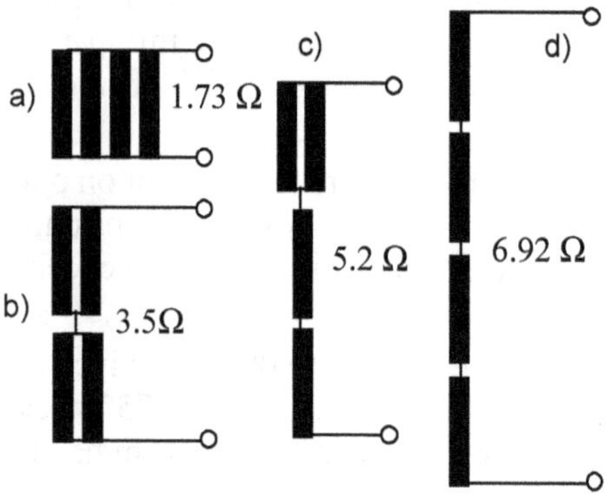

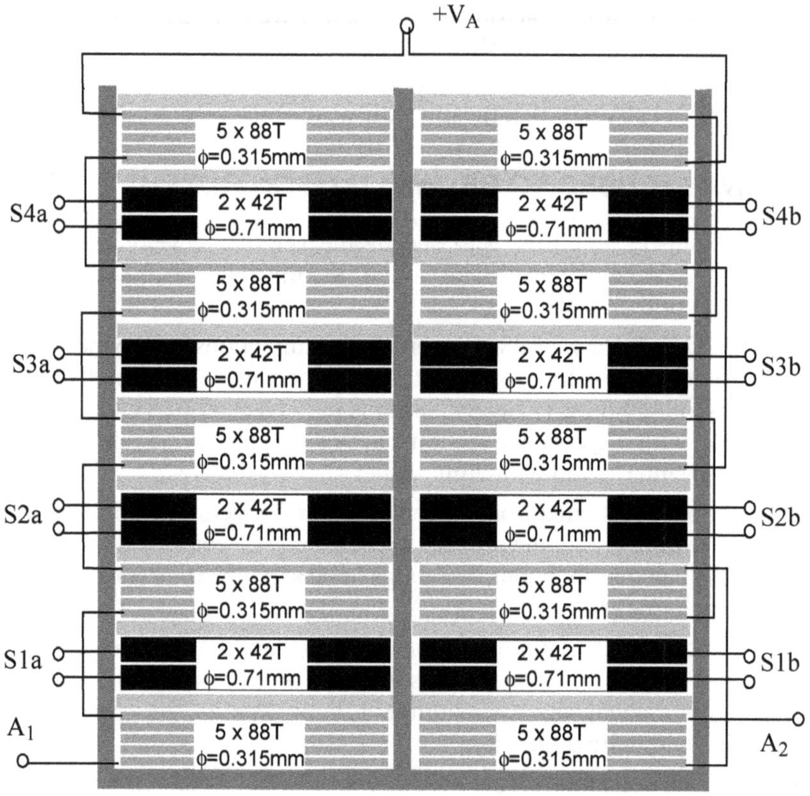

COMMERCIAL CASE STUDY: ERTOCO TOROIDAL OUTPUT TRANSFORMERS

Ertoco, a transformer making business in Vilnius, Lithuania, started selling on eBay late in 2015 so we got a pair of their toroidal SE output transformers. Rated at 10 Watts with a 2k5 primary, they are suitable for 2A3, 300B and many other triodes and pseudo-triodes of around 30-40 Watts anode dissipation, such as F2a and EL153 strapped as triodes.

The square wave reproduction is quite good, but the primary inductance is only 6H. With a relatively high leakage inductance of 26mH, the quality factor is only 6/0.026 =250.

If you are on a tight budget and/or want to mount output transformers under a relatively thin chassis, you may consider these or similar low profile transformers.

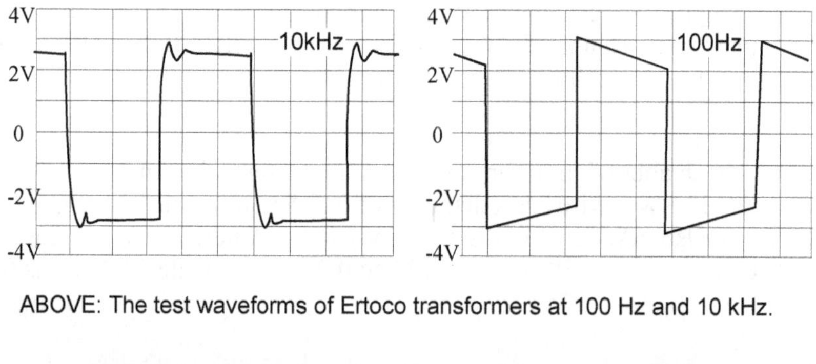

ABOVE: The test waveforms of Ertoco transformers at 100 Hz and 10 kHz.

MANUFACTURER'S SPECIFICATIONS:

- Input / output impedance: 2,500 Ω / 8 Ω
- BW: 10Hz - 130 kHz (-3dB)
- Primary / secondary DC resistance: R_1=50 Ω, R_2=0.3 Ω
- Maximum primary DC current: 150 mA
- L_P=6.5 H @ 100 mA, L_L= 26mH
- Rated power: 10W
- Diameter 105mm, height 40mm, weight 1.6 kg

AUDIO CLASSICS: DYNACO ST70 ULPP OUTPUT TRANSFORMER (A-470)

Dynaco A-470 is a highly acclaimed output transformer used in iconic ST-70 amplifiers with two EL34 tubes in Class AB_1 push-pull per channel. The 4.3 kΩ push-pull EI transformer is rated at 35 Watts, with simple 3/2 sectionalizing (3 primary sections and two secondary sections).

EI84 laminations were used with a 45mm thick stack, resulting in A=2US= 2*1.4*4.5= 12.6 cm^2 center leg cross-section.

WL=42, Coil Length CL=WL-5=38 mm.

The primary wire is AWG33. While the maximum turns-per-layer is TPL_{1MAX}= CL/d_1= 38/0.2mm = 190, the actual primary turns-per-layer used is TPL_1=180, meaning the horizontal fill factor is HFF_1=180/190 = 95%.

The total number of primary turns is N_1= 2*(360+720) = 2*1,080 = 2,160 turns.

The secondary wire is AWG21, wound in a bifilar fashion (two wires wound in parallel). TPL_{2MAX}=CL/d_2= 38/(2*0.72) = 26.4, with the actual TPL_2 =23, so HFF_2= 23/26 = 88.5%. With bifilar winding, the HFF is inevitably lower, below 90%, due to lost space between each turn and the next one and between the two parallel wires in each turn itself.

Since N_2 (8Ω)=92 turns, the turns ratio (8Ω) is TR= 2,160/92= 23.48 and the impedance ratio is ZR= 551. This means that an 8Ω speaker impedance would reflect back to the primary as a Z_P=8*551= 4,408 Ω load.

The white arrows on the winding diagram indicate the rotational (winding) sense.

A single layer of paper insulation is used between all layers, while triple insulation is used between sections, a single paper layer sandwiched between two plastic insulation layers.

ABOVE: Our redesigned & improved ST-70, with new fascia and bias test points in front of the power transformer. The on-off rotary switch (left knob) and volume control (right knob) are at the front!

Dynaco's David Hafler's 1957 patent simply and cryptically titled "Transformers" explains the reasons and benefits of this winding scheme. For more information visit:

http://www.google.com/patents/US2815408

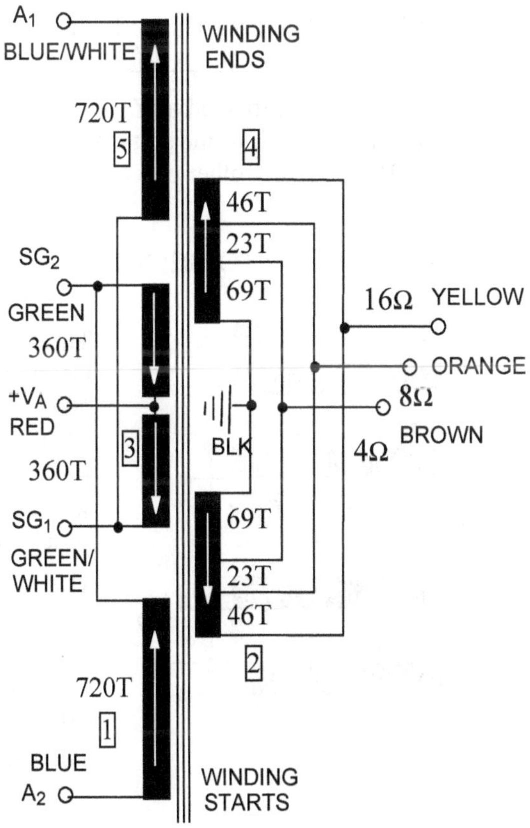

RIGHT: The winding diagram and connections of Dynaco A-470 push-pull output transformers

DIY DESIGN: HI-END 9k ULPP TRANSFORMER (FOR EL84, 6V6)

These low voltage secondary mains transformers sold under the Powertran brand name can be rewound as output transformers.

Since a=2.86 cm and S=4.2 cm (EI85.8 laminations), A= aS = 12 cm^2, so the rough estimate of their power rating at 50Hz is $P=A^2$ = 144 VA.

Electrically, the rating is 24V*5A=120VA, so they look promising for push-pull transformers of up to 50VA of RMS power.

Design parameters

Detailed design steps were outlined in many other designs in this book, so they will not be repeated here; let's look at the final winding parameters.

Window length: WL=43mm, coil length CL=39mm

The 0.22mm wire is actually 0.24mm in diameter (with insulation) so TPL_{1MAX}= CL/d_1= 39/0.24 = 162T. We will use TPL_1= 150, so each half of the primary will have a 600 turn section (4 layers) and a 900 turn section (6 layers).

TPL_{2MAX}= CL/d_2= 39/0.66 = 59T, we will use TPL_2= 45.

The coil height is CH=PH+SH+IH = 2*10*0.24 + 4*0.66 + 8*0.2 = 4.8+2.7+ 1.6 = 9.1 cm or 9.1*1.15= 10.5 cm with a 15% bulging allowance.

The window height is 14.3cm, and the maximum coil height is 2mm less (for bobbin wall thickness and top clearance) or 12.3cm, so we have 10.5/12.3= 85.4% vertical fill, perfect!

P: 600+900+900+600= 3,000T wire d = 0.22 mm

S: (45T + 45T) in parallel with (45T + 45T) = 90T|| 90T, wire d= 0.60 mm

TR=3,000/90 = 33.33 and $IR=TR^2$ = 1,111

With an 8Ω load, the primary impedance is Z_{AA} = 1,111 x 8 = 8,888Ω, which is perfect for 2x EL84 or 2x6V6 tubes.

A 4Ω load would be reflected as Z_{AA} = 4,444Ω impedance to the primary side, which would suit a pair of EL34, 6L6, 6AR6, 7027A, 7591, and many other power tubes.

INFORMATION ON THE ORIGINAL MAINS TRANSFORMER

- Powertran M2190L
- Primary: 240V Secondary: 24V, 5A
- VA rating: 120VA
- Weight: 2.2kg
- Magnetizing current: <85mA
- Temperature rise: <65°C
- Core dimensions: EI85.8
- 0.35mm thick laminations

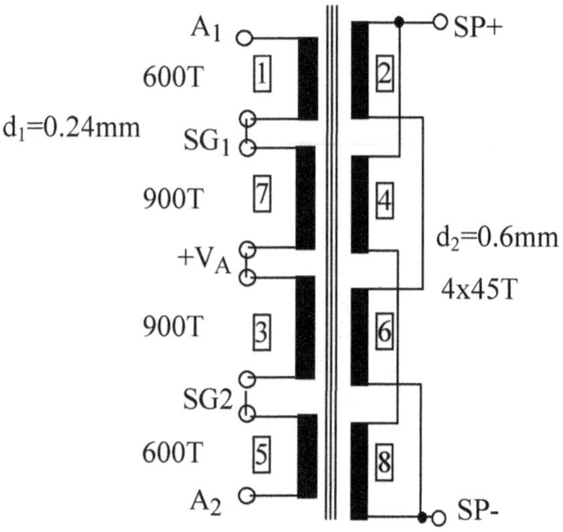

MEASURED RESULTS:

- R_{AA} = 175 Ω L_{AA}= 68H
- L_L=8.1 mH
- C_P=2.1 nF, C_S= 938 nF, C_{PS} = 1.5 nF
- BW @12W: 11Hz - 46kHz (-3dB)

THE WAY IT USED TO BE: VINTAGE GERMAN 50W PP OUTPUT TRANSFORMERS

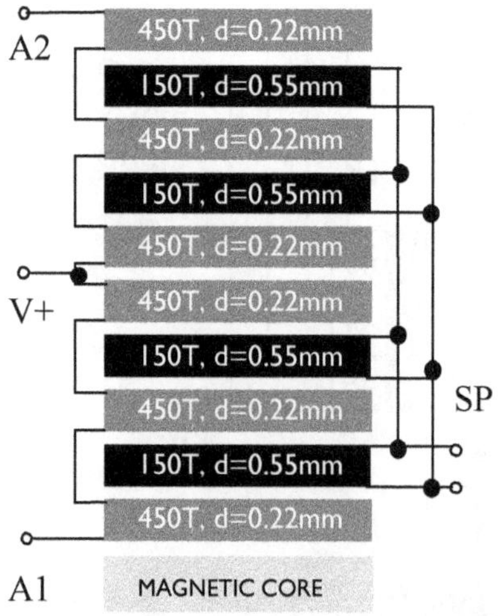

These push-pull output transformers from 50W-rated EL34-based vintage German amplifiers used M102A laminations. Core dimensions were 103x103x38 mm. Such transformers are easy to spot: if a transformer has a square footprint and rounded corners of laminations, it uses M-laminations.

Primary: 6 sections with 450 turns each (wire d=0.22mm), the secondary is in 4 sections of 125 turns each, wire d=0.55mm. All primaries were connected in series, all secondaries in parallel.

The nominal primary impedance was declared as Z_{PP}= 4 kΩ (plate-to-plate impedance with an 8Ω load). The turns ratio is TR=2,700/125 = 21.6 and $IR=21.6^2$ = 466.6, so the 8Ω load is reflected as Z_{PP}= 8*466.6 = 3,733Ω primary plate-to-plate impedance. Close, but not quite the precision you'd expect from normally pedantic Germans.

The primary DCR was R_P=115 Ω.

DIY DESIGN: SPLIT COIL C-CORE PP OUTPUT TRANSFORMER FOR EL84, 7189 OR 6V6

Even if you don't have or don't want to use C-cores, you can choose the same topology for implementation on EI laminations. You will not have two completely separate bobbins but one bobbin split vertically in half. The DC plate currents flow from the common CT (Center Tap) towards anodes A1 and A2. The magnetic fluxes produced by these currents (as per the right-hand rule) need to oppose each other, so if the currents are equal, the two fluxes will cancel one another, and the magnetic core will not be magnetized.

Choosing the lamination size and stack thickness

Two EL84 in push-pull as pentodes can easily produce 12 watts of power. Z_{PP}=8,800 Ω is the most common plate-to-plate load used for EL84 in push-pull. If you are happy with the lower -3dB frequency of 20 Hz, we can calculate the required cross-sectional area of the center leg: $A_{EF}=k\sqrt{(P/f_L)} = 20*\sqrt{(12/20)} = 15.5\ cm^2$.

EI114 laminations with 2U=38mm need a stack thickness of 15.5/3.8= 4.1 cm. The standard bobbins are 44 mm, which is ideal. The stack thickness versus center limb ratio is S/2U= 44/38= 1.16, which is very good.

This ratio should never exceed 2.0 for output transformers. A square winding (where S=2U) would be ideal, and this ratio is almost there.

The new A= 3.8*4.4= 16.72 cm^2 and A_{EF}= 0.96*16.72= 16 cm^2.

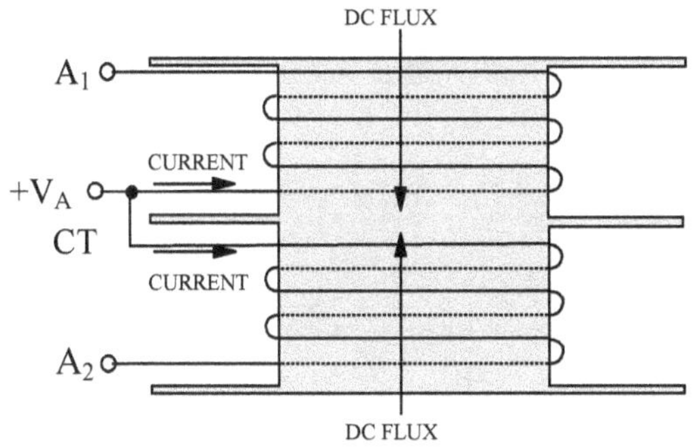

ABOVE: The winding direction stays the same for the two sections of a push-pull output transformer with vertically-split bobbin.

Primary and secondary turns

Primary AC signal voltage (plate-to-plate) $V_{PP}=\sqrt{(PZ_{PP})} = \sqrt{(12*8{,}800)}$= 325V.

Primary turns: $N_1=V_{PP}10^4/(4.44f_LB_{MAX}A_{EF}) = 325*10^4/(4.44*20*0.8*16) = 325 * 8.8 = 2{,}860$.

With Z_{PP}=8,800 Ω and Z_S=8 Ω, we get the impedance ratio of IR= Z_P/Z_S = 8,800/8= 1,100, so the turns ratio TR=$\sqrt{(IR)} = \sqrt{1{,}100}$= 33.16.

Many designers use 3.2 Ω instead of 4 Ω and 6.4 Ω instead of 8 Ω in this step, since most speakers' average impedance is lower than the "nominal" 8 Ω ,and is closer to 6 Ω. Plus, there are lots of 6 Ω speakers on the market. In that case, with Z_P=8,800 Ω and the load Z_S=6.4 Ω the impedance ratio changes to IR= Z_P/Z_S =8,800/6.4 = 1,375 and turns ratio TR= $\sqrt{(IR)} = \sqrt{(1{,}375)}$= 37.1 and $N_2=N_1/TR$ = 2,860/37.1 = 77.1 turns.

Primary and secondary currents and wire sizing

The DC current through primary will be between 60 and 80 mA. The AC component is $I_{AC}=\sqrt{(P/Z_P)} = \sqrt{(12/8{,}800)}$= 37 mA, so the total maximum primary current is $I_1= I_0+I_{AC}$ = 80+37 = 117 mA.

Secondary current at maximum power $I_2=\sqrt{(P/R_L)} = \sqrt{(12/6.4)}$= 1.37A.

Primary wire: $d_1=0.71\sqrt{(I_1)} = 0.71\sqrt{(0.117)} = 0.243$ mm Since we chose the worst case scenario (high bias current and low load impedance), let's stick with the 0.24 mm wire.

Secondary wire: $d_2=0.71\sqrt{(I_2)} = 0.71*\sqrt{(1.37)} = 0.83$ mm.

We are not going to automatically select 0.8mm wire. Since secondary windings or their sections have a small number of turns, we need to consider how many we can fit into the available bobbin length. Plus, the calculated size is the minimum we should use. To minimize the output resistance and increase the damping factor, use the largest secondary wire that can fit.

Primary and secondary layering

Maximum number of turns per layer: $TPL_{MAX}= WL/d_1$= 26/0.24 = 108T.

We have to divide 2,860 primary turns into 10 sections, so each section will have 2,860/10=286 turns.

Since 286T/108T= 2.65 layers, we will have 3 layers per section. TPL=286/3= 95.3, so let's use TPL=96, or 288 turns-per section.

The horizontal fill factor HFF is 96/108= 88.9%, meaning we have an 11% margin, which is enough, providing winding is done carefully and professionally. This complex winding job is not for beginners anyway!

The winding length of one-half of the bobbin is WL=(57-5)/2= 52/2=26mm, so if we distribute 86 turns over two rows, we will have 43 turns in each row.

Since 26mm/43 turns = 0.6 mm, that is the maximum diameter wire we can use, so all 43 turns will fit in one layer. Referring to the connection diagram below right, we will have two halves connected in parallel to form one secondary, and then four such secondaries S1-S4 (or 8 sections) will be paralleled again (for 3.2Ω output), illustrated in a). For 6.4Ω load, two and two secondaries will be paralleled again, as in b) and then connected in series. For the 16Ω output, the four secondaries will be connected in series, as in d).

We have 8 sections, so the current through each section will be 1.37A/8 = 171mA. The 0.3mm diameter wire can take up to 181mA, so we could use such wire, but then our winding length would not be fully filled.

Also, the secondary resistance would be higher and the damping factor lower. The largest standard wire size we can comfortably fit is 0.6 mm, which is good for 322 mA. So, let's settle on 0.6 mm secondary wire!

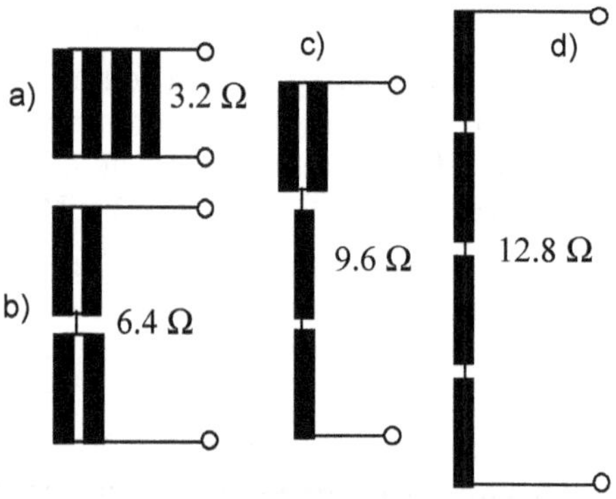

ABOVE: Four separate sections give us four choices, of which a) would be used as a 4Ω output, b) as an 8Ω output and d) as a 16Ω output.

Estimating the window fit

Method 1 (by area):

The gross window area for each vertical half is $F=1.9cm*2.6cm = 3.9\ cm^2$. We use German "Fenster" for "window", thus the capital "F"!

The primary area is $F_P=2{,}860/2*0.24^2*\pi/4=65mm^2 = 0.65cm^2$. The area of each secondary section is $F_S=86*0.6^2*\pi/4 = 24\ mm^2$, or $1cm^2$ for all four of them. Total copper area is $F_C= F_P+F_S=0.65+1cm^2 = 1.65\ cm^2$

The ratio of copper area to gross window area is 1.65/3.9 = 42.3%, and since it is higher than 40% (border case), we may not be able to fit our windings. Let's use a more precise method to see if we should proceed with this design.

Method 2 (by winding height):

The window is 19mm tall, but we lose 2mm on bobbin thickness and another 1mm on top for clearance, so our total winding height must not exceed 16mm!

Primary height: H_1=5 sections*3 layers*0.24 mm = 3.6 mm, secondary height: H_2=4 sections*2 layers*0.6 mm = 4.8 mm.

With two layers of 0.15mm Mylar® insulation between sections, the number of insulation layers is nine. The total insulation thickness is H_I=0.3 mm*9 = 2.7 mm

Total winding height: CH = PH+SH+IH =3.6+4.8+2.7 = 11.1 mm. Allowing for the "bulging factor" of 15%, we multiply H by 1.15 and get 11.1mm*1.15 = 12.8mm. We have filled 12.8/16 = 80% of the window height, pretty close to optimum!

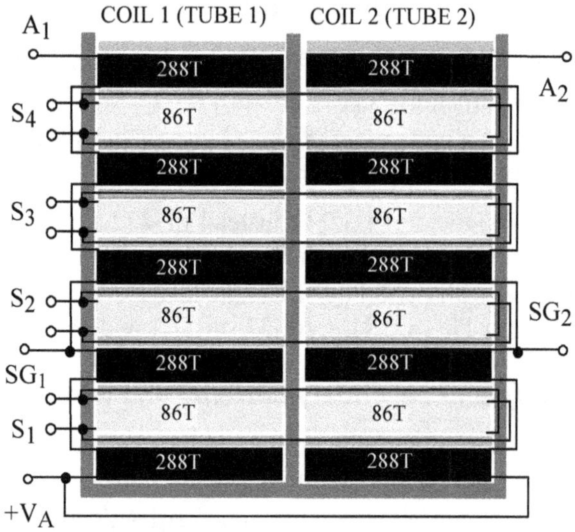

ABOVE: The winding diagram for EL84 PP output transformer with ultralinear taps for screen grids.

DIY DESIGN: LOW IMPEDANCE TRIODE PP OUTPUT TRANSFORMER (6C33C-B and 6080)

The beauty of low impedance triodes

We have already seen how the f_L (the lower +/- 3 dB frequency of the output transformer) depends primarily on the L_P (primary inductance) and the source resistance (R_I or the internal resistance of the tube). Tubes with low R_I, such as the triodes mentioned above, require low values of L_P to achieve low f_L, compared to high internal impedance tubes such as 6BQ5 pentode or high-impedance directly-heated transmitting triodes such as 211, 811, 813, 845.

This means that transformers for these low impedance triodes are the easiest and fastest to make since the number of primary turns N_1) can be relatively low (L_P is also low, and L_P depends on N_1).

We often get inquiries from DIY novices who want to make their transformers for an Ongaku clone, and we have to dissuade them from such a project since 211 tubes operate on 1,200 V_{DC}, meaning the insulation requirements are very stringent, and such high Z_P transformers are hard to wind.

A beginner should start with winding a transformer of this kind, which operates on lower voltages (typ. 200-300 V_{DC}) and requires fewer turns.

Key design parameters

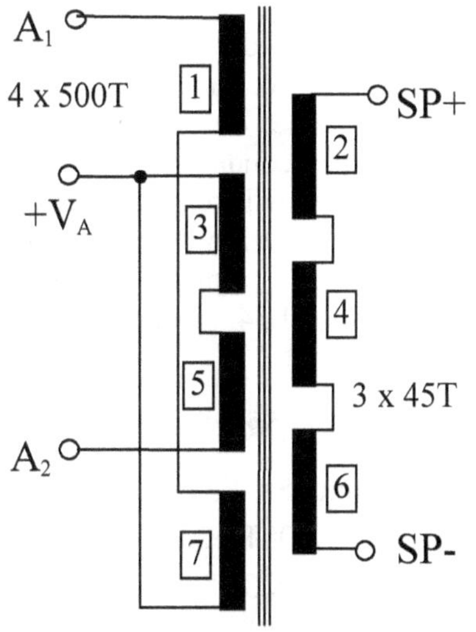

EI114 laminations: a=38.1 mm S= 4.5 cm A= aS = 17 cm^2

Primary: 500T + 500T + 500T + 500T = 2,000T wire d_P = 0.25 mm

Secondary: 3 x 45T = 135 Turns, wire d_S = 1.0 mm

0.315mm wire would be OK, but we want the output impedance as low as possible, meaning the secondary wire diameter should be as high as possible. For mechanical reasons (the strength and tightness of the wound coil) and to achieve a uniform height of the windings, we want those 45 turns to fill a whole layer, and a larger wire will help us do just that.

I_0 = 200mA I_{MAX}=280 mA per tube

TR=2000/135 = 14.8

ZR=TR^2 = 220, $Z_P(8\ \Omega)$ = 220 x 8 = 1,760 Ω plate-to-plate

Sectionalizing

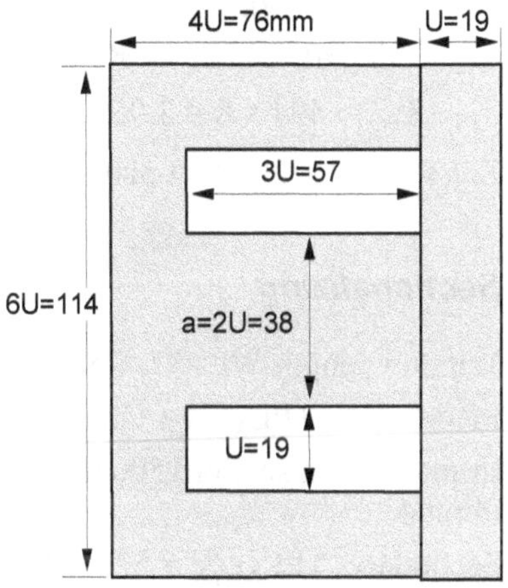

EI114 laminations

WL=57 mm, useable window length or coil length CL=52 mm

Primary sectionalizing:

TPL_{MAX}=CL/d_1= 52mm/0.25mm = 208 T, 500T/208T = 2.4 layers, so we will use 3 layers for each of the 4 primary sections.

500 turns / 3 layers = 166 (actual TPL)

Check: 166/208 = 79.8%, perfect!

Secondary sectionalizing:

Theoretical maximum TPL (turns per layer): 52mm/1.0 mm = 52 turns.

We need only 45 turns, so each secondary section will fit in one layer!

Check: 45/52 = 86.5% HFF, almost perfect.

Winding

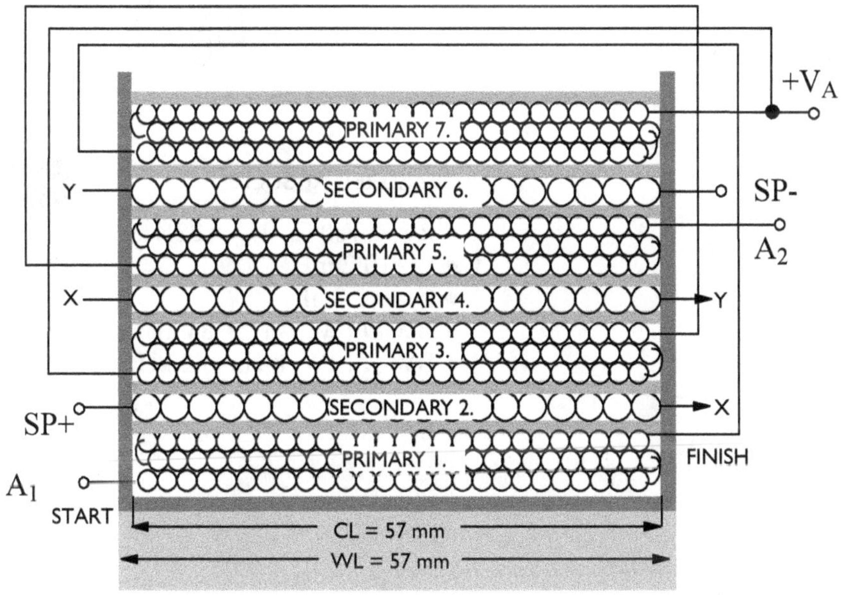

As per previous discussions, to balance the DC and AC properties of the two halves of the primary winding, the four sections are not connected consecutively A_1-V_A-A_2, but the top half is split (Section 1, the first half of the A_1 winding is wound first and its second half, section 7, is wound last) and two bottom sections (3 and 5) are inserted in between them.

This way, section 1 will have the lowest R_{DC}, section 7 will have the highest R_{DC}, and combined, their total R_{DC} will be approximately the same as the R_{DC} of sections 3 and 5 together!

DIY DESIGN: 4 kΩ PP OUTPUT TRANSFORMER (2A3, 300B, EL34, 6L6)

Manufacturers' datasheets from 1938, reproduced on the next page, for 2A3 (upper) and 300B triodes (lower), specify 3kΩ push-pull impedance of a fixed-biased 2A3 output stage and 5kΩ plate-to-plate impedance for cathode- or self-bias. Interestingly, 300B tubes are specified at 4 kΩ for both. Instead of designing two different transformers, we can use the same 4 kΩ design for both tubes.

It is the same design (same lamination size, EI114, and sectionalizing) as the previous example. The only difference is a lower number of secondary turns (higher impedance ratio), so we will only give you the summary here; the design process will not be repeated.This primary impedance is also suitable for EL34, 6L6, and 7027 power tubes.

Values are for 2 tubes	Fixed Bias	Cathode Bias	
Plate Voltage	300	300	volts
Grid Voltage	-62	-	volts
Cathode-Bias Resistor	-	780	ohms
Peak AF Grid-to-Grid Voltage	124	156	volts
Zero-Signal Plate Current	80	80	ma.
Max.-Signal Plate Current	147	100	ma.
Load Resistance (plate to plate)	3000	5000	ohms
Total Harmonic Distortion	2.5	5.0	%
Power Output	15	10	watts

PUSH-PULL AMPLIFIER – CLASS A_1 Values are for 2 tubes

Plate Voltage	300	350 volts
Grid Voltage	−61	−67.5 volts
Peak A-F Grid-to-Grid Voltage	122	135 volts
Zero Signal Plate Current	100	170 milliamps
Maximum Signal Plate Current	150	200 milliamps
Effective Load Resistance (Plate-to-Plate)	4000	4000 ohms
Maximum Signal Power Output	10	20 watts
Total Harmonic Distortion	4.5	2 per cent

Key design parameters

a=38.1 mm, S= 4.5 cm, A= aS = 17 cm^2

Primary: 4x500T= 2,000T wire d = 0.25 mm

Secondary: 3 x 30T = 90 Turns of 2 x 0.8 mm (bifilar winding)

I_0 = 170mA, I_{MAX}=200 mA for 2 tubes

TR=2,000/90 = 22.2, ZR=TR^2 = 494

Z_{PP} (8 Ω)= 494 x 8 = 3,952 Ω plate-to-plate

Z_{PP} (6.5 Ω) = 3,236 Ω plate-to-plate

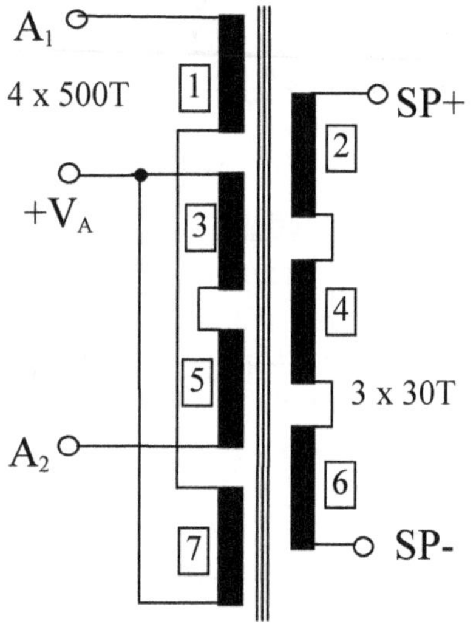

Sectionalizing

Window length WL=57, Coil Length CL=52mm

PRIMARY: TPL_{MAX}= CL/d= 52mm/0.27mm = 208 T

Primary, each section: 500 turns / 3 layers = 166 TPL, 12 layers in total

Secondary: TPL_{MAX} = 52mm/1.6 mm = 32.5 turns

We need 30 turns, so with careful winding each section will fill one layer!

MEASURED RESULTS:

- R_{AA} = 160 Ω L_{AA}= 22H
- L_L=10.2 mH
- C_{PS} = 2.2 nF
- BW @1W: 4 Hz - 84 kHz (-3dB)
- BW @12W: 8 Hz - 42 kHz (-3dB)

SPECIAL MAGNETIC COMPONENTS: LOW POWER INPUT, PREAMP OUTPUT & DAC OUTPUT TRANSFORMERS, TRANSFORMER VOLUME CONTROL

- MOVING COIL STEP-UP TRANSFORMERS
- LINE-LEVEL INPUT TRANSFORMERS
- DAC OUTPUT TRANSFORMERS
- THE WAY IT USED TO BE: NEUMANN PREAMP OUTPUT TRANSFORMER 60kΩ : 600Ω
- PROJECT #1: NEUMANN PREAMP OUTPUT TRANSFORMER CLONE USING EI41 LAMINATIONS
- PROJECTS #2 & #3: PREAMP OUTPUT TRANSFORMERS USING EI57 & EI66 LAMINATIONS
- PROJECT #4: PREAMP OUTPUT TRANSFORMER (LINE DRIVER) 5kΩ :150Ω
- TVC (TRANSFORMER VOLUME CONTROL)

9

MOVING COIL STEP-UP TRANSFORMERS

To use moving coil cartridges with standard phono preamps designed for higher moving magnet signals, we must amplify the voltage by using an additional voltage amplification stage.

Alternatively, we need to transform the high current, low voltage MC signals into the low current, high voltage signals needed at the input of the MM preamps by using a transformer.

MC transformers are always used with secondaries terminated, usually with the standard input resistance of MM phono stages of R_I=47kΩ. We are mostly interested in two parameters: the step-up ratio(s) and the frequency range.

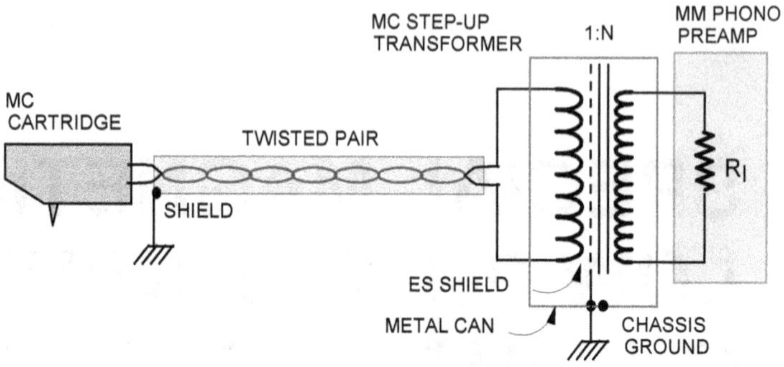

The wiring of a typical signal chain: MC cartridge - shielded twisted pair - MC transformer - MM phono stage

How to evaluate the specifications of commercial MC transformers

Designing and winding small-signal step-up transformers is not a job for the faint-hearted, so we better leave it to the specialists. We do need to know, though, how to evaluate the specifications of commercial units. They can be fully assembled ones, such as our first example, which are usually very expensive, $500 and up to thousands of dollars, or the transformers themselves, which are much cheaper to buy and which you would need to mount into some kind of chassis or enclosure, together with the input and output sockets.

Example #1: Ortofon ST-80

Ortofon ST-80 MC transformer is specified as "Power ratio: 27dB (47k ohm at 1kHz)" and also "Input impedance: 2 - 6 Ω, output impedance: 47 kΩ ". How can the output impedance be a fixed 47 kΩ if the input impedance varies from 2 to 6 Ω? Of course, it can not! They are saying that the cartridge's resistance should be in the 2-6W range and that the transformer should be terminated by 47kΩ.

What is the voltage ratio VR? Ortofon did not specify it, but from the specified 27dB "power ratio" we know dB = 20log(1/VR), so we can calculate log(1/VR) = 27/20 = 1.35, so 1/VR=$10^{1.35}$ = 22.39, or VR=0.0447. So, a 47k secondary load will be reflected to the primary as 47k/VR^2 = 47,000/22.4^2 = 47,000/501.8 = 93.7 ohms.

Example #2: UTC O-3

This vintage UTC O-3 input transformer can be used with low impedance microphones or moving-coil cartridges. It has two primaries, which can be connected in series or in parallel. The secondary impedance is specified as 50kΩ. With a 7.5Ω primary (parallel) connection, the impedance ratio is IR=Z_P/Z_S = 7.5/50,000 = 0.00015, meaning that the Voltage Ratio or Turns Ratio is VR= √(IR) = 0.01225.

If 30Ω primary is used (two coils in series), IR= 0.0006 and VR= 0.0245. This is a step-up transformer with few primary turns and lots of secondary turns; that is why both IR and VR are lower than 1!

500V HIPOT is the insulation test voltage, and 30-20 kc (kilocycles or kHz) means the frequency range of the transformer is from 30 Hz to 20 kHz.

Notice that dB levels are not specified, making the frequency range specification meaningless! Are they -0.5dB points, -1dB points, -2dB points or -3 dB points? Usually, when the dB level is not specified, -3dB is assumed.

Now we get to V_{SMAX}, the maximum signal level on the secondary, specified as +9 dBm. We have seen earlier that dBV= dBm-2.218, so +9 dBm means 9-2.218 = 6.78 dBV. With some basic maths, dBV=20logV_{SMAX}, 6.78 = 20logV_{SMAX}, logV_{SMAX}= 6.78/20 = 0.339 or V_{SMAX}= $10^{0.339}$ = 2.183V.

That is the maximum allowable secondary voltage, presumably before the transformer goes into saturation and starts severely distorting the waveform.

Multiply the V_{SMAX} with VR (voltage ratio) to calculate the maximum primary voltage. If the lower step-up ratio is used, the maximum primary voltage is V_{PMAX}= 2.183*0.0245 = 53.5 mV.

With the higher step-up ratio, the maximum primary voltage is V_{PMAX}=2.183*0.01225 = 26.95 mV.

The amplitude - frequency response characteristic of UTC O-3 transformer

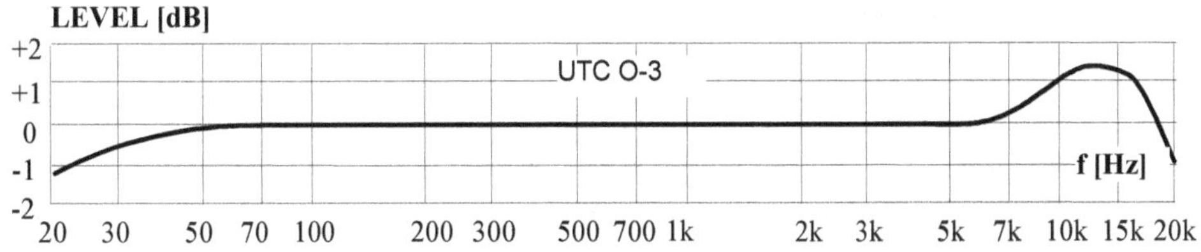

The level vs. frequency curve of the UTC O-3 transformer is shown above. The low-frequency attenuation is the standard 6dB/octave or 20 dB/decade drop, while the high-frequency response shows a resonant peak at around 12 kHz, after which the curve steeply drops to -1dB at 20 kHz. The peak is only 1.4 dB above the 0 dB level, so it may not be audible. The steep drop afterward is akin to the addition of another time constant (attenuation) after 21 kHz. As such, it does not need to be implemented as part of the RIAA filter if an MC transformer such as O-3 is used.

Example #3: Harman-Kardon XT-2

This vintage pair of microphone input transformers by Harman Kardon has the best frequency response of all MC transformers we have tried so far, from 10Hz to 41 kHz (-3 dB). Unfortunately, they use some odd plug, requiring a socket similar in size to the octal (8-pin) tube socket but with nine pins. These sockets are very difficult to source and expensive. Luckily, the pins can be soldered in place permanently.

Specs: 120 or 250Ω primary, 50 kΩ secondary, so two step-up ratios can be used, $\sqrt{(50/0.25)} = \sqrt{200} = 14$ or $\sqrt{(50/0.12)} = \sqrt{417} = 20.4$.

ABOVE: Harman-Kardon XT-2 transformers

Example #4: Vintage Russian microphone transformers

Claimed data: made in Soviet Union in early 1980s, step up-ratio 1:2, 1:10 or 1:20, dimensions: H 50 mm, Ø 40 mm.

The DC resistances are marked on the wiring diagram (right). We will evaluate this and a few other MC transformers further soon.

BELOW: DC winding resistances of a vintage Russian microphone transformer

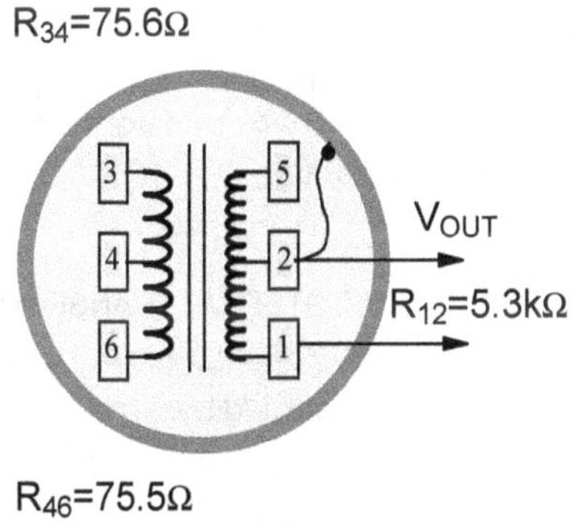

Example #5: Raphaelite PM30Ω :7k

Claimed data: made in China, primary impedance 30Ω , Secondary impedance: 7k , turns ratio: 1:13.5, 5Hz - 61kHz (-3dB), weight: 1 kg each, applications: MC with DCR 3-50 Ω .

NOTE: With 7k and 30Ω ,the impedance ratio IR would be 7k/30 = 233, and TR would be $\sqrt{(233)} = 15.3$, not 13.5 as claimed! Also, the weight is 207g each, not 1 kg!

Simple impedance tests using a multimeter and LCR meter

Although AC signals are required for proper testing, you can get an approximate idea of the step-up ratio of an unknown input or MC transformer by measuring its DC resistances. For the Russian transformer, 5.3k/75Ω = 70, $\sqrt{70} = 8.4$ for half of the primary winding and 5.3k/151Ω = 35, $\sqrt{35} = 5.9$ for the whole primary winding.

DC measurements were done with a digital multimeter, the AC measurements with a digital LCR meter at 1kHz. The step-up ratios of the Russian transformer are 7.3 and 14.75.

If you divide the standard 47k input resistance of MM phono stages by the IR, you will get the reflected resistance onto the primary side, the one the MC cartridge will "see". The Russian transformer would reflect 47k load as 47k/218 = 215Ω (1:14.75 ratio), the Chinese transformer as 47k/179 = 263Ω (1:13.4 ratio).

RESISTANCES AND IMPEDANCES OF MC STEP-UP TRANSFORMERS

	R_{P1} [Ω]	R_{P2} [Ω]	R_P [Ω]	Z_{P1} [kΩ]	Z_{P2} [kΩ]	Z_P [kΩ]	R_S [kΩ]	Z_S [kΩ]	IR	TR (VR)
Russian	75.6	75.5	151	9.32	9.31	37.6	5.3	2,028	54/218	7.3/14.8
Raphaelite	N/A	N/A	11.4	N/A	N/A	4.06	0.624	728	179	13.4

Testing the frequency range and voltage ratio of MC transformers

Most oscilloscopes' vertical sensitivity goes down to only 5 mV/division, and if a 0.1mV signal is used for this test, it will not be possible to measure the signal amplitudes on the screen (the amplitude will be too small). So, although MC cartridges produce 0.1 mV order of magnitude signals, it is more practical to perform this measurement at a 1mV input level. The 10x higher input signal will not overload most MC transformers.

Notice the apparent relative size difference of the four input TXs tested. The NO233BK is not shielded at all, so it is most likely the same size as the Chinese and Russian transformers, which are shielded in large cans.

NOTE 1: Since the Russian transformers have a CT primary, the measurements were taken between terminals 3 and 6, referring to the whole primary winding.

NOTE 2: C_{PS} A and B columns refer to the capacitance measurements between the primary and secondary windings taken between different ends of the primary and one end of the secondary winding.

RIGHT: Test setup for measuring moving coil and microphone input transformers' voltage ratio, frequency range and fidelity of waveform reproduction.

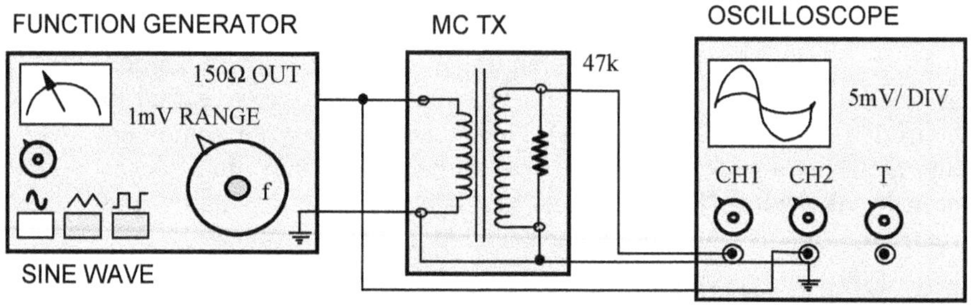

The C_{PS} parasitic capacitance of the Russian transformers is roughly half of the Chinese ones, and their leakage inductance is nine times lower! T

he Russian transformers' primary inductance at 120Hz (important for bass reproduction) is 15 times higher, and their leakage inductance is nine times lower. Despite all that, the frequency ranges of the two transformers were very similar.

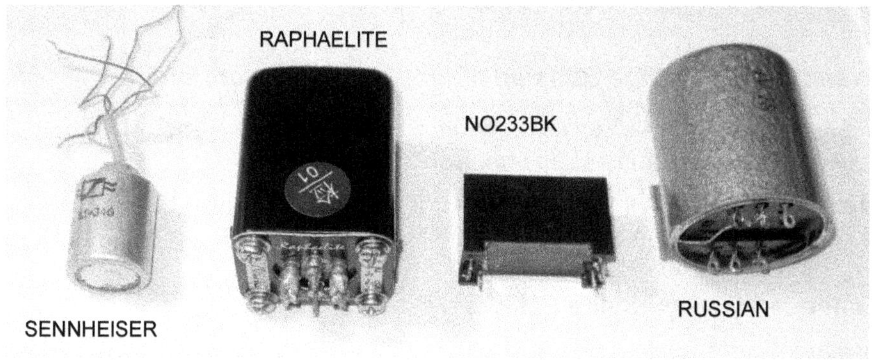

TESTING MC STEP-UP TRANSFORMERS

	C_{PS} A [pF] (at 1 kHz)	C_{PS} B [pF] (at 1 kHz)	Lp 120Hz [H]	Lp 1 kHz [H]	L_L 120Hz [mH]	L_L 1 kHz [mH]	f_L (-3dB) [Hz]	f_U (-3dB) [kHz]
Russian	119	395	15	2.2	4.5	4.2	9	22
Raphaelite	258	660	1.23	0.2	31	35	12	20
Sennheiser	27.8	27.8	0.46	0.1	0.23	0.24	30	26
NO233BK	16.5	17.2	35	9.8	47.3	16.8	40	90

NO233BK had by far the highest primary inductance of 35H at 120Hz, yet it had the worst f_L (40 Hz). It had the lowest parasitic primary-secondary capacitance and as a result, had the highest f_U of all, by far, a whopping 90 kHz!

These measurement results depend on the output impedance of the signal source. Although the 150Ω Z_{OUT} of Radford precision oscillator used for this test is quite low compared to other generators (typically 600Ω), it is relatively high compared to a very low impedance of moving coil cartridges, and that contributed to the treble roll-off.

You can insert a voltage divider at the function generator's output, say 5kΩ and 5Ω, which will ensure that the transformers' primaries see a very low source impedance of around 5Ω. With a ratio of 1,000:1, the divider will give you 1mV out for 1V input. Our measurements were mostly concerned with relative comparisons between the tested four transformers and not absolute accuracy.

LINE-LEVEL INPUT TRANSFORMERS

Assuming there is no grid current flowing in the first stage of a tube amplifier, input transformers are invariably of a very low power rating, their main function being to galvanically decouple the source from the amplifier.

Some can be used in a step-up mode to increase the voltage signal's amplitude and provide impedance matching. Most can be configured to be used with both balanced (XLR) and single-ended (RCA) inputs, in which case they can provide a phase splitting function and convert a single-ended (unbalanced) signal into a balanced one.

The topology of a hi-end Marantz T1 amplifier is fully balanced from its input to the output, with transformer coupling throughout. However, notice one coupling capacitor at the output of the first stage to prevent DC current from flowing through the primary of the first interstage transformer (not shown on the partial diagram on the right).

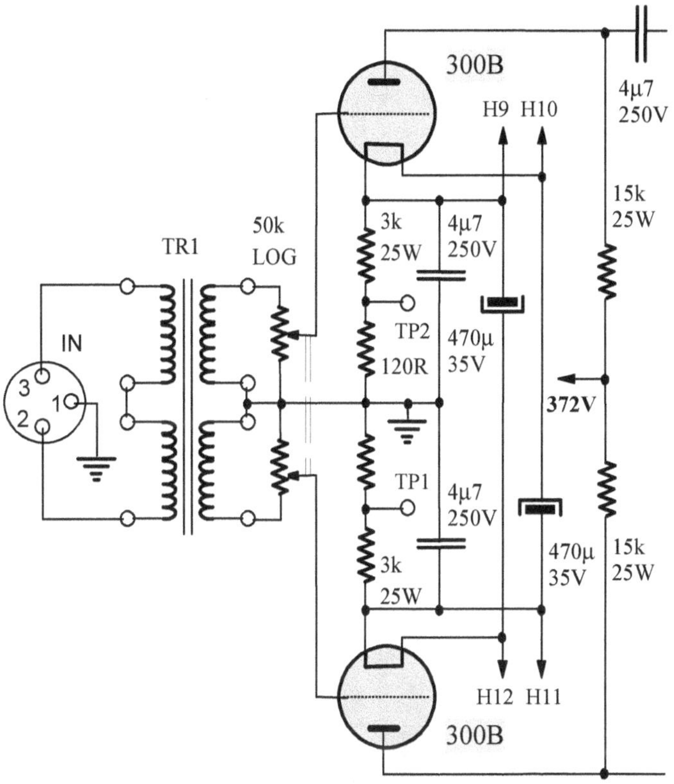

ABOVE: The input stage of Marantz T1 amplifier, © Marantz

The way it used to be: UTC A-20

A20 is a vintage multi-impedance input transformer. Salvaged units are still available on eBay. There are also similar vintage models by Sescom and Triad. This is by no means a hi-end transformer, the winding is often sloppy and haphazard and the lamination size is minuscule.

In audio dBm (also called dBmW) is referenced to the 1 mW of power that a sine source of 0.775 V_{RMS} dissipates on a 600Ω resistive load. Don't confuse it with dBV (or simply "dB"), the unit used for voltage ratios such as amplification and attenuation factors.

The referent power level for dBm is P_0=0.001 W, so P [dBm] = 10log(P/0.001). Since UTC A-20 maximum power level is +15dBm, this means 15 = 10log(P/0.001). We can calculate P in Watts from here, log (P/0.001) = 15/10 = 1.5, or P/0.001=$10^{1.5}$ = 31.62, so finally P=31.6mW (0.0316 Watts). This is truly a flea-sized transformer!

RIGHT: 500Ω – 500Ω and 125Ω – 50Ω connections

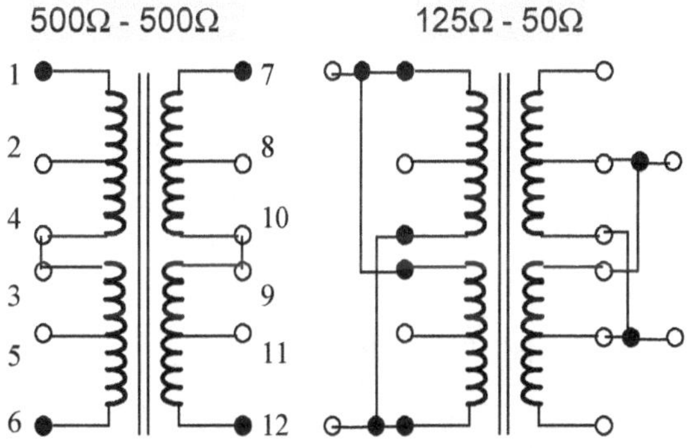

UTC A-20 SPECIFICATIONS:

- Made in USA by United Transformer Corporation
- Primary: 50, 125/150, 200/250, 333, 500/600 Ω
- Secondary: 50, 125/150, 200/250, 333, 500/600 Ω
- Frequency range: 10 Hz - 50 kHz
- Max. level: +15 dBm

The way it used to be: Triad HS-66 transformer

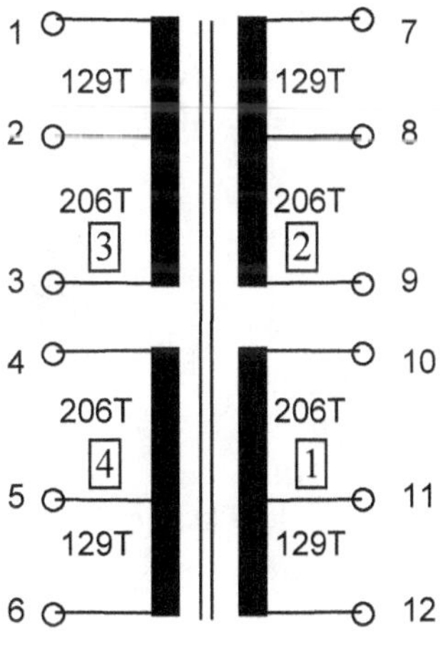

ABOVE: The winding diagram of Triad HS-66 input transformer

Triad HS-66 is similar to UTC's model A-20 transformer, with a 600:600 Ω impedance ratio (for the whole primary & secondary), compared to 500:500 Ω for A-20. The frequency range was specified as 10 Hz - 50 kHz.

Its 3/8" by 3/4" stack is made up of 27 EE laminations , the material is 29 gauge M6 GOSS steel. Its maximum power level is +15 dBm or log (P/0.001) = 20/10 = 2, or P/0.001=10^2 = 100, so finally P = 100 mW.

AWG36 magnet wire was used for all windings. There are two copper-foil electrostatic shields, one between windings 1&2, the other between sections 3&4.

The sections with 206 turns have an inductance of around 1.1H and the smaller sections, with 128 turns, measured between 0.5 and 0.6H, for a total inductance of around 2.35H per winding. The DCR varies, approx. 24Ω for winding #1, 26Ω for winding #2, 29Ω for winding #31, and 31Ω for winding #4.

If connected in series, the inductance of the two windings forming the primary (terminals 1&6) is around 7.8H, as is the inductance of the whole secondary (terminals 7&12).

DAC OUTPUT TRANSFORMERS

From the cheapest to the most expensive, CD players and digital-to-analog converters (DACs) use only a dozen or two common D/A converter chips. The weakest link is usually the output stage for which cheap integrated circuits or discrete transistors are used. Even many "tubed" players still have a solid-state output stage, with a single triode "tacked on" often as a cathode follower to bring the output impedance down. Although manufacturers charge a hefty premium of hundreds of dollars for such a tube "output stage, this is no improvement.

However, replacing the semiconductor output stage with a quality transformer will significantly improve its sound. Clarity, microdynamics, the level of detail, and "musicality" will be significantly better.The transformer's secondary current will feed the load resistor R_X and thus perform the current-to-voltage (I/V) signal conversion. Of course, two transformers are needed for stereo output.

There are additional benefits of this conversion. Distortion and noise can be reduced by isolating the digital (on the primary side) and analog ground (on the secondary side).

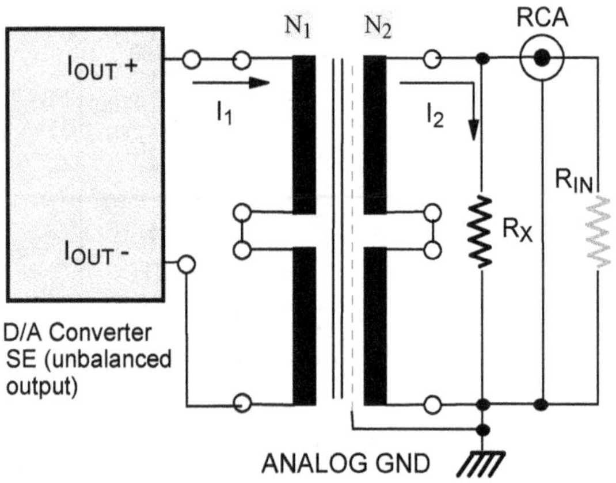

ABOVE: transformer output stage for a DAC with a single-ended (unbalanced) current output

BELOW: transformer output stage for a DAC with a balanced current output

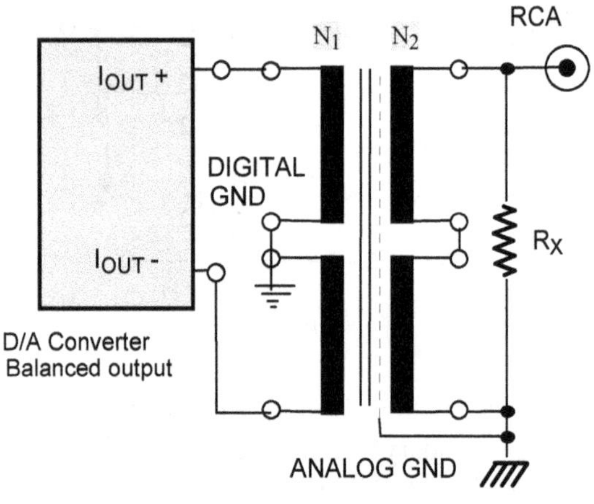

Also, the transformer's limited upper-frequency range will filter out higher harmonics of the quantization noise. The transformer would not only serve as an I/V converter and analog filter but can also provide signal amplification!

The principle is the same for DAC with balanced and DAC with unbalanced outputs.

As an example, Burr Brown PCM1738 produces 2.48 mA_{PP} of output current. The effective value is 2.92 times lower than the peak-to-peak figure so I_1=2.48mA/2.82 = 0.88 mA. If the turns and voltage ratio of the transformer is say 1:5 (or TR=V_1/V_2=1/5), the secondary voltage will be five times higher than the primary voltage, but the secondary current will be five times lower, so I_2=I_1*TR = 0.88*0.2=0.176 mA.

We will assume that all that current will pass through resistor R_X and that none will enter the preamplifier's input stage (connected to the secondary).

If the preamp's input impedance is not infinite, it needs to be added in parallel with R_X, and such a value is used in calculations.

Say we need the voltage signal at the RCA output to have a value of $1V_{RMS}$. $V_{OUT}=I_2R_X$, so $R_X=V_{OUT}/I_2$ = 1/0.176 *10^{-3}= 5.68 kΩ.

We have assumed an ideal transformer, which is far from reality. Real transformers of this kind have a significant insertion loss. Increase the value of R_X up to 20%, or 1.2*5.6 = 6.72 kΩ, which would make the standard value of R_X=6k8 perfect for the job.

If the input impedance of the following tube stage (or even a transistor preamplifier) is R_{IN}=47kΩ, the total resistance of R_X and in parallel must be R_P=6k8, so R_X= $(R_PR_{IN})/(R_{IN}-R_P)$ = 6.8*47/(47-6.8) = 7.95kΩ, so use 8k2 standard value.

Commercial benchmark: Sowter 1465 transformer

Wound using multi-section winding techniques to increase the bandwidth, 1465 and other Sowter DAC transformers use mumetal (76% Nickel) cores enclosed in mumetal enclosures.

1465 has twin bifilar primary windings, which can be connected in parallel, in series, or driven by twin DACs supplying signals of the opposite phase. Its electrostatic shield should be connected to analog ground only, never to digital ground.

SOWTER 1465 TRANSFORMER SPECIFICATIONS:

- Can dimensions 45 mm (dia.) x 52.5 mm high
- Primary inductance (each coil) 5.0 H typ.
- Primary DC resistance (each coil) 6.7Ω typ.
- Secondary resistance (two coils in series) 1.29 kΩ typ.
- Voltage ratio 1+1:5+5
- Frequency response: 5 Hz to 100 kHz (+/- 1.5 dB, paralleled secondary windings), 5 Hz to 50 kHz 5 Hz to 50 kHz (+/- 1.5 dB, secondary windings in series)
- £105.89 each (2018)

THE WAY IT USED TO BE: NEUMANN PREAMP OUTPUT TRANSFORMER 60kΩ : 600Ω

BV30 output transformers by Georg Neumann, from their 1961 studio preamplifiers, were used in a parafeed (LC) mode with no primary DC current. Wound on M42 laminations of 0.35mm thickness, 0mm gap (no air gap), 43 pieces, stack thickness around 15mm. The coil was vertically sectionalized in two sections to reduce interwinding capacitances, and each vertical section was horizontally sectionalized in a 2/1 manner to reduce leakage inductance.

SPECIFICATIONS: Neumann BV30

- Frequency response: 10 Hz to 50 kHz +/- 0dB
- Input impedance: 60 kΩ, output impedance 600Ω or 2.4kΩ
- IR: 100 (paralleled secondaries) or 25 (secondaries in series)
- VR: 1:10 or 1:5
- Primary 4*850T (3,400), 0.1 mm wire
- Secondary 2*340T, 0.22 mm wire
- Insulation: 2*0.1mm lacquered paper
- Weight: 260 g

	a = b	c	e = g	f	t (mm)
M42	42	6	30	12	0-0.3-0.5-1

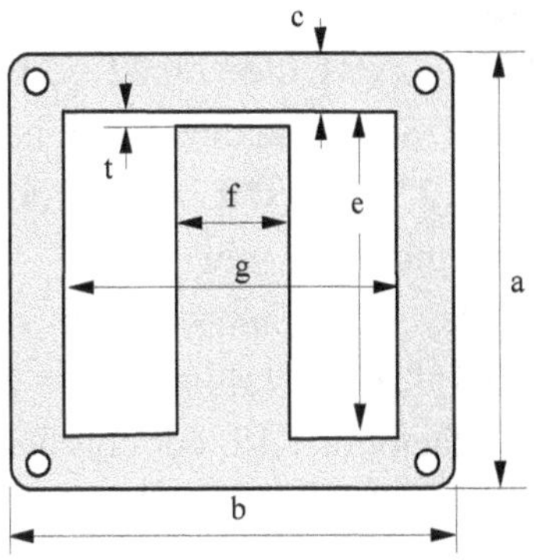

Magnetic properties

The magnetic material used was Hyperm766. The high percentage (75-81%) of nickel resulted in high initial permeability (35,000), high maximum permeability (90,000), and very low losses of 0.025 Watts/kg. The saturation flux density was also low, only 0.8 Tesla. Transformers were shielded in a Mumetal shielding case.

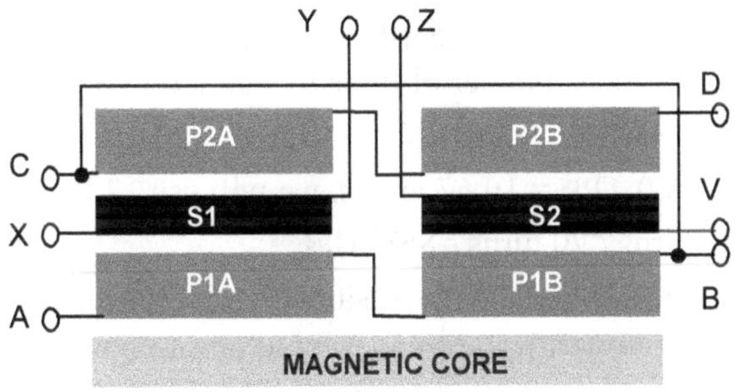

Winding diagram

Based on Neumann's finished coil, there is a 3-4 mm clearance between the top of the winding and the core, so they didn't need all of the 14mm window height for their windings.

Thus, in the design of your output transformer along the Neumann lines, should you wish to eliminate the output capacitor and have 5-15mA of DC anode current flowing through its primary, introduce an air gap and increase the number of turns to the maximum that would fit.

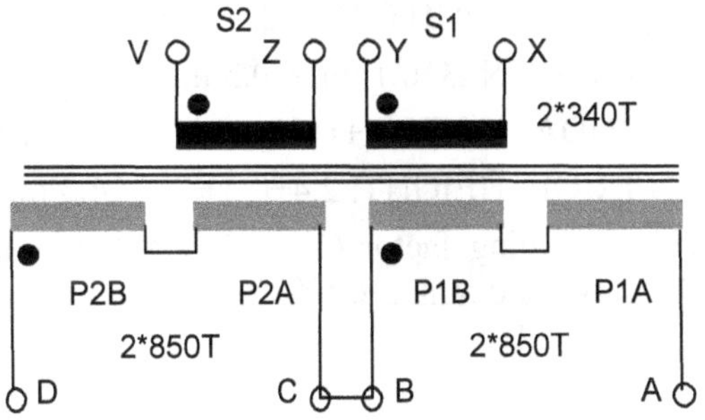

ABOVE RIGHT: How Neumann designers sectionalized the windings in BV30 transformers.

RIGHT: The winding diagram of Neumann BV30 transformers

PROJECT #1: NEUMANN PREAMP OUTPUT TRANSFORMER CLONE USING EI41 LAMINATIONS

Although M-cores can still be bought, let's replicate Neumann's design using standard EI laminations. We don't know all the parameters of such a design, but a bit of detective work can yield some close estimations.

Nickel alloy laminations

These nickel alloy (49%) EI41 laminations were sold on eBay for US$49.- (2 sets), including transport (2016). EI-4118 bobbins were included (18.5mm Stack, 5pin + 5pin). The laminations were 0.35mm thick.

The window area is 21*8 mm or 168 mm^2. Neumann's window was 30*14 = 420 mm^2. Here you see the beauty of M-cores that Neumann used, where for similar outside dimensions, you get a 250% larger window area! Their stack was 15 mm thick, ours is 18.5 mm; their A=1.2*1.5= 1.8 cm^2, ours is A=1.3*1.85= 2.4 cm^2.

Determining the number of turns

Again, we start with $N_1=V_1 10^4/(4.44 f_L B_{MAX} A_{EF})$. We don't know Neumann's V_1, B_{MAX} or f_L, but we can safely assume that their B_{MAX} is below 0.75 Tesla since the saturation value of their material is 0.8 T!

Our cross-sectional area A is 2.4/1.8 = 1.33 or 33% bigger and our B_{MAX} can be 1.5X higher (1.2T), so assuming the same V_1 and the same f_L, our N_1 can be up to 1.33*1.5 = 2 times lower. Thus gives us the minimum values, N_{1MIN}=3,400/2 = 1,700, and $N_{2MIN} = N_1/5$ = 170!

Sectionalizing, first attempt

Primary sections: 1,700/4 = 425 turns, secondary sections: 170/2 = 85 turns

With 3*2 = 6mm for the bobbin thickness, each coil length is (21-6)/2 = 15/2 = 7.5mm so we will work with 7mm.

PRIMARY: TPL_{PMAX}= CL/d=7/0.1 = 70, 425/70 = 6.1 layers, we will use 7 layers, TPL_P = 425/7 = 61

SECONDARY: TPL_{SMAX}= CL/d=7/0.22 = 32, 64/32 = 2.66 layers, we will use 3 layers, TPL_S = 64/3 = 21

Window height: WH = 8mm, Maximum coil height: CH= 8-2= 6mm (2mm BT)

PRIMARY HEIGHT: 2*7 layers*0.1mm = 1.4 mm

INSULATION: 3*0.1mm = 0.3 mm

SEC. HEIGHT: 1*3 layers*0.2mm = 0.6 mm

TOTAL COIL HEIGHT: CH = 1.4+0.3+0.6 = 2.3 mm

Even with a bulging factor of 1.15, our coil height is only 2.65mm, less than half of the available window height of 6mm! So, we can safely increase the number of turns, thus reducing B_{MAX} and lowering distortion!

Revised design

Let's keep the same B and instead of halving the Neumann's number of turns reduce it only by 25%!

PRIMARY: 3,400*0.75 = 2,550 or 2,550/4 = 640

640/60 TPL = 10.67 layers, we will use 12 layers

60 turns / 70 turns =86% HFF

SECONDARY: $N_1/5$ = 640*4/5 = 512 turns in total

512/2 = 256 turns in each section and 256/32 = 8 layers, we will use 8 layers of 0.2 mm wire

PRIMARY HEIGHT: H_1 = 2*12*0.1mm = 2.4 mm

INSULATION: 3*0.1mm = 0.3 mm

SECONDARY HEIGHT: H_2= 8*0.2mm = 1.6 mm

NETT COIL HEIGHT: 2.4+0.3+1.6 = 4.3 mm

With a bulging factor (1.15) CH = PH+SH+IH = 4.3*1.15 = 5.0 mm, 5/6 = 83% vertical fill factor.

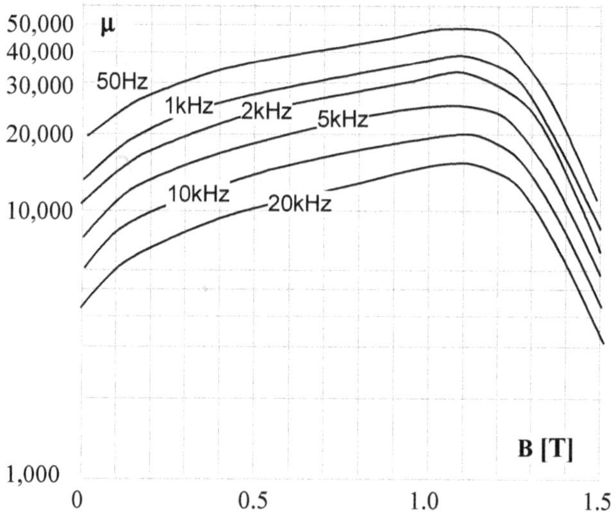

ABOVE: Permeability curves for 49% nickel alloy. Despite the proclaimed B_{MAX} of 1.5T, a significant drop in permeability starts at around 1.21T, so keep B_{MAX} under 1.1T!

PROJECTS #2 & #3: PREAMP OUTPUT TRANSFORMERS USING EI57 & EI66 LAMINATIONS

Small (30VA and 15VA) mains transformers with grain-oriented laminations (the ones already discussed and featured as chokes and guitar output transformers) can be rewound into preamp output transformers as well. They usually have vertically-split bobbins, exactly what is needed for this type of design (Neumann clone).

A_{30} =aS = 2.2*3.1 = 6.8 cm^2 (EI66 transformer), A_{15} =aS = 1.9*1.9 = 3.6 cm^2 (EI57 transformer)

Neumann's number of turns:

N_1=3,400 (4*850T), 0.1 mm wire, N_2 = 680 (2*340T), 0.22 mm wire

Sectionalizing for EI66 lamination stack

Allowing 3*2 = 6mm for the thickness of the three bobbin walls, each of the two coil lengths is (33-6)/2 =27/2 = 13.5 mm. We will work with 13 mm, use 0.16mm primary wire, whose diameter is actually 0.18mm with insulation, and 0.22mm secondary wire.

PRIMARY: TPL_{PMAX}= CL/d= 13/0.18 = 72, 850/72 = 11.8 layers, we will use 12 layers, TPL_1= 850/12 = 70

SECONDARY: TPL_{SMAX}= CL/d=13/0.22 = 59, 340/59 = 5.76 layers, we will use 6 layers, so TPL_2= 340/6 = 57

Window height: WH = 11 mm, maximum coil height: CH= 11-2= 9 mm.

PRIMARY HEIGHT: PH=2*12 layers*0.18mm = 4.32 mm

Insulation: 3*0.1mm Mylar = 0.3 mm

SECONDARY HEIGHT: SH=1*6 layers*0.22mm = 1.32 mm

TOTAL COIL HEIGHT: CH = 4.32+0.3+1.4 = 6 mm

Max. coil height including the bulging factor is around 7mm, so 7mm/9mm = 78% vertical fill, which means the coil will fit comfortably into the window.

Sectionalizing for EI57 lamination stack

Allowing 6mm for the bobbin walls' thickness, each of the two coil lengths is (28.5-6)/2 =11.25 mm, so we will work with 11 mm and use 0.16mm wire, which is actually 0.18mm diameter with insulation, and 0.22mm secondary wire.

PRIMARY: TPL_{PMAX}= CL/d= 11/0.18 = 61, 850/61 = 13.9 layers, we will use 14 layers, TPL=850/14 = 60.

SECONDARY: TPL_{SMAX}= CL/d=11/0.22 = 50, 340/50 = 6.8 layers, we will use 7 layers.

Window height: WH = 9.5 mm, Maximum coil height: CH= 9.5-2=7.5 mm.

PRIMARY HEIGHT: PH= 2*14 layers*0.18mm = 5.04 mm, Insulation: 3*0.1mm Mylar = 0.3 mm, SECONDARY HEIGHT: SH=1*7 layers*0.22mm = 1.54 mm.

TOTAL COIL HEIGHT: CH=CH = PH+SH+IH = 5.04+1.54+0.3 = 6.88 mm.

Max. coil height, including the 15% bulging factor is 6.88*1.15=7.8 mm.

We have 7.5mm available, so tight winding is needed to reduce the bulging factor, and the coil should just fit. It would be safer to reduce the wire diameter, primary wire from 0.16 mm down to 0.12 and secondary from 0.22 to 0.2mm.

PROJECT #4: PREAMP OUTPUT TRANSFORMER (LINE DRIVER) 5kΩ :150Ω

The way it used to be: vintage benchmark by Tamura

Tamradio A-4714 are vintage audio transformers by Tamura, Japan, with a 5kΩ primary and 600Ω CT secondary. The frequency range was specified as 30Hz - 20kHz (-1 dB), the insertion loss was 0.2 dB, and the maximum primary DC current was 60mA. ust as with all vintage transformers, they rarely come up for sale due to their limited supply and desirability and are priced accordingly (read: extremely expensive).

Typical application circuit and the required specifications

One possible line-level preamplifier stage with Tamura A-4714 is illustrated. The secondaries are in series for single-ended operation, but balanced outputs are possible by grounding the CT instead. The DC current through the primary is around 18mA.

Let's settle on the following specs: I_1=20mA, f_L=15 Hz, Z_P=5kΩ, Z_S=150 Ω ! There is no need to allow 60mA in the primary as Tamura did; 20mA is just right for driver tubes such as 5687, 6SN7, 6CG7 or 12BH7. The smaller gap will give us higher effective permeability and higher primary inductance. Since the DC current is relatively small, the required air gap is only around 0.05 mm.

We will reuse the laminations and the bobbin from surplus Taiwanese low voltage mains transformers since they use GOSS laminations. The bobbin is vertically split in two (primary on one and low voltage secondaries on the other half), but that divider can be cut off.

The EI66 laminations with 50mm stack give us A_{EF}=0.95*A= 0.95*2.2*5.0 = 10.45 cm^2

With Z_P=5,000 Ω and Z_S=150 Ω, we get the impedance ratio IR= 5,000/150 = 33.3, so the turns ratio is TR=√IR =√33.3 = 5.77!

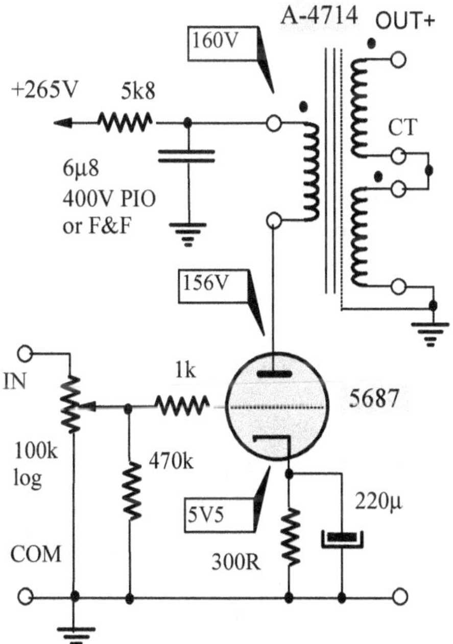

ABOVE: High current (18mA) single-ended preamplifier (line stage) with 5687 triode and Tamura A4714 or similar output transformers

Window sizing and maximum coil height

We have chosen 4:3 sectionalizing - 4 primaries in series and 3 secondaries in series. Now we have to determine the maximum number of turns we can fit into the available window, which is 33x11mm.

Window length is WL=33mm so coil length is CL=33-(2*2)-1= 28mm (B=2mm for the bobbin thickness on each side plus 1 mm clearance for insulation). Window height: WH = 11mm, Coil height: CH= WH-B-1= 11-3= 8mm

Normally, with power and output transformers for power amplifiers, we would calculate the minimum required wire diameter based on the primary and secondary current levels, but in this case, the current is only up to 20 mA, so 0.1 mm wire would be sufficiently large. However, we have chosen a larger diameter wire, 0.135 mm (with insulation), to reduce copper losses and make the winding job easier. Thin wires are fiddly to wind and can break.

We will use a reverse process in this calculation, similar to that used for chokes. The thickness of insulation will be determined first, and the leftover height will then be used to determine the maximum total number of turns that can physically fit into the rest of the window height.

INSULATION THICKNESS d_I: There are 7 layers of insulation, including the final (outside) layer, total insulation thickness is d_I=7*0.25 = 1.75 mm

GROSS COPPER HEIGHT (GCH):

GCH= CH-d_I= 8-1.75= 6.25 mm

Factoring in 15% bulging factor we get the Net Copper Height of NCH=GCH/1.15=6.25/1.15 = 5.4 mm

INITIAL LAYERING:

Maximum number of layers NL=NCH/d=5.4/0.135= 40

Maximum TPL (turns per layer): TPL_{MAX}= CL/d = 28/0.135 = 207 turns

Actual TPL=0.9*207 = 186 (assuming 0.9 horizontal fill factor)

The total number of turns we can fit is TT= NL*TPL = 40*186 = 7,440

Since TT = $N_1+N_2 = N_1+N_1/TR = N_1+N_1/5.77 = N_1(1+0.173)$, we can determine N_1 = TT/1.173 = 7,440/1.173 = 6,343 and N_2= 7,440-6,343 = 1,097.

ABOVE: Mains transformers with GOSS laminations can be rewound as audio transformers

BELOW: Winding diagram. The framed numbers indicate the winding order of transformer's sections.

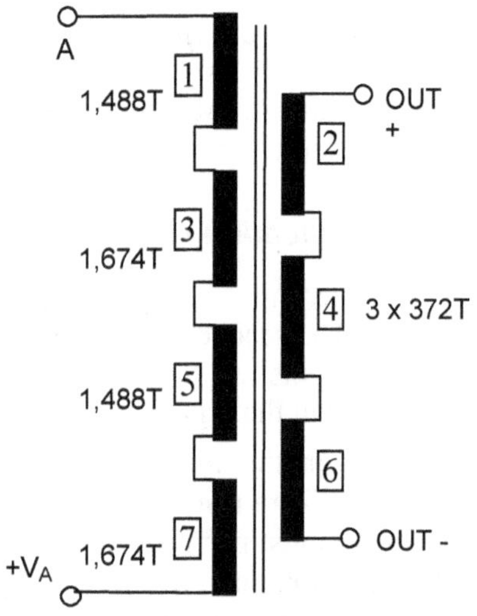

Final layering and final winding data (after adjustments)

The number of primary layers: 6,343/186 = 34.1, so we will use 34 layers.

Since 34/4 = 8.5 two of the sections will have 8 layers and two will have 9 layers (2*8+2*9=34).

The number of secondary layers: 1,097/186 = 5.9, so we will use 6 layers in total, or 2 layers per section.

<u>FINAL DATA:</u>

PRIMARY: 2*8*186 + 2*9*186 = 2*1,488 + 2*1,674 = 6,324 turns

SECONDARY: 3 * (2*186) = 3*372 turns = 1,116 turns

The actual Turns Ratio: TR = 6,324/1,116 = 5.667, so IR= TR^2 = 32.11

The initial goal was IR of 33.3, but due to our fitting adjustments, we got 32.1, which is close enough. Instead of 5k:150Ω impedance ratios, we will have 5k:156Ω, making no practical difference.

VINTAGE EXAMPLE: STANCOR WF-35 AND UTC A-25 LINE OUTPUT TRANSFORMERS

STANCOR WF-35 or A-25 by their UTC competitors are interstage transformers with 15kΩ primary impedance and various output impedances, from 600Ω down to 50Ω, depending on which secondary terminals are used. The DC resistance of the primary is R_P=1.6kΩ.

These transformers could take up to 8mA of DC current, but their frequency range only extends down to 40 Hz due to the air gap. That is due to a much lower primary inductance, which is a consequence of the DC magnetization of the core and lowered primary inductance.

In the design example of a capacitorless line stage, we have 7.7 mA of E86C triode's anode current flowing through the transformer's primary winding.

The un-bypassed 220Ω cathode resistor provides local negative feedback and also enables a mild global negative feedback to be brought in from the output transformer's secondary via the 5k6 resistor. NFB improves the bass response. If removed, the lower -3dB frequency would increase significantly, and you would lose the bass frequencies below 40-50 Hz!

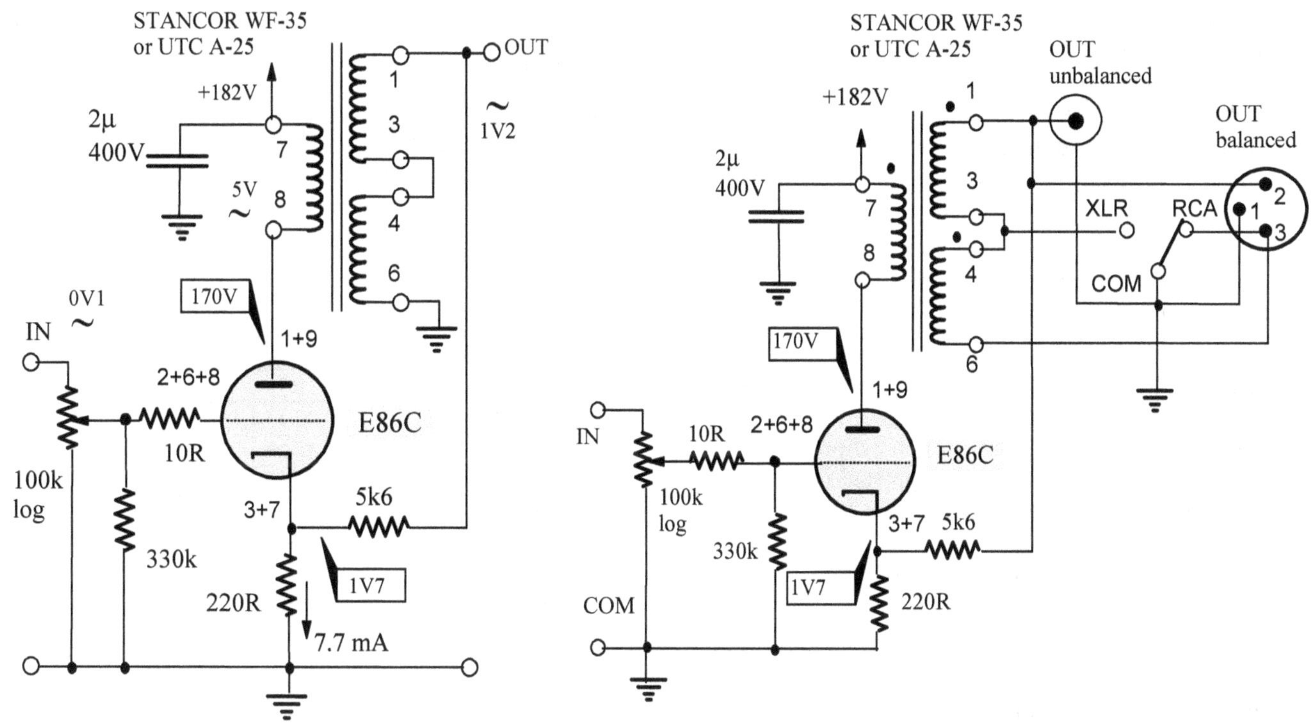

An example of a single-stage line-level audiophile preamplifier using E86C triode and STANCOR WF-35 or UTC A-25 transformers with a single-ended (LEFT) and balanced/unbalanced (RIGHT) output.

15kΩ to 600Ω means the impedance ratio of the output transformer is IR=25, and the voltage ratio is 5. The triode circuit amplifies around 50 times, and the overall amplification factor is 12, so with 1V input, up to 12V are available at the output.

Dual secondaries on output transformers lend themselves to a balanced preamplifier output or if used in a power amplifier, as a phase splitter for driving push-pull grids! If both single-ended (RCA connector) and balanced (XLR connector) outputs are needed, a simple switch makes selecting between the two output modes easy.

Pin 1 of the balanced output is always grounded, the positive (pin 2) is wired to secondary terminal #1, and the negative output (pin 3) is wired to secondary terminal #6. When the switch is in the "RCA" position, the center tap (terminals 3&4) of the output transformer is disconnected from the ground, and a full secondary (between terminals #1 and #6) is connected across the unbalanced output.

TVC (TRANSFORMER VOLUME CONTROL)

These volume control transformers are from a Malaysian-made Promitheous Audio preamplifier. The preamp can work as a passive TVC control unit or an active line stage with a 6DJ8 duo-triode per channel. The TVC is operational in both options.

The transformer is actually an autotransformer, EI66 laminations (0.35mm thickness) with a 35mm stack. The measured values of DC resistance, secondary (output) inductance at 1 kHz, the impedance at 1 kHz (all measured between the common and output terminals), and the attenuation from the previous step are tabulated (next page).

The attenuation figure for each position is referenced to the previous step and was calculated from the measured input and output signal RMS voltages.

The attenuation is not uniform or regular, at least not according to our measurements; it ranges from -2.51dB to -6.22 dB. The total attenuation from position 23 (maximum signal) to position 1 (minimum) is around -96 dB (the sum of the rightmost table column), making the average attenuation -4.17 dB per step.

The inductance of the whole winding is around 2.2H due to the low-grade magnetic material used, which equates to the input impedance of around 37kΩ at 1 kHz.

You may have already identified the significant weaknesses inherent in transformer volume control. First, due to its magnetic core's highly nonlinear hysteresis curve, TVC distorts significantly.

Secondly, its inductance forms an LC circuit with its parasitic capacitances. The input capacitance of the following triode, so resonant peaks are present in its amplitude vs. frequency characteristics, resulting in an oscillatory nature of its response to step and square-wave signals. The Promitheous Audio's line stage oscillogram tells you all you need to know (next page).

Thirdly, the input impedance of a TVC-based control is not constant; it varies with the signal's frequency since it is a vector sum of a relatively constant resistive and a widely varying inductive component.

There is also a capacitive component, but let's simplify things a bit; the aim here is not to impress you with my knowledge of vector algebra or to complicate things unnecessarily but to illustrate the pitfalls of using TVC.

That's exactly what TVC is - an unnecessary overcomplication, a "solution" that creates at least three new problems. All the newly-created problems are much more serious than the original one it is supposed to solve, namely the thin conductive plastic film of potentiometers in the signal path.Unfortunately, the audio field is bursting full of complex solutions to minor or nonexistent problems.

Let's look at a couple of frequencies. The reactance or the inductive impedance component is $X_L=\omega L=2\pi fL$, meaning it increases with rising frequency f.

Since L isn't constant but also varies with f, that raise isn't even linear. The modulus of the impedance is $Z=\sqrt{(R^2+X_L^2)}$. With 2.2H of input inductance and 146Ω of input resistance, the input impedance at 20 Hz is $Z=\sqrt{[146^2 + (2*3.14*20*2.2)^2]} = \sqrt{[146^2 + 276^2]} = 312\Omega$!

Notice that the order of magnitude of the resistive and inductive component of the input impedance at such low frequency is roughly the same, 146Ω versus 276Ω.

If such a TVC were used at the input of an integrated power amplifier, the load on a preamplifier would be below 300Ω at 20Hz, so most preamplifiers would struggle to drive such a low impedance load, and the distortion would increase significantly as a result.

At say 10kHz, the situation is different. $Z=\sqrt{[146^2 + (2*3.14*10{,}000*2.2)^2]} = \sqrt{[146^2 + 138{,}230^2]} = 138{,}230\Omega$!

The resistive component is now negligible compared to 138kΩ of inductive impedance, almost 1,000 times higher.

MEASURED RESULTS:

SP	DCR [Ω]	L	Z [Ω]	ΔA [dB]
1	0.0	0.9 μH	0.58	
2	0.3	0.3 μH	0.98	-4.56
3	0.7	14.8 μH	1.40	-3.10
4	1.1	30.9 μH	1.87	-2.51
5	1.5	81.3 μH	2.67	-3.09
6	2.0	152 μH	3.74	-2.93
7	2.6	293 μH	5.88	-3.93
8	3.1	512 μH	9.76	-4.40
9	3.8	863 μH	16.14	-4.37
10	4.6	1.33 mH	26.02	-4.15
11	5.5	2.01 mH	40.79	-3.91
12	6.6	3.2 mH	66.26	-4.21
13	7.9	5.14 mH	107	-4.16
14	9.5	8.17 mH	172	-4.12
15	12.3	16.7 mH	341	-5.94
16	16.2	36.7 mH	683	-6.03
17	21.7	82.8 mH	1,398	-6.22
18	29.2	202 mH	2,764	-5.92
19	40.1	409 mH	5,392	-5.80
20	55.7	782 mH	10,640	-5.90
21	78.3	1.41 H	18,110	-4.62
22	111.4	2.01 H	27,610	-3.66
23	146.2	2.22 H	37,300	-2.61

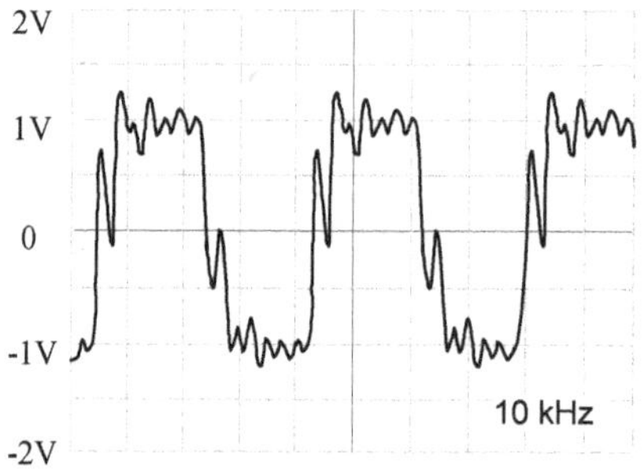

SP = switch position

L= inductance (at 1kHz)

Z = impedance at 1kHz

ΔA = attenuation form the previous position

LEFT: The fidelity (or rather "lack of") of a 10kHz square wave signal reproduced by the Promitheous Audio line stage with TVC

BELOW: The inside view of one transformer.

BELOW: The autotransformer TVC is the simplest configuration with only one winding.

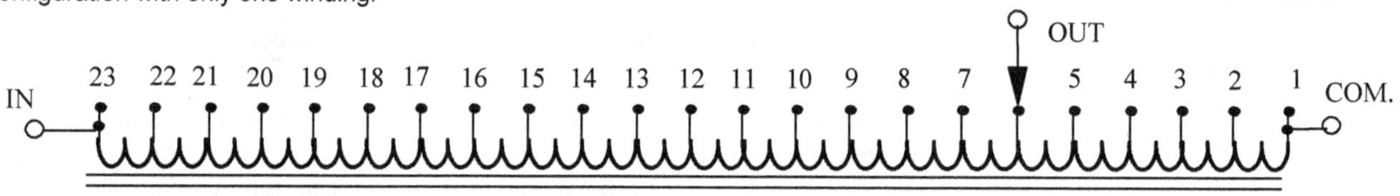

INTERSTAGE TRANSFORMERS, GRID & ANODE CHOKES

- IMPEDANCE COUPLING, ANODE- AND GRID-CHOKES
- TRANSFORMER-COUPLED STAGES
- COMMERCIAL INTERSTAGE TRANSFORMERS
- DIY DESIGN: THE UNIVERSAL INTERSTAGE & PHASE SPLITTER TRANSFORMER
- DIY DESIGN: SINGLE-ENDED DRIVER TRANSFORMER FOR 2A3 & 300B TUBES
- DIY DESIGN: 1:1.5 SE TO PP DRIVER & PHASE SPLITTER TRANSFORMER
- DIY DESIGN: PP-PP INTERSTAGE TRANSFORMER FOR A MCINTOSH-TYPE CLASS B AMPLIFIER

10

IMPEDANCE COUPLING, ANODE- AND GRID-CHOKES

RC interstage coupling is the cheapest and easiest to implement. However, resistors have the same impedance for DC and AC currents, and to maximize the amplification factor of the stage, the anode resistance R_A should be as high as possible, ideally infinite. However, the DC voltage drop on such a resistor would be very high.

An inductor has low resistance for DC (typically 1-3 kΩ), and very high impedance for AC currents and seems ideal for an anode load; there is very little DC voltage drop across it.

In our LC coupled version of the ECC40 stage, the DC voltage drop on the anode choke is $I_A R_A$ = 3mA*2k3 = 6.9 V_{DC}, while the DC voltage drop on the 47k anode resistor was 3mA*47kΩ = 155V! Instead of a 400V_{DC} anode supply, we now only need 242V_{DC}.

This means a lower voltage power supply can be used, or, using the same available power supply, the tube can work with a much higher DC voltage on the anode than with an anode resistor.

Since the reactance X_L of the choke is very high, the amplification factor or voltage gain of the stage approaches that of μ of the tube. The voltage gain of the common-cathode stage is $A = -\mu R_L/(r_I + R_L)$, or, with anode choke, $A = -\mu X_L/(r_I + X_L)$, and if X_L is much larger than r_I, as it is, at least at higher frequencies, we get $A \approx -\mu$.

There is a special treat for those who like chokes and dislike resistors: double-impedance coupling! This concoction involves using an anode choke in the driver stage and a grid choke in the driven stage.

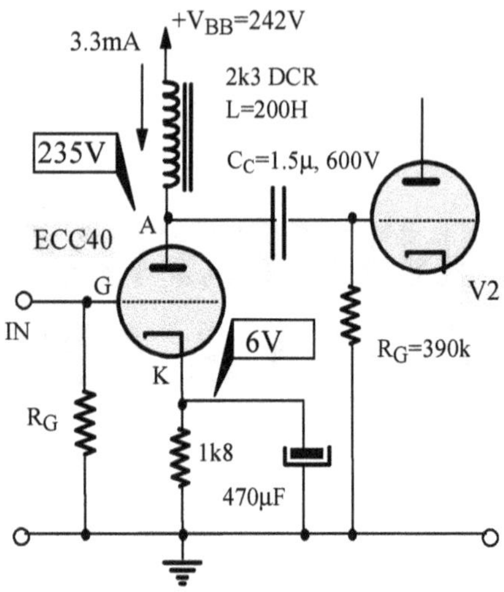

ABOVE: Impedance or LC coupling

BELOW: Double-impedance or LCL coupling

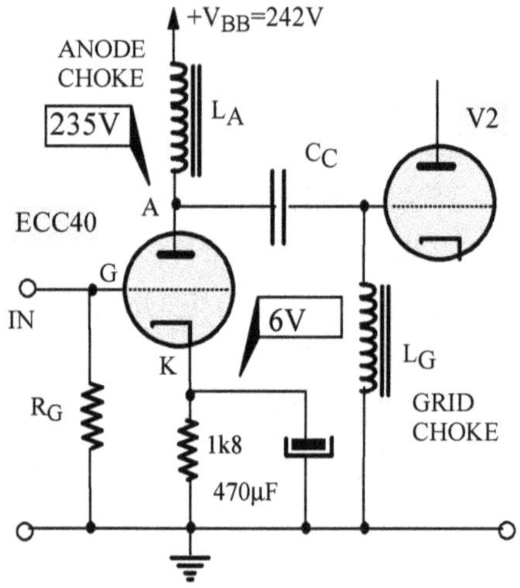

The low- and high-frequency models

The full model can be simplified for three frequency bands. C_A, C_C, C_G and R_G have the same meaning as with RC-coupling. L is the inductance of the anode choke, C_L is its distributed capacitance, and R_F is the choke's core loss resistance, which is mostly its iron loss. Please do not confuse it with its DC resistance R_L.

The two chokes and the coupling capacitor form a series tuned circuit resonant at a certain low frequency.

This will increase the gain in the bass region, but since two chokes are in parallel for AC signals, it will also double the value of the parasitic distributed capacitance of the chokes. This increased shunt capacitance will reduce the gain at high frequencies, improving the frequency response of an amplifier at one frequency extreme (bass in this case) makes its response at the opposite end (treble) worse.

In the mid-band, all capacitances can be neglected, the shunt capacitances are open circuits, and the coupling capacitance is a short circuit. At low frequencies, the reactance of CC becomes significant, but in contrast with the RC-coupling, where it plays the dominant role, with LC-coupling, it can be neglected without a significant error.

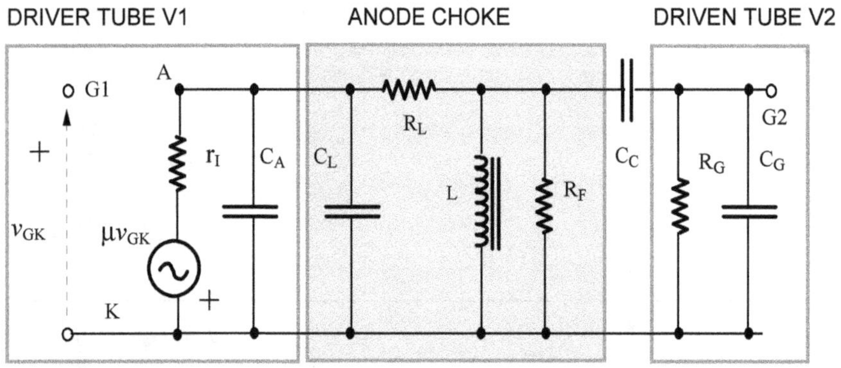

ABOVE: The full model of impedance-coupled tube stages

This is because the decreasing reactance of the anode choke has a much more dominant effect on the low-frequency response than the capacitor's increasing reactance.

At high frequencies, the three shunt capacitances can be agglomerated into one, and the coupling capacitance can again be considered a short circuit. The mathematical analysis is very similar to RC-coupled stages, so let's look at a practical amplifier instead.

PRACTICAL EXAMPLE: The LC-coupled ECC40 stage

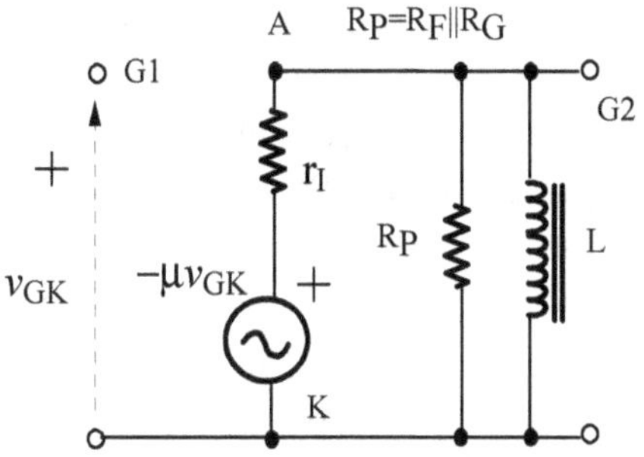

ABOVE: Low frequency model of impedance - coupled stages

BELOW: High frequency model of impedance - coupled stages

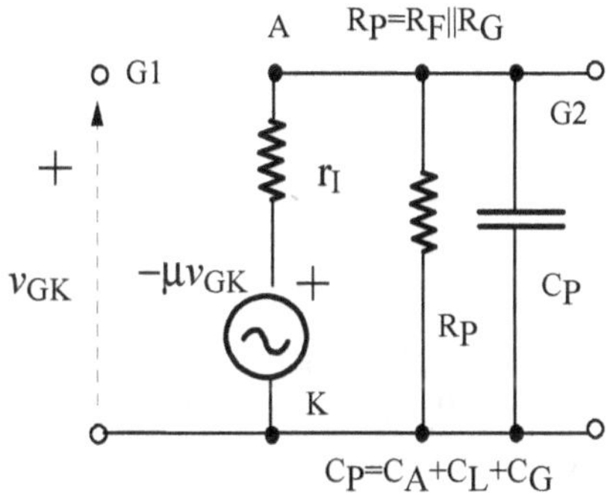

Since the impedance of an anode choke raises linearly with frequency, HF response of LC-coupled stages isn't of concern, but LF response is because, at low frequencies, the choke's impedance may drop to a value that is too low for the required LF response of the stage.

At a midrange frequency of 1,000Hz, the 200H choke's reactance is $X_L=2\pi fL = 1.257M\Omega$! The anode load Z_L is X_L in parallel with grid resistor R_G of the output stage; in this case, $R_G=390k\Omega$, so $Z_L=X_L||R_G = 298k\Omega$.

The voltage amplification of the stage with one triode is $A_1=-\mu Z_L/(Z_L+r_I) = -32*298/(298+11) = -32*0.96 = -30.86$.

Connecting two triodes in parallel does not increase μ, but halves the internal resistance, so $A_2=-\mu Z_L/(Z_L+r_I/2) = -32*298/(298+5.5) = -32*0.98 = -31.42$. The gain is slightly higher, but look what happens at low frequencies.

At f=20Hz, the choke's reactance is only $X_L=25k\Omega$! The anode load in this case is $Z_L=X_L||R_G = 25k || 390k = 23.5k\Omega$, a drop from almost 300k at midrange! Using one triode, the amplification factor has dropped to $A_I=-\mu Z_L/(Z_L+r_I)=-32*23.5/(23.5+11)= -32*0.68 = -21.8$, or $21.8/30.86 = 70.6\%$ of the midrange gain.

By pure coincidence, we have stumbled upon the lower -3dB cutoff frequency of the stage, f_L, which with one tube is around 20Hz.

Using two paralleled triodes, $A_2= -\mu Z_L/(Z_L+r_I) = -32*23.5/(23.5+5.5) = -32*0.81 = -25.9$, a much better result, a drop to $25.9/31.42 = 82.5\%$, a significant improvement over 70.6% with one triode.

To reduce the drop of amplification (gain reduction) at low frequencies, the driver triodes used in LC-coupled stages should have a low internal resistance r_I and be paralleled to reduce such resistance as much as possible. At the lowest desired frequency of operation, the reactance of the anode choke should be at least twice the internal resistance of the tube. Pentodes (due to their high r_I) should never be used as drivers!

LC-coupling compared to RC-coupling

For impedance and transformer-coupled stages, the static load line will be very steep, almost vertical, since the DC resistance of the choke or transformer's primary winding is low (in our case 2k3), while its AC impedance is very high. At 1kHz, $X_L=2\pi fL = 2*3.14*1{,}000*200 = 1.257M\Omega$, so the parallel with 220k grid resistor presents a load of $1.257M\Omega||220k\Omega = 187k\Omega$!

The voltage amplification factor of the LC coupled stage can be determined from the curves. $V_{MIN}=80V$ and $V_{MAX}=375V$, so $\Delta V_A= 375-80= 295V$. The grid swing is the same as for RC-coupling, $\Delta V_G=10V$, so voltage amplification is $A_V= -\Delta V_A/\Delta V_G= -29.5$ or 29.4 dB.

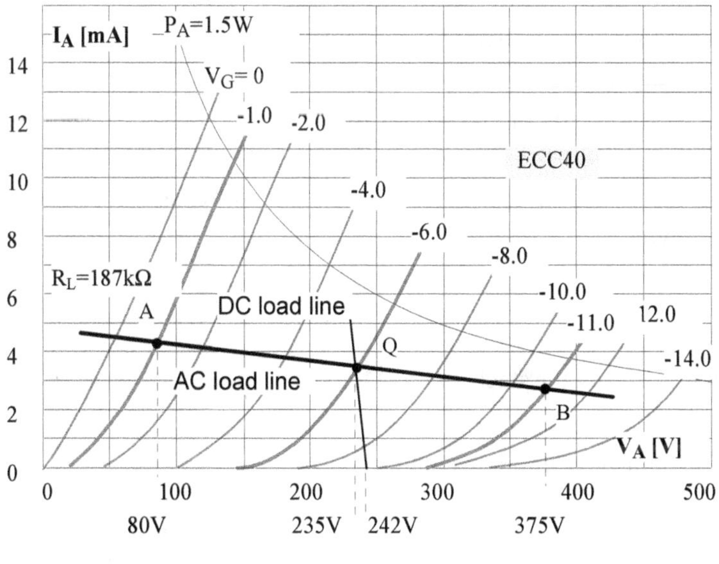

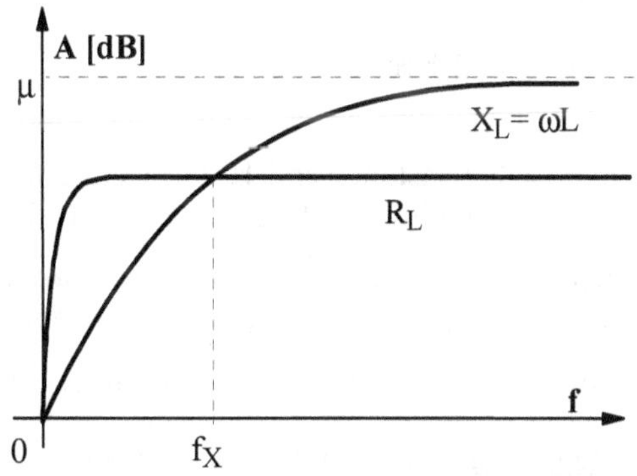

ABOVE: The frequency dependence of the amplification factors of a RC and LC-coupled common cathode stage.

LEFT: The AC and DC load lines for LC-coupled ECC40 stage at 1 kHz.

While inductive coupling is advantageous at higher frequencies, it is inferior at lower frequencies because X_L is frequency-dependent and R_L isn't! At frequency zero (DC), XL drops to zero, and so does the amplification factor, so inductive coupling cannot be used in DC amplifiers!

Depending on the RL and L values, you can find at what frequency the inductive coupling results in the same AV as that of the resistive coupling: $R_L=X_L$ or $R_L=2\pi fX_L$, so $f_X=R_L/2\pi L$. For our anode choke used in a real amplifier (L=200H), instead of using a resistor $R_L=47k\Omega$, we get $f_X= 47{,}000/400\pi = 37$ Hz!

This is bad news; you have to think carefully if the higher amplification of a certain stage is worth the price paid - a compromised low-frequency response. The midrange gain with impedance coupling is higher than with RC coupling, but it drops off more rapidly at frequency extremes, so the bandwidth is narrower.

The graph (previous page) illustrates the low-frequency response of a common cathode stage with a resistive load R_L and inductive load X_L.

TUNING THE IMPEDANCE AND DOUBLE-IMPEDANCE COUPLED STAGES

CALCULATION

The anode choke in the ECC40 stage has L_A=200H. Find the value of the coupling capacitor needed to tune the resonant circuit to 20Hz, to improve the bass response. How would that value change if a grid choke L_G=5,000H is used instead of R_G (double impedance coupling)?

At resonance, the reactive impedance of the capacitor and the choke are equal: $X_C=X_L$ or $1/\omega_R C=\omega_R L_A$, so we have $C=1/(\omega_R{}^2 L_A)$. In this case $\omega_R=2\pi f_R=2*3.14*20=125.7$ rad/s and $C=1/(\omega_R{}^2 L_A) = 0.317\mu F$, so we would use the nearest standard value, which is $0.33\mu F$!

If a grid choke is also used, the effective inductance is $L=L_A+L_G=5{,}200$ H. In Class A_1 operation, there is practically no current through grid chokes, thus wound using a very fine wire to get a high number of turns. They have ten or more times higher inductance than anode chokes, which must pass 5-20 mA of current and are wound using a thicker wire, meaning fewer turns. The required coupling capacitance is now much lower: $C=1/(\omega^2 L) = 0.0122\ \mu F = 12.2$ nF

Since $\omega=1/\sqrt{(LC)}$ or $f=1/[2\pi\sqrt{(LC)}]$, using a larger value capacitor will shift the resonant frequency lower. For instance, if a 22nF coupling capacitor is used instead of 12.2nF just calculated, the resonant frequency would drop from 20 Hz to 14.9 Hz!

COMMERCIAL EXAMPLE: Valab 50H/100mA plate choke

Valab AP50-100 plate choke was designed to be used with power tubes with internal resistance R_I under $2k_W$, for instance, 300B, 2A3, and most pentodes and beam power tubes strapped as triodes (6L6, EL34, 7027, and many more). Z6 silicon steel laminations were used (the lowest grade of GOSS material).The inductance was specified as L= 50H @ 120Hz and 74H@20Hz, DC resistance as 455Ω , AC impedance as 790 kΩ @120Hz, and 3.2 MΩ @1kHz.

Dimensions are 75 x 70 x 65 mm, weight 1,600g. For educational reasons and to practice numerical manipulation skills, let's try and reverse-engineer this plate choke by estimating the number of turns from inductance and DCR.

The laminations seem to be EI74 with 38 mm stack: A= aS = 8.75 cm^2. From the formula for inductance $L= N^2\mu_{EF}\mu_0 A_{EF}/\ell_{MP}10^{-8}$ we could estimate the number of turns as $N=\sqrt{[10^8 L\ell_{MP}/(\mu_{EF}\mu_0 A_{EF})]}$.

We know $\ell_{MP}=2(b+e-d) + \pi c = 2*(14+51-14)+\pi*11.5 = 138$ mm, but we need to know the effective permeability μ_{EF}, which we don't (since we don't know the width of the air gap used).

The 455Ω DC resistance and 100 mA current point out that the winding wire has a diameter of between 0.2 mm and 0.236 mm, depending on the current density J chosen by its designer. Let's try to find N from the primary resistance.

$N=\ell_{TOT}/MLT$ and the Mean Length of Turn is $MLT=2(a+S+4B)+\pi W$, B=bobbin thickness, W=winding thickness: MLT = 2*(23+38+4*2) + 3.14*(14-3) = 172mm = 17.2 cm

Now we need to find the winding length ℓ_{TOT} from the DC resistance!

<u>First attempt (assuming d=0.2mm magnet wire used):</u>

$R_{DC}=r\ell_{TOT}/S$ (S is the cross-sectional area of the winding wire, not the stack width of the core), so $\ell_{TOT}=SR_{DC}/\rho$, where $\rho = 1.678*10^{-8}\ \Omega m$ (resistivity of copper)

$S=d^2\pi/4 = 0.00022*\pi/4 = 3.1416*10^{-8} m^2$ (here we use diameter without insulation)

$\ell_{TOT}=S*R_{DC}/\rho = 3.1416*455/1.678 = 851.86$ m

$N=\ell_{TOT}/MLT$= 85,186 cm/17.2cm = 4,952 Turns

CHECK: Could they have physically fit almost 5,000 turns? Again, we assume 2mm bobbin thickness and allow for 1 mm clearance on top of the winding, so Net Winding Area WA_{NETT}= (14-3)*(34-4)= 11*30 =330 mm^2.

Square area taken by wire d=0.22 mm (with insulation) $A_{SQ}=d^2$ = 0.0484 mm^2.

$N_{MAX}=WA_{NETT}/A_{SQ}$= 330/0.0484 = 6,818 Turns. The window utilization ratio is 4,952/6,818 = 72.6%, which is plausible although a bit low for a professional job, so let's repeat the same exercise for a thicker wire of d=0.212 mm and see what we get.

<u>2nd attempt (assuming d=0.212mm magnet wire used):</u>

$S=d^2\pi/4 = 0.000212^2*\pi/4 = 3.53*10^{-8}m^2$. Now we can calculate $L_{TOT}=SR_{DC}/\rho$ = 3.53*455/ 1.678 = 957 m

$N=L_{TOT}/MLT$= 95,700 cm/17.2 cm = 5,564 turns.

CHECK: Square area taken by wire d=0.232 mm (with insulation) is $A_{SQ}=d^2$ =0.053824mm^2. $N_{MAX}=WA_{NETT}/A_{SQ}$= 330/0.053824 = 6,131 turns, and 5,564/6,131= 90.8%, the most likely scenario. So, our estimate is of N around 5,500 turns!

Let's see what μ_{EF} we get from the simplified formula $L= N^2\mu_{EF}\mu_0 A_{EF}10^{-2}/\ell_{MP}$. Since $\mu_0 = 4*\pi*10^{-7}$ Vs/(Am) = $1.257*10^{-6}$ T*m/A we can write $L= 1.257N^2\mu_{EF}A_{EF}*10^{-8}/\ell_{MP}$.

Now we can calculate $\mu_{EF} = L\ell_{MP}10^8/(N^2 1.257A_{EF})$ =50*13.8*10^8/(5,564^2*1.257*8.75*0.95) = 213 at 120Hz.

Using the more precise formula, assuming a 0.1 mm gap "g", we get $\mu_{EF} = L\ell_{MP}/(N^2 1.257A_{EF}10^{-8} - Lg)$ = 252. We could have estimated μ_{EF} from the curves on page 19 (graph in the upper right corner). For 1,000*K= 1,000*g/ℓ_{MP}= 1,000*0.1/138 = 0.725, depending on B we get the range of μ_{EF} from 200 at 1T to 250 at 0.4T!

COMMERCIAL EXAMPLES: Silk and Valab grid chokes

Our personal experience confirms the claims of many audiophiles: tube amps using grid-chokes instead of grid resistors sound better, more dynamic, and musical.

The arrangement below shows a simple yet very good sounding amplifier topology with an SRPP input stage and a choke-driven grid of a single-ended output stage. The end of the choke connected to the ground can instead be taken to a negative DC voltage source, -70 to -90 V_DC, for an even larger swing of the driving voltage on the output tube's grid and higher output power.

A grid choke wound with a center tap (or any other % tap) can be used as a step-up autotransformer, which increases gain or in a step-down mode. In this case, the step-up ratio is $V_{OUT}/V_{IN}= (N_1+N_2)/N_1$.

In 2017, grid-chokes by SILK Thailand sold for US$176 a pair. An inductance of more than 7,000 H at 12Hz is claimed, DCR of 1.3 kΩ, AC impedance at 1kHz of more than 5 MΩ (meaning that $L_{1,000}$=796H), and parasitic capacitance of less than 20pF. The dimensions are 37 x 32 x 32 mm (LxWxH).

Chokes have a center tap for phase splitting applications and use Supermalloy material.

Permalloy is an alloy with about 20% iron and 80% nickel content. Supermalloy has 5% molybdenum, 79% Ni and 15% Fe. It belongs to the 80% nickel family of materials.

It saturates early, at 0.7 to 0.8 Teslas. The usable induction (the flux level at which the incremental permeability has substantially decreased) is 0.65 to 0.7 Tesla, much lower than silicon steel at around 1.6 T.

Let's try to deduct the basic design parameters from the known data. From the 32mm height, we can guess that the laminations used are EI35, where e+d=29.5mm, and the 2.5mm difference to the overall dimension is twice the thickness of the metal frame. The stack seems to be around 10 mm.

Since A=aS=0.96*1=1cm^2 and ℓ_{MP}=2(b+e-d)+ πc= 2*(7.7+24.5-5) + 3.14*5 = 70.1mm = 7cm, from $L= N^2\mu_R\mu_0A/\ell_{MP}10^{-8}$ we can express N as $N=\sqrt{[10^8 L\ell_{MP}/(\mu_R\mu_0A)]}$.

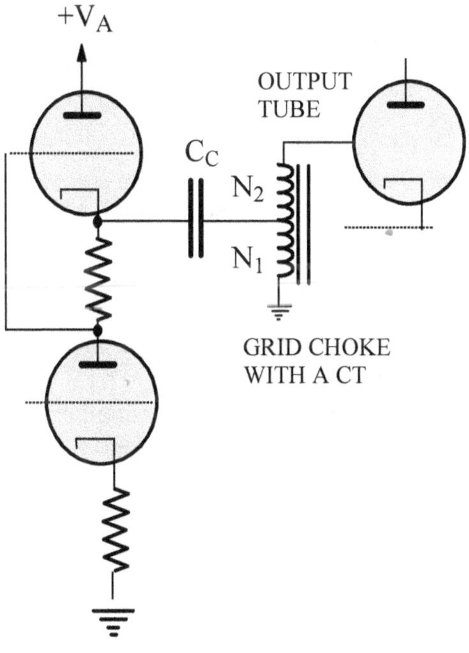

ABOVE: A grid choke with one or more taps can be used to step the voltage up

Depending on the lamination thickness (which we don't know without seeing the actual choke), the permeability for Supermalloy is μ_R= 20-50,000 and $\mu_0 = 1.257 * 10^{-6}$ Vs/Am. For μ_R= 20,000 we get N=1,784 turns.

The Mean Length of Turn is MLT=2(a+S+4B)+πW, B=bobbin thickness, W=winding thickness, MLT = 2*(9.6+10+4*2) + 3.14*5.0 = 2*27.6+15.7 = 71 mm. For 1,784 turns the total wire length is ℓ_{TOT}=126,664 mm.

$R_{DC}=\rho\ell_{TOT}/S$, so $S=\rho\ell_{TOT}/R_{DC}$ where $\rho = 1.678*10^{-8}$ Ωm From here S= $1.635*10^{-9}$ m^2. Since $S=d^2\pi/4$, we get wire diameter $d=\sqrt{(4S/\pi)} = 4.56*10^{-5}$ m = 0.0456 mm, or AWG45.

Supermalloy is expensive, and winding such small-sized chokes is a fiddly task. To get similar results, you can use larger GOSS laminations such as EI48 up to EI66, where you can fit a larger number of turns or the same number of turns of a larger diameter wire.

Valab GC640-5 grid chokes are also sold on eBay, but are cheaper, only US$50 +US$10 shipping per pair (in 2017). Z11 lamination are used, 20,000 turns, L=640H @ 120Hz and 2,500H @ 20Hz. Z11 is an old Nippon steel specification, equivalent to the current 35Z155 material, M6 equivalent, the lowest grade of GOSS.

TRANSFORMER-COUPLED STAGES

Believing that capacitors, even of paper-in-oil (PIO) or film&foil kind, should be avoided, audiophiles advocate the merits of transformer coupling between stages instead of the cheaper and simpler RC coupling. Transformer coupling, they claim, results in a more natural, less colored, or stringent sound.

Interstage transformers also make the precise matching of impedances between the stages possible. They can serve as phase splitters, converting a single-ended signal into two signals of the opposite phase to drive the output tubes in a push-pull configuration.

Nothing is free in electronics, and the price to pay here is high. Quality interstage transformers are even more difficult to design and build than output transformers and are costly, especially if there is a primary DC current. For that reason, a parafeed arrangement can be used where a coupling capacitor prevents DC current from entering the primary winding. Much wider bandwidth can be achieved without the air gap, but we are back to capacitive coupling, and the nasty capacitor sound, meaning all the benefits are lost, a typical exercise in futility.

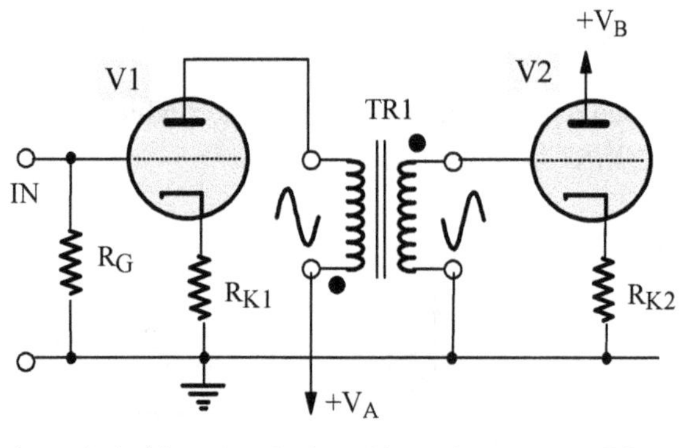

Single-ended driver to single grid, series output, DC current flowing through transformer's primary, phase inverting

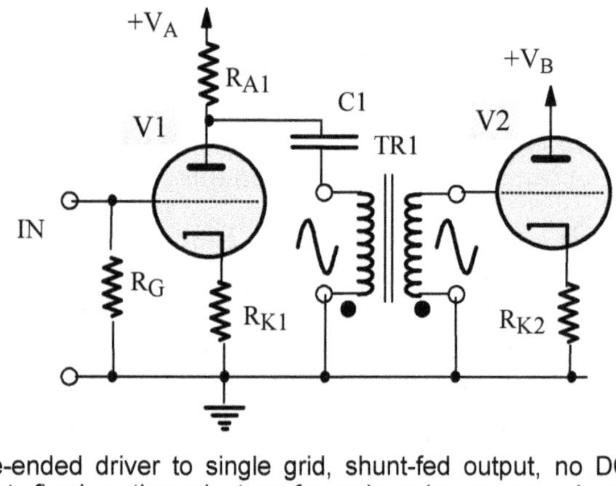

Single-ended driver to single grid, shunt-fed output, no DC current flowing through transformer's primary, no phase inversion

Low frequency response of an interstage transformer

The low-frequency model for interstage transformer coupling is a simple RL high pass filter. The inductive impedance or reactance is $Z_L = \omega L_P = 2\pi f L_P$, where L_P is the primary inductance. The total resistive component of the filter is equal to the series combination of the tube's internal resistance rI and the transformer's primary resistance R_1, which we will call R_S, where $R_S = r_I + R_1$.

Of most interest is the frequency at which the inductive reactance of the transformer primary inductance L_P becomes equal to the resistive component of the filter ($X_L = R_S$): $f_L = R_S(2\pi L_P) = (r_I + R_1)/(2\pi L_P)$ so $L_{PMIN} = (r_I + R_1)/(2\pi f_L)$. This is why high μ tubes such as 12AT7, 6SL7, or 12AX7 should not be used as drivers. Due to their high internal resistance r_I, their low-frequency response would be poor in this situation.

Primary inductance can be increased by using larger laminations and a higher number of primary turns, but then parasitic capacitances also increase, and the HF response deteriorates. Primary inductance can also be increased by using magnetic materials of increased permeability, such as Permalloy and Hiperm, but they are expensive and saturate at much lower levels of B, so no luck there either.

LF RESPONSE OF INTERSTAGE TRANSFORMERS

$f_L = (r_I + R_1)/(2\pi L_P)$

$L_{PMIN} = (r_I + R_1)/(2\pi f_L)$

Choosing the driver tube and its anode current

These graphs from the Tango NC-20 spec-sheet illustrate two important points. In the upper graph, if you compare the two curves for the same primary current (10 mA in this case), the higher the internal impedance, the faster the low-frequency response drops and the earlier such a drop starts.

At 5 Hz and 10 mA, the response is down only -1dB with a 1kΩ driver tube, compared to -6dB for a 5kΩ driver tube. Primary DC current also affects the low-frequency response. The higher the current, the stronger the low-frequency attenuation and the higher the f_L!

Looking at the graph for a 5kΩ driver tube, at 10 mA, the -3dB f_{LA} is 9 Hz, which is an acceptable result, while at 30 mA, f_{LB} is around 21 Hz, too high for a hi-fi amp.

So, choose driver triodes with as low internal resistance rI as possible, and minimize the DC current through the transformer's primary.

However, the internal impedance of a tube varies inversely with its plate current, which is very unfortunate here. The graphs for 6SN7 triodes show rI dropping from about 35 kΩ at 1 mA to between 7 and 8 kΩ above 10 mA. This is an obvious contradiction, so a compromise is needed.

Minimizing plate current through transformer's primary winding will cause the driver tube's internal resistance to go up, and vice versa.

A designer should never go below the current where μ drops, in this case, below 5 mA. Don't go too high, either, since rI does not drop much further (in this case, rI is constant from 7 to 17 mA). Depending on the transformer used, the optimal range for 6SN7 would be 7 - 13 mA.

We could connect two or even more triodes in parallel and substantially reduce the overall internal resistance. At 9mA, one 6SN7 triode has internal resistance of 8 kΩ, so two in parallel would have half that, which would lower f_L and improve the LF response of the amp!

However, the current through the interstage transformer's primary would then double, so it would need to be wound with thicker wire, necessitating a larger size EI laminations or C-core.

The bad news is that the inter-electrode capacitances of two paralleled tubes will double in value and start shunting or bypassing high frequencies to the ground earlier, meaning that the upper -3dB frequency of this stage will be reduced will drop significantly, up to half of the single-tube value.

So, if the upper -3dB frequency with one tube was, say, 120 kHz, now it will drop to about 60 kHz!

As is always the case with transformers and amplifiers, whatever you do to improve the bass (LF response) makes the HF response (treble) worse!

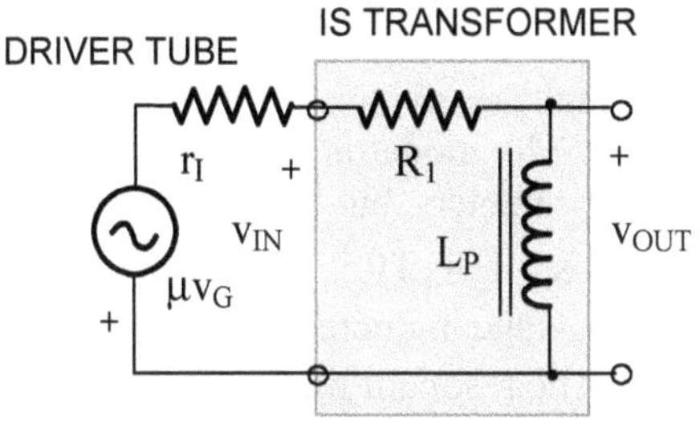

ABOVE: Low frequency model for Class A_1 and AB_1 interstage transformers

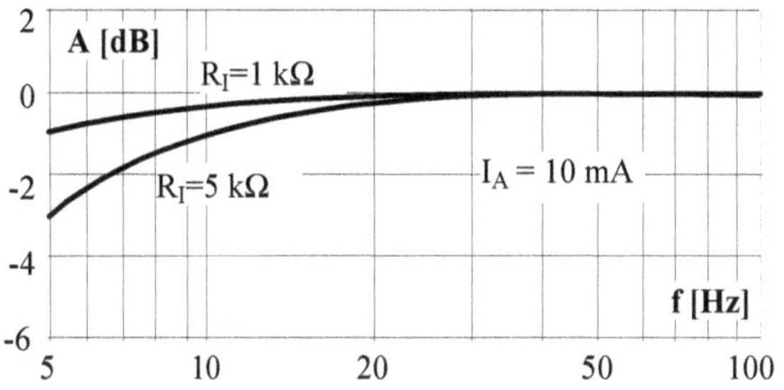

ABOVE: Low frequency response of Tango NC-20 interstage transformer for r_I=1 kΩ and 5 kΩ

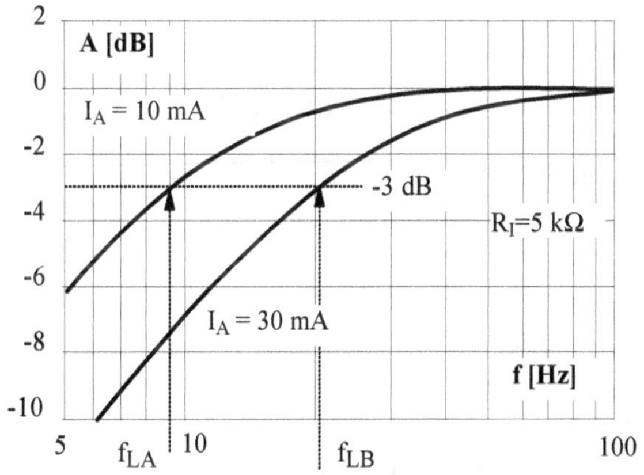

ABOVE: Low frequency response of Tango NC-20 interstage transformer for r_I=5 kΩ and two different values of primary DC current

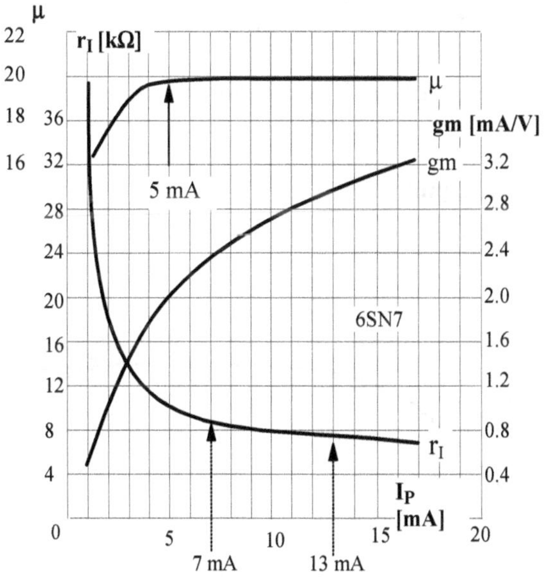

ABOVE: The three main parameters of 6SN7 triode as a function of anode current

High frequency model for interstage and output transformers

At high frequencies, the transformer behaves like a series resonant circuit formed by leakage inductances and parasitic capacitances. The model is simplified, of course, since we are using lumped capacitance, while in reality, they are distributed parameters, but the results are very close to measured figures.

$R_{SER} = r_I + R_1 + R_2TR^2$, $L_L = L_{LP} + L_{LS}TR^2$, and $C = C_1 + (C_2 + C_{IN})/TR^2$. L_{LP} and L_{LS} are the primary and secondary leakage inductances. At the upper frequency limit (-3dB) $X_C = R_{SER}$ or $1/(\omega C) = R_{SER}$, so $\omega_U = 1/(R_{SER}C)$. The resonant frequency of the LC circuit is $\omega_R = 1/\sqrt{(L_LC)}$. The optimal HF response is achieved when the resonant frequency f_R equals the upper -3dB frequency f_U: $\omega_R/\omega_U = 1$.

The primary inductance is $L_P = \mu_{EF}\mu_0 N_1 A_{EF}/(g + \ell_{MP}/\mu_{EF})$, where ℓ_{MP} is the length of magnetic path, and g is the air gap. Once we determine the L_P required for the low frequency response, we need to find N_1.

We know A_{EF} and ℓ_{MP}, but don't know the size of the gap g or the effective permeability μ_{EF}. There are ways to estimate these, but basing designs on assumptions can be grossly inaccurate - even a small discrepancy in such parameters has a huge impact on the calculated number of turns, requiring two, three, or even more iterative steps.

The alternative approach is similar to "designing" chokes. Simply fit as many turns as possible, make the coil, assemble the transformer and adjust the air gap; measure its performance and see what you've achieved.

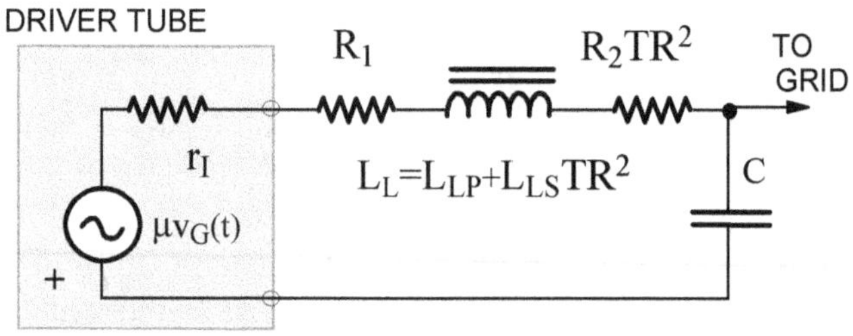

ABOVE: The HF model of an interstage transformer in class A_1 (infinite grid resistance, no grid current).

The step-up transformer problem

A line-level input transformer has a ratio of 1:10; the leakage inductance is 10mH, and primary and secondary capacitances are roughly the same, around 40pF. It drives the grid of EF86 pentode with an input capacitance of around 15pF. What is the circuit's resonant frequency, and what upper -3dB frequency can we achieve?

$C = C_1 + (C_2 + C_{IN})/TR^2 = 40 + (40+15)/(1/10)^2 = 40 + 55*100 = 5{,}540$ pF = 5n5

When reflected to the primary, C_2 and C_{IN} are divided by the square of the TR, and since TR, in this case, is smaller than 1 ($N_1/N_2 = 0.1$), the capacitances are multiplied by 100! For 1:20 ratio, they would be multiplied by 400!

The resonant frequency is $\omega_R = 2\pi f_R = 1/\sqrt{(L_LC)}$ so $f_R = 1/\sqrt{(L_LC)}/2\pi = 21{,}382$ Hz, barely enough for a hi-fi amplifier. To make things worse, this is one of the best cases using a low leakage transformer and a low input capacitance pentode. With a triode at the input with C_{IN} of say 115 pF, we would get $C = 40+155*100 = 15{,}540$ pF = 15nF and $f_R = 1/\sqrt{(0.01*15.54*10^{-9})}/2\pi = 4{,}037$ Hz.

Assuming $f_R = f_U$, an upper frequency around 4 kHz would not be enough even for a guitar amp, let alone hi-fi. Now you understand why quality wideband IS transformers never go above the 1:2 ratio!

The impact of Q-factor on the high frequency response

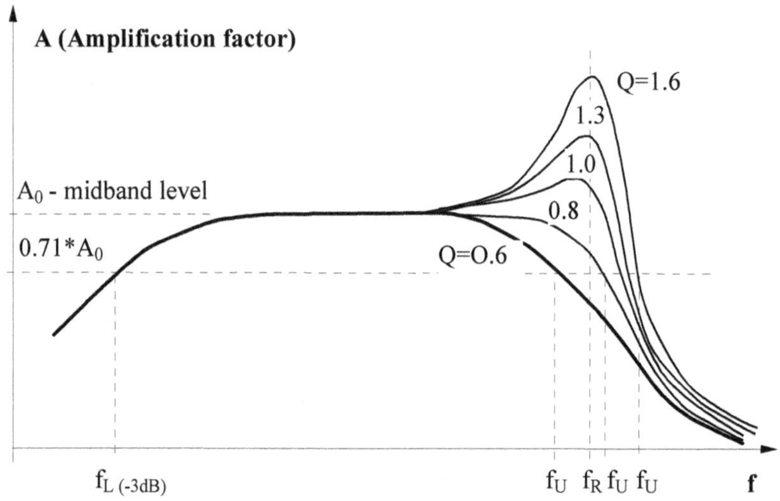

LEFT: The frequency response of an interstage transformer depends heavily on its Q-factor

The high-frequency resonant peak is an issue with interstage transformers, where the load impedance is very high (the grid impedance of the following tube). For step-down interstage and output transformers, quality factor Q is usually below 1, and there is no peak.

The optimum Q is about 0.7-0.8, resulting in the flattest and widest response curve. In the first approximation, we can assume that the upper -3dB frequency f_U is the same as the resonant frequency f_R. The curves are not to scale; the differences are smaller than illustrated.

Interstage transformers to drive Class A_2, AB_2 or Class B_2 output stages

If the grid of power tubes is to be driven into the positive region, as in classes A_2, AB_2, and B_2, the grid current will start flowing, and the input impedance of the output stage will drop to a very low value. That will cause the loading of the driver stage and a large distortion. Therefore, RC coupling should *never* be used for Class A_2, AB_2, or Class B_2 amplifiers.

This is illustrated in the equivalent diagram below right. Driver tube is the voltage source with its internal impedance R_I, R_A is its anode resistor, and R_G grid resistor of the output stage.

The voltage divider ratio is constant, and there is no distortion. The flow of grid current is equivalent to switching a very low resistance R_{GK} between the grid and the cathode of the power tube, in parallel with R_A and R_G.

Secondly, a driver tube should have as low internal resistance as possible. A single 6SN7 triode section has r_I=6,800 Ω. Some designers parallel two sections together and get half that impedance, r_{IP}=3,400Ω.

Instead of step-up action, the best choice would be a step-down interstage transformer, no matter how counterintuitive it sounds. The grid driving voltage fed to the output stage would be reduced, but two advantages would be achieved.

First, the secondary winding would have a much lower DC resistance, eliminating the danger of bias change due to grid current flowing. Second, the step-down ratio would reduce the "internal" impedance of the driver stage (as seen by the grid of the power tube).

For instance, with one 6SN7 triode and a 2:1 transformer, the source resistance, reflected onto the secondary side, as seen by the power tube) is cut from 6,800Ω to 1,700Ω or 4 times, since the impedance ratio $IR=TR^2 = 2^2 = 4$.

The good news is that despite halving the driving voltage's amplitude by the transformer, we can still realize the same overall gain since the voltage drop due to internal source impedance r_I is also cut to 1/4 of the previous value.

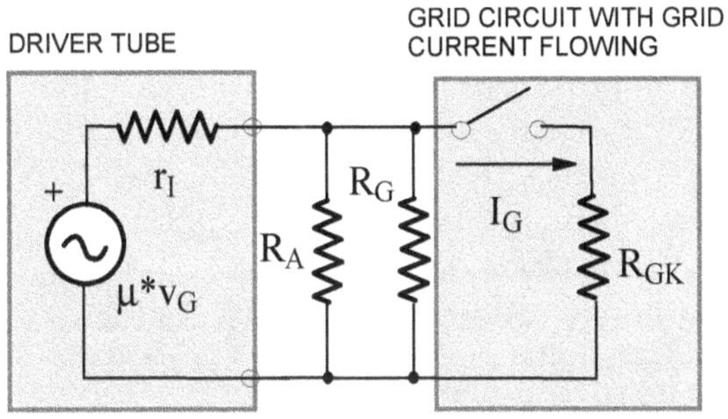

ABOVE: The simplified model of a SE driver tube stage and power tube driven into class A_2, AB_2 or B_2 with grid current flowing (once the "switch" is closed).

BELOW: Control grid current as a function of positive grid voltage

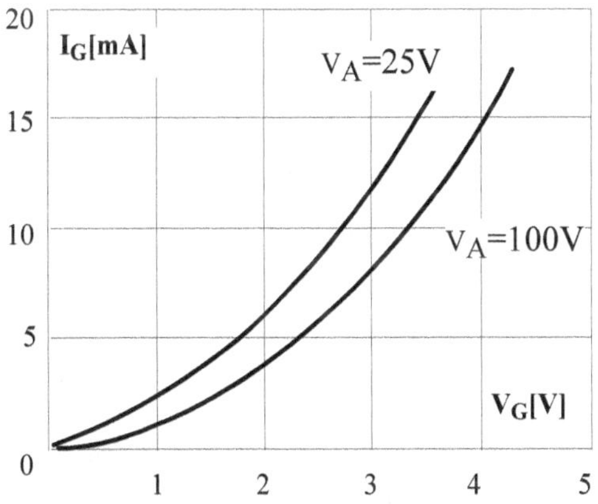

RC- versus transformer coupling

In conclusion, it is impossible to obtain the same bandwidth with transformer coupling as can be achieved with capacitive coupling. The flatness of the frequency response is usually also inferior, with transformer coupling having a pronounced resonant peak in the 15-60 kHz range. Technically, transformer coupling cannot be justified.

Listening tests, however, show that transformer coupling sounds different from capacitive coupling, but does it sound better or just different is the ultimate question, one that divides audiophiles and causes endless debates and disagreements.

BELOW: Stage gain (log scale) versus frequency of a typical RC-coupled (A_{RC}) and transformer-coupled triode stage (A_{TX}).

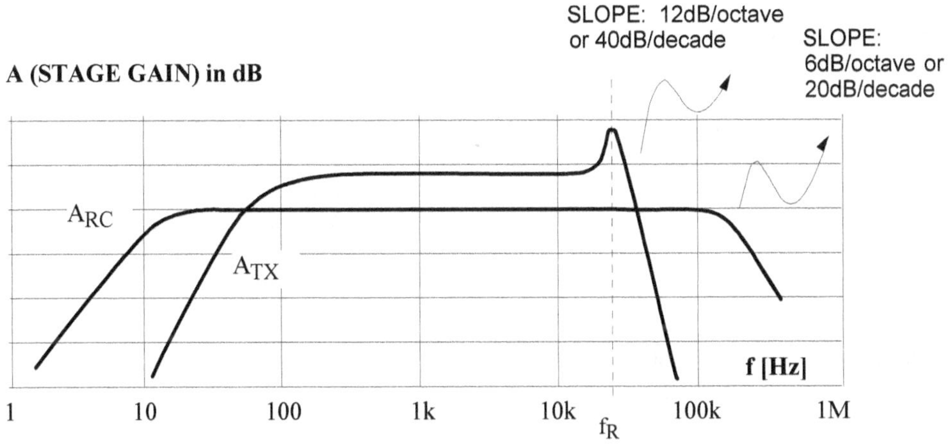

COMMERCIAL INTERSTAGE TRANSFORMERS

CASE STUDY: Hashimoto A-105 and A-107

Made in Japan, Hashimoto A-105 is a versatile albeit expensive interstage transformer; both the primary and the secondary can be connected in series or parallel. The secondary windings can be connected in series, parallel, or configured in center tap push-pull connection, with 1:1, 1:0.5, or 1:0.5+0.5 voltage ratio, respectively.

It is wound on a pair of tiny C-cores, confirmed by only 0.8 kg of total weight (including the steel case). In 2017, a pair cost US$596.00 + US$42 transport!

The manufacturer's amplitude-versus-frequency graphs (below right) show this transformer's high-frequency response for 1:2 ratio (secondaries in series) and 1:1 ratio (secondaries in a phase-splitting arrangement). Notice how f_U increased from 47kHz for the 1:2 ratio to 75kHz for the 1:1 connection.

The lower -3dB frequency limit f_L depends on the parallel combination of R_1 and R_2: $R_P=(R_1R_2)/(R_1+R_2)$.

For interstage transformers with no secondary current (no grid current), the load impedance can be considered infinite, so at the -3dB frequency $Z_L= \omega L=2\pi f L=R_1$ or $f_L=R_1/(2\pi L)=(R_I+R_P)/(2\pi L)$.

For primaries in series the lower -3dB frequency is $f_L= (5000+350)/(2\pi 60) = 5{,}350/377 = 14.2$ Hz (-3 dB) which agrees with the declared -2dB f_L of 25 Hz.

Hashimoto model A-107 has a 7kΩ primary, 25-25,000Hz (+-2dB) frequency range and primary inductance of L_P=60H.

MANUFACTURER'S SPECIFICATIONS:

- Hashimoto A-105, made in Japan
- Dual primary & dual secondary driver transformer
- US$596.00 for a pair + $42 transport on ebay (in 2015)
- Three voltage ratios are possible, 1:1, 1:2 and 1:0.5
- Z_P (series) 5 kΩ, parallel 1.2 kΩ
- L_P = 30H each half, 60H (series), 15H (parallel)
- Maximum primary current: 15mA in series or 30mA in parallel
- Primary resistance: 350 Ω, secondary resistance 450 Ω
- Frequency range: 25-35,000Hz (+/-2dB)
- Dimensions: 52 (W) x 58 (D) x 85 (H) mm, weight 0.8 kg

BELOW: High frequency response of Hashimoto A-105 interstage transformer for two different voltage ratios

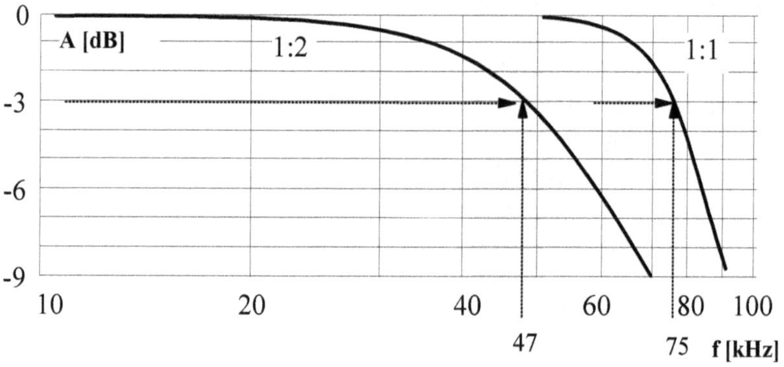

A-107 is suitable for a triode driver stage with a maximum of 10 mA primary current (20 mA with two primaries in parallel). The secondaries can be used in series, parallel, or push-pull connections (1:2, 1:1, or 1:1+1 ratio).

A-107's fU increased from 46 kHz for a 1:2 single-ended connection to 67 kHz for a 1:1 PP secondary connection (as a phase splitter). The higher the step-up ratio, the lower the bandwidth of an interstage transformer!If you want to experiment after winding your own version, make the primary and secondary windings "modular," as in this case.

If you are a real fanatic, you can go even further and have 4 identical primaries and 4 secondaries if you need so many voltage and impedance ratios. Two & two primaries can be paralleled and then connected in series, the same with the secondaries, all four primaries in series or parallel, ditto for the secondaries, and so on.

COMMERCIAL CASE STUDY: Sowter and Audio Note IS transformers

Sowter Transformers, based in Ipswich, England, was established in 1941. A wide transformer range and useful technical information make their website a worthwhile place to visit.

Sowter type 8423 is a 1:1+1 transformer on EI78 laminations with 5k nominal impedance of each of its three windings. The primary inductance is 25H (frequency not specified?), the maximum primary current is 50mA, and the bandwidth is modest, 30Hz-30kHz (-3dB). In 2016 it sold for £133.57 (approx. US$190.-) each.

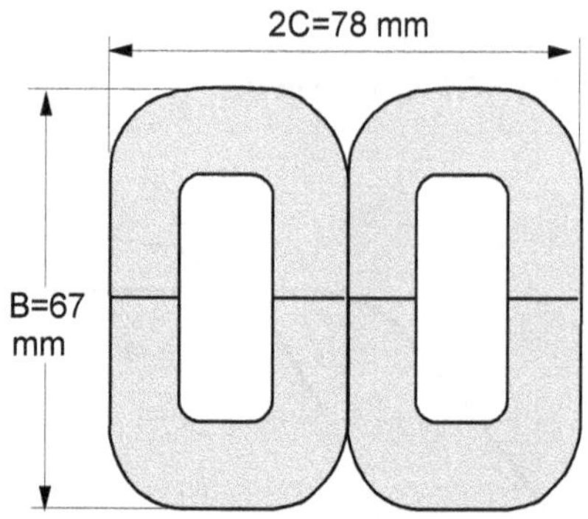

RIGHT: The specified overall dimensions of 78mm and 67mm, indicate this size of Audio Note IS transformer cores.

Audio Note UK offers 33 (!) different interstage transformers, grouped into three types (11 models in each group) by the maximum primary current (10, 20, and 30mA). In each group, the cheapest model uses EI laminations (M4 material) and costs £155.40 each. The next three models also use copper wire, but their C-cores are made of better grade magnetic materials, from "Improved HiB" (£353.60) through "Super HiB" (£424.00) to "ultra HiB" (£635.20 each).

If you thought those prices were steep, the next group uses copper on the primary side and Audio Note silver lead-out wires on the secondary side. It is unclear if that includes silver secondary windings or only the lead-outs? Depending on the magnetic material used, they range in price from £789.60 to £2,438.00 each.

The last three models are fully wound with silver wire and cost from £2,109.80 to £3,392.76 each, excluding the VAT ("Value-Added Tax"). The most expensive units use "Super Perma" 50% or 55% nickel cores. The impedance and voltage ratios are the same for all models, 1:1, and all are of the same physical size.

CASE STUDY: Hammond 126A

Hammond 126A is a 1:1 interstage transformer with a 5k impedance. The maximum primary current is 15 mA, and the bandwidth is specified as 20Hz-20kHz (-1dB). The laminations are 76x63 mm. The magnetic material is specified as 29M6. The stack thickness is 38 mm, which means the cross-sectional area of its center leg is $A = aS = 2.6*3.8 = 9.9\ cm^2$. Each transformer weighs 3.125 lb. (around 1.4 kg). In 2015 these were reasonably priced at US$79.- each.

While the claimed primary inductance was 59H (the test frequency and the measurement method were not specified), we measured only 11-12H with a digital LCR meter. Since LCR meters measure inductance without any DC bias, with 15 mA of DC current magnetizing the core, the primary inductance of such transformers can only go down!

There was also a significant difference in the inductance between the two transformers. The inductance of SE transformers varies with the airgap size, and simply tightening one transformer's bolts can cause such differences.

Hammond claims that "all models use bifilar wound windings for exceptional coupling and bandwidth", so one would expect no DCR differences between primary and secondary windings. However, one transformer had R_P= 190Ω and R_S=186Ω, and there was also a difference in test results between our two samples. The other transformer's R_P was only 177Ω, meaning a (190-177)/177 =0.073 or 7.3 % difference. Such a difference cannot happen with automated winding, where five or ten transformer coils are wound simultaneously, and the number of turns is controlled by a computer; thus, this could indicate hand winding and sloppy counting of turns.

In their reply to our question about such resistance variations, Hammond attributed that difference to magnet wire, which has "a resistance tolerance of +/-15%" and "then, during the winding operation, with winding tension, there may be stretching that adds an additional 5%". In other words, they claim that DCR variations of up to 20% are normal. I am not convinced!

Due to a single secondary winding, so they cannot be used as phase splitters in push-pull amplifiers. We can do better; our universal IS transformer will have a split primary and a split secondary winding for both SE and PP applications.

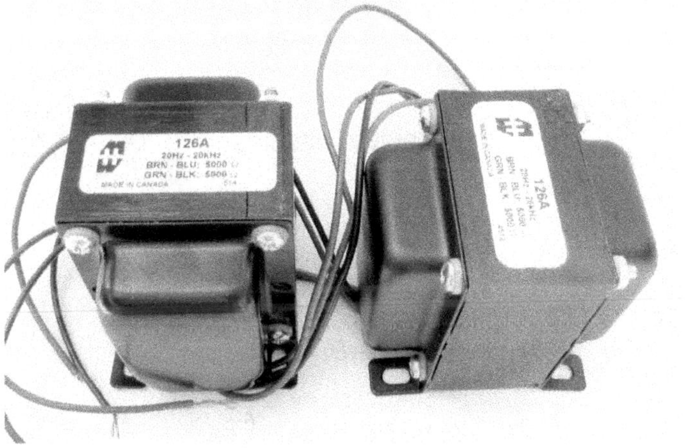

MEASURED RESULTS:

- TRANSFORMER #1:
- R_P= 177Ω, R_S= 176 Ω
- L_{P120}= 12.4H, L_{P1000}= 11.3 H
- L_L=0.40 mH, C_{PS1} = 64 nF, C_{PS2} = 2.2 nF
- TRANSFORMER #2:
- R_P= 190Ω, R_S= 186 Ω
- L_{P120}= 19.5H, L_{P1000}= 13.6 H
- L_L=0.42 mH, C_{PS1} = 64 nF, C_{PS2} = 2.4 nF

VINTAGE EXAMPLE: Triad HS-35 interstage transformer

Triad HS-35 interstage transformer appears in their 1961 catalog and is specified as a single plate to single or push-pull grids. The primary impedance is 15 kΩ, and the total secondary impedance (both secondaries in series) is 111 kΩ, resulting in a 1:2.72 voltage ratio. The frequency range at the 26 mW power level is 20 Hz - 20 kHz.

The DC resistance of the primary winding is 1.67 kΩ, and its inductance is above 200H. The secondary DC resistance is 3,940 Ω, and each secondary's inductance is above 200H. 49% nickel alloy laminations were used, size 26-27 EE. The center leg is 3/8" wide (9.525mm), and the stack is 13/32" thick (10.32mm).

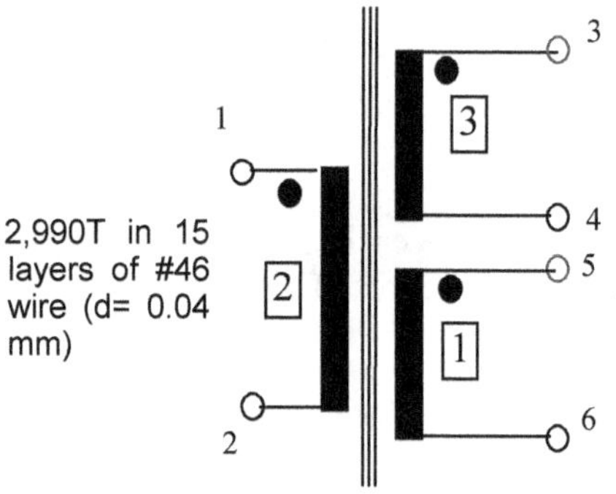

LEFT: Triad HS-35 interstage transformer winding diagram

RIGHT: 26-27 EE laminations

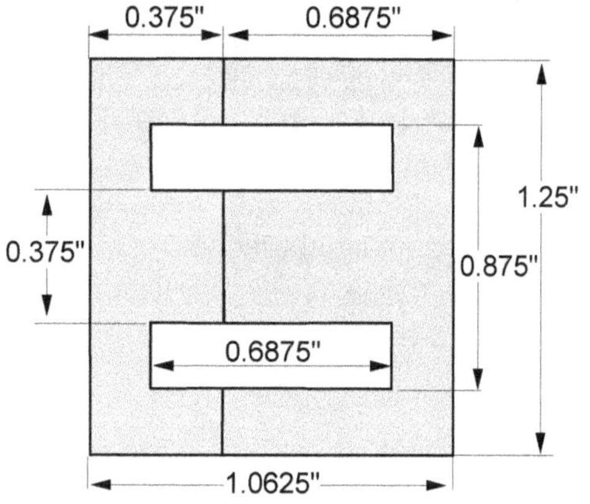

A simple 1/2 sectionalizing is used; secondary 5-6 is wound first, then the primary, then the secondary 3-4. Let's check the horizontal and vertical fit. The window length is WL=0.6875" = 17.5 mm and coil length CL=WL-4 mm = 13.5 mm. The 0.04mm diameter is the dimension of the bare wire; with heavy insulation, its diameter is around 0.055mm.

PRIMARY: TPL_{MAX}= CL/d = 13.5/0.055 = 245 turns, and with an assumed horizontal fill factor of HFF=0.8 we get TPL= TPL_{MAX}*HFF = 245*0.8 = 196.

Indeed, the dissection of this transformer by one keen enthusiast showed 15 layers of between 196 turns per layer and 204 TPL, so our theoretical estimate is spot-on.

The window height is WH=0.25"= 6.35mm, which reduced by a 2mm bobbin gives us net maximum coil height of CH_{MAX}=4.35mm. Now we can check the vertical fit: PH= 15*0.055 = 0.825 mm, SH= 40*0.047 = 1.88 mm, total coil height is thus CH = PH+SH = 2.7 mm

Since CH_{MAX} - CH = 4.35 - 2.7 = 1.65 mm, there is plenty of insulation and bulging factor allowance.

CASE STUDY: Raphaelite interstage transformers

These "5k to 6k6" interstage transformers are made in China and have an impedance ratio of 1:1.25. The maximum allowed primary current is 30mA. The -3dB frequency range with 1k source impedance was declared as 11Hz to 45kHz and the core size as 19mm x 30mm, meaning that EI57 laminations were used, with 19mm center leg width and a 30mm stack thickness.

Normally sold on eBay for US$82 each (in 2018), they could be found on other websites for half that price, a true bargain. The powder-coated steel covers are very elegant, and the connections are easy to solder onto.

The DCR figures for the two transformers (R_P and R_S) were almost identical, but, as always, the primary inductance varied (17.3-16.1)/16.1*100 = 7.45%!

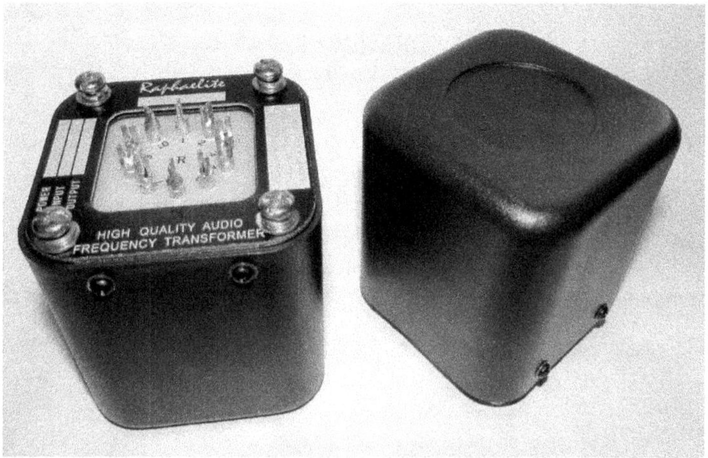

MEASURED RESULTS:

	TR. #1:	TR. #2:
R_P [Ω]	434	433
R_S[Ω]	546	543
L_{P120} [H]	17.3	16.1
L_{P1000} [H]	16.5	14.8
L_{S120} [mH]	26.1	24.5
L_{S1000} [mH]	26.8	22.7
L_{L1000} [mH]	2.80	2.85
C_{PS1000} [nF]	2.2	2.4

CASE STUDY: ElectraPrint interstage transformers

A "5k to 5k" interstage transformer with a maximum allowed primary DC current of 15mA and a 1:1 impedance ratio, but no model number or markings. An eBay seller from the USA was selling these regularly (ElectraPrint's owner itself?) for US$150 a pair (2018) and accepted our offer of US$125.-

EI76 laminations were used with a 26mm stack. The -1dB frequency range was specified as 29Hz to 32kHz, from which the -3dB frequency range can be estimated from the "dB graph" as 1.96 times wider, or 14.5 Hz to 64 kHz.

The inductance at 1kHz is 50% higher than at 120Hz, which is not a good sign. That wasn't the case with Raphaelite transformers. Likewise, the leakage inductance is almost an order-of-magnitude (ten times!) higher, meaning the quality factor of ElectraPrint transformers is approx. ten times lower.

In the bass region (measured at 120Hz) the primary inductance varied (18.6-17)/17*100 = 9.4%!

MEASURED RESULTS:

	TR. #1:	TR. #2:
R_P [Ω]	515	512
R_S[Ω]	516	512
L_{P120} [H]	12.5	11.5
L_{P1000} [H]	18.6	17.0
L_{S120} [mH]	12.5	11.5
L_{S1000} [mH]	18.7	17.0
L_{L1000} [mH]	22.8	22.4
C_{PS1000} [nF]	0.37	0.38

The loaded versus unloaded transformers

While output transformers of power amplifiers are always loaded with the speaker impedance, interstage and preamplifier output transformers can have various loads on their secondaries. One option is to leave them unloaded, as we have done in the E86C single-stage line-level preamplifier (page 113). However, in that case, there will always be a power amplifier connected to the secondary of the line preamp's output transformer. Typically the input impedance of tube amplifiers is in the 47k to 100k range, thus loading the preamp's output transformer.

When an interstage transformer drives a grid of the output stage the input impedance of such a stage is very low, so the secondary is practically unloaded. The drawbacks of this situation are best identified by comparing its A-f curve to those of loaded transformers. Unloaded secondary yields the highest midrange amplification, but low frequencies are rolled off very early, so the lower -3dB frequency f_L is relatively poor (high).

Also, due to low damping, there is a pronounced peak at the transformer's resonant frequency, which for lesser quality units may fall into the audible range (at 20kHz as illustrated). This is a general discussion and the curves illustrated are not for the amplifier described above!

As the secondary load increases, the midrange gain of the stage drops (from A=40 to A=30 in the illustrated example), but f_L is lowered, f_U is raised slightly and the resonant peak flattened.

Larger loads (R_S>$R_{NOMINAL}$) completely remove the resonant peak and widen the frequency range further, but at the expense of a significantly reduced amplification factor (halved in this case).

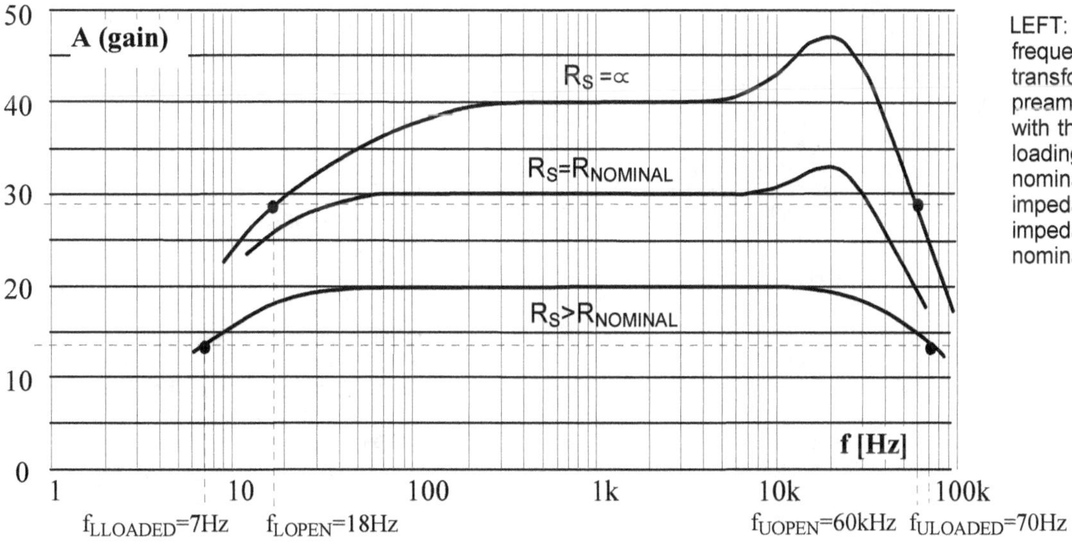

LEFT: The amplitude vs. frequency characteristics for a transformer coupled preamplifier or a driver-stage with three different secondary loading: open secondary, nominal secondary load impedance R_S and load impedance much higher than nominal.

DIY DESIGN: THE UNIVERSAL INTERSTAGE & PHASE SPLITTER TRANSFORMER

Now that we understand some of the issues and problems arising from the use of interstage transformers let's look at a few practical designs.

The transformers described here were installed into PSET (Parallel Single-Ended Triode) monoblocks with 300B tubes (pictured below) and sounded very good. When Hammond 126A transformers were temporarily substituted, the amplifier's lower -3dB frequency worsened from 22 Hz with our transformers to 29 Hz, resulting in weaker bass. This directly results from Hammond's much lower measured primary inductance, 12 H, versus our 36 H.

We adopted a pragmatic approach by taking both the laminations and the winding wire we had at hand and designing an IS transformer around those two givens. The wire was 0.11mm in diameter, and the laminations were made of GOSS, reused from a pair of small but good quality vintage push-pull output transformers.

The window length is 30 mm, and the winding or coil length is CL= 30 - (2 x 2) mm = 26 mm.

The same wire is used for the primary and secondary, so the maximum number of turns per layer for all windings is TPLMAX = CL/d = 26mm/0.11mm = 236. Assuming a Horizontal Fill Factor of HFF= 80% (fast bulk winding without careful layering), we can fit 236*0.8 = 188 turns in each layer.

The window is 10 mm tall, but 2mm are lost on bobbin thickness and 1 mm for top clearance, so the total winding height must not exceed 7 mm! Since we have four identical windings, there are four insulation layers of 0.2mm or 0.8mm total. After subtracting 0.8mm for insulation, we are left with a net winding height (copper height) of 6.2mm.

We have to consider the bulging (bowing) effect, which increases the height of the winding in the middle of the bobbin by up to 15%, so 6.2/1.15 = 5.4mm.

The 0.11 mm wire is 0.14 mm thick (with insulation) so we can fit a maximum of 5.4/0.14 = 38 layers. Since we have four windings, the total number of layers must be divisible by 4, so let's settle on 32 layers. 32 layers x 188 turns = 6,016 total turns or 1,500 turns in each winding.

The primary and secondary inductances are around 36 Henry, a very good result for such a small core.

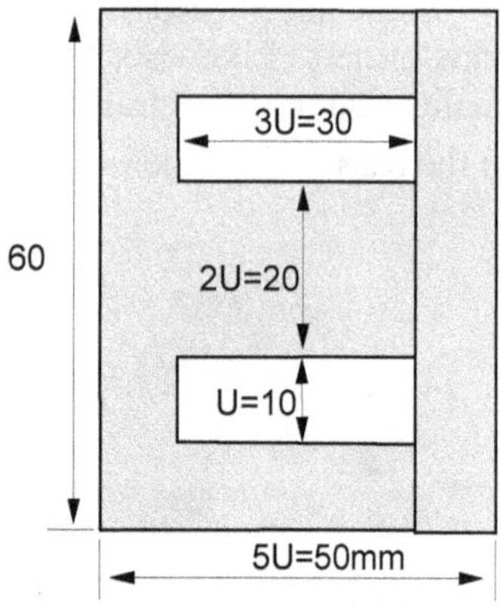

ABOVE: EI60 laminations

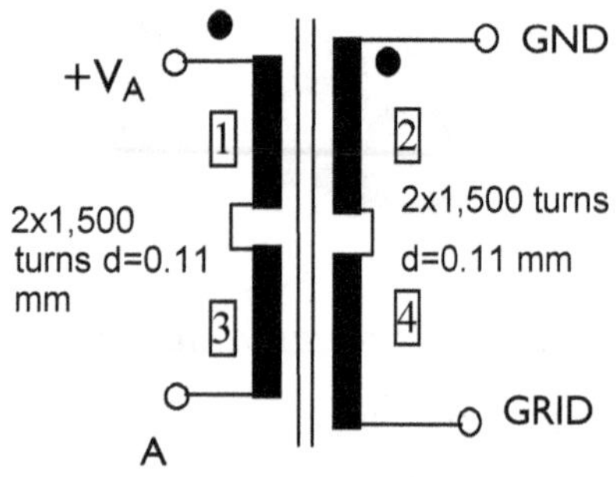

ABOVE: The winding diagram

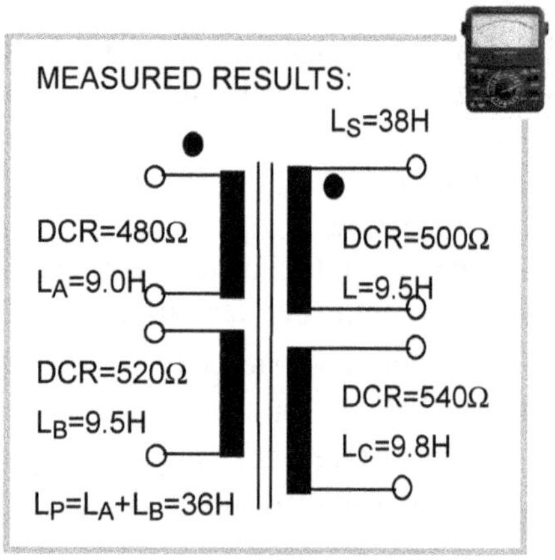

The DC resistance of the two primary and two secondary halves are not identical since we didn't use a bifilar winding nor sectionalized the winding vertically, but the difference is fairly small, (520-480)/480 = 40/480 = 0.083 or 8.3%, which is acceptable.

LEFT: Our parallel SE 300B monoblocks, with power and output transformers under the transformer covers. The filtering choke and interstage transformer are under the chassis.

Thanks to Class A_2 operation, the maximum output power was measured at 40.5 Watts!

6EM7 - 6DN7- 6SN7 driver stage transformer-coupled to PSET 300B otuput stage

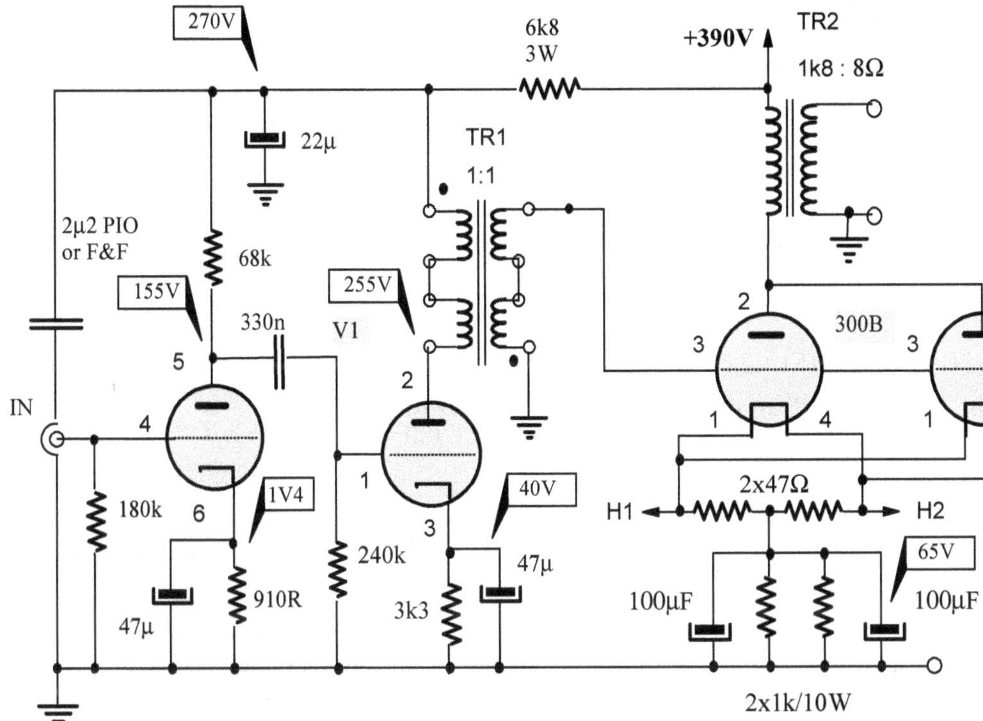

NOTE: 6EM7, 6DN7 and 6SN7 duo-triodes can be used for V1. The marked DC voltage are with 6EM7 tubes.

DIY DESIGN: SINGLE-ENDED DRIVER TRANSFORMER FOR 2A3 & 300B TUBES

We needed a 1:1.15 interstage transformer between a triode-connected 6AQ5 or 6V6 driver at 20 mA anode current and a 2A3 or 300B power tube.

The Noval 6AQ5 power tube has a plate dissipation of 12 Watts as a beam tube or 9 Watts when triode connected. Its performance is similar to the octal 6V6-GT. Connected as a triode, it has an amplification factor of around 10 and a low internal impedance of under 2 kΩ (1,970Ω), 3-4 times lower than the audiophile favorite 6SN7.

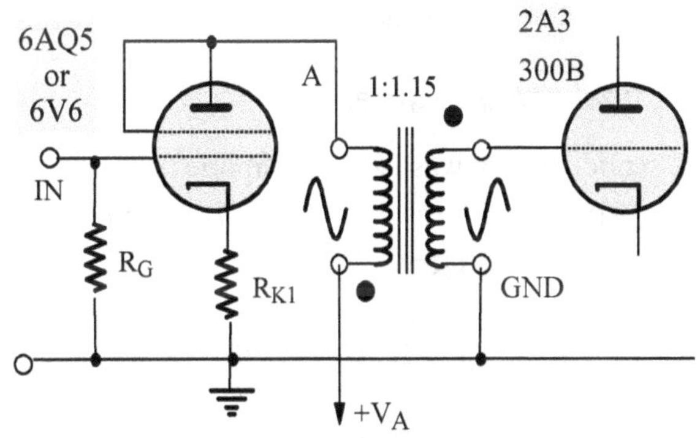

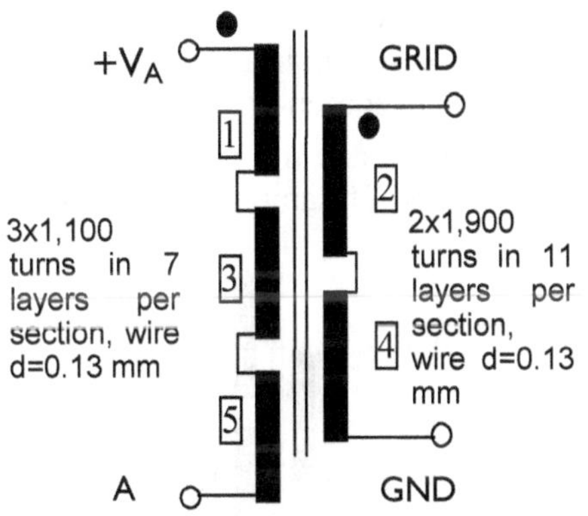

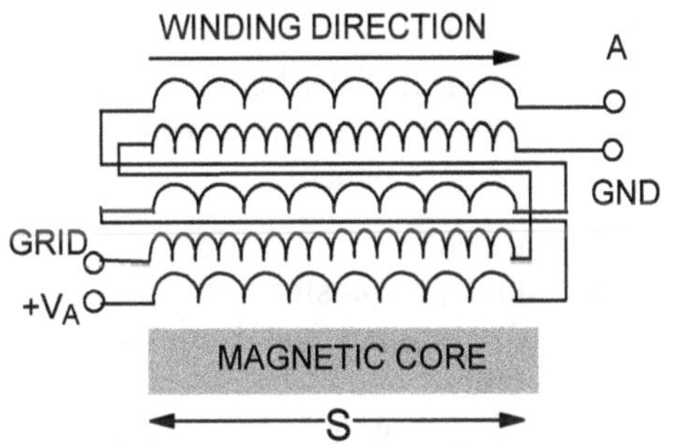

MEASURED RESULTS:

- L_P=19.5 H
- R_l=750 Ω
- I_0=20 mA
- air gap g= 0.1mm

Sectionalizing and the number of turns

The EI66 laminations (reused from a 70V line transformer) have the length of the magnetic path ℓ_{MP}=6*2U= 6*2.2= 13.2 cm and gross cross sectional area A=aS=6.82 cm^2.

We have chosen 3:2 sectionalizing, 3 primaries in series and 2 secondaries in series. We have to determine the maximum number of turns we can fit into the available window, which is 33x11mm. Since the window length is: WL=33mm, subtracting the thickness of two bobbin walls (sides) and 1mm for end clearance, we get coil length CL= 33-(2*2)-1 = 28mm. TPL_{MAX}= CL/d= 28/0.135 = 207.

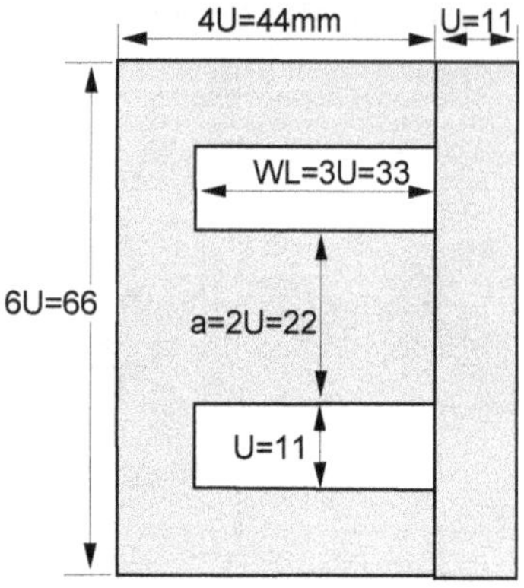

ABOVE: laminations EI66

Window height is WH = 11mm and coil height: CH= WH-B-1= 11-3= 8mm.

There are 5 layers of insulation, including the final (outside) layer. The total insulation thickness is d_I=5*0.25 = 1.25 mm.

Subtracting the insulation height from coil height, we get the gross copper height: GCH= CH-d_I = 8-1.25= 6.75 mm.

Factoring in 15% bulging factor we get the Net Copper Height of NCH = GCH/1.15 = 6.75/1.15 = 5.85 mm, so the maximum number of layers we can comfortably fit is NL= NCH/d = 5.85/0.135 = 43.

Total number of turns (maximum): TT_{MAX}=NL*TPL_{MAX}= 43*207 = 8,901

Assuming 0.85 horizontal winding factor we get TT=8,901*0.85 = 7,565.

Actual TPL=207*0.85=176.

TT=N_1+N_2 = N_1+1.15*N_1 so N_1=TT/(1+1.15) = 7,565/2.15 = 3,518.

Since we are splitting N_1 into 3 sections, let's round the total turns TT down to 3,300; that way, each section will have 1,100 turns and N_2=1.15*N_1=1.15*3,300 = 3,800, nice and round numbers.

The number of layers in each primary section NL_P=11,100/176 = 6.25, so we will use 7 layers per primary section. 1,900/176 = 10.8, so we will use 11 layers per secondary section. Total number of layers is NL = NL_P + NL_S = 3*7 + 2*11 = 21+22 = 43, which is exactly our maximum allowed number of layers!

The lower -3dB frequency can be estimated by the formula f_L= (r_I+R_1)/(2πL_P), where r_I is the driver tube's internal resistance, R_1 is the transformer's primary DC resistance, and L_P is the transformer's primary inductance. In this case, we get f_L = (1970+750)/(2*π*19.5) = 2,720/122 = 22.3 Hz, very close to the measured f_L of 24 Hz. To lower the f_L, we would need larger laminations and a thicker stack.

Connect the primary start lead to the anode supply (+V_A) and its end to the anode (A). The secondary winding's start point (marked with the dot on the winding diagram above to indicate the "in phase" points) should be connected to the output triode's grid. This leaves adjacent turns in both windings at zero audio potential, and the effective primary-secondary capacitance is zero. That way, a much wider frequency range is achieved.

The final design figures

3:2 sectionalizing was used, 3 primaries in series and 2 secondaries in series.

P_1=P_2=P_3=1,100 turns d=0.13mm, 7 layers in each section, S_1 =S_2=1,900 turns, d=0.13mm, 11 layers per section.

The turns ratio is TR=V_1/V_2 = N_1/N_2 = 3,300/3,800 = 0.8684, so V_2 = 1.15V_1.

DIY DESIGN: 1:1.5 SE TO PP DRIVER & PHASE SPLITTER TRANSFORMER

We bought a Chinese-made stereo amp on eBay ("Soundtrack" brand), attracted by the large transformers (or rather large transformer *covers*), the 300B tubes, and the engraved tops on the two smaller cases, indicating interstage transformers. Despite its good looks, the amp turned out to be the biggest lemon we have ever seen or heard.

The output transformers were tiny, less than 1/2 of the height of the mostly empty covers, and the two "interstage transformers" flanking the center-mounted power transformer were actually filtering chokes! The amp sounded so bad that we took it apart and reused some of its components.

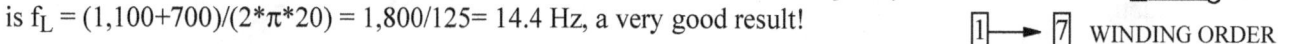

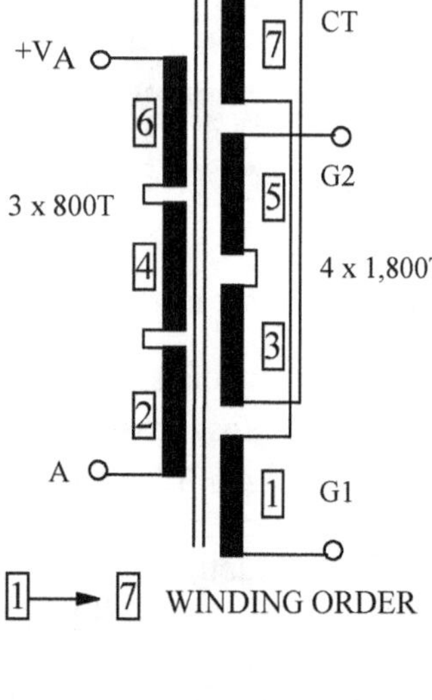

There was one piece of good news; the said chokes were made of 0.35mm GOSS laminations which we made into true interstage transformers. The laminations were EI76.2, with a 1" center leg (25.4mm). The stack was 15mm thick, resulting in a cross-sectional area of 2.54*1.5=3.75 cm^2.

We wound it with 3 primaries in series, each with 800T of d=0.11mm wire and 4 secondaries in CT configuration, each with 1,800T of d=0.11mm wire.

The turns ratio primary to half of the secondary is N_1/N_2=2,400/3,600 = 0.67 = V_1/V_2 so V_2 = 1.5V_1 (for each grid). Grid-to-grid voltage is twice that or V_{GG}=2V_2=3V_1!

L_1=20H at 120 Hz and 40H at 1 kHz, DC resistance of the primary is R_1=700 Ω For I_0=15 mA, and the air gap needs to be g=0.1mm!

For EL86 with r_I=1.1 kΩ (triode connection) as a driver, the lower -3dB frequency is f_L = (1,100+700)/(2*π*20) = 1,800/125= 14.4 Hz, a very good result!

DIY DESIGN: INTERSTAGE TRANSFORMER FOR A MCINTOSH-TYPE CLASS B AMPLIFIER

The goal is to design an interstage transformer for a McIntosh - style class B push-pull amplifier with a 6SN7 duo-triode driving the output stage of a 6L6 PP amp.

The plate-to-plate impedance of the primary should be 10 kΩ, and the ±3 dB frequency (f_L) should be under 10Hz. A voltage boost is not needed, so a 1:1 voltage ratio should be used to get a reasonably wide bandwidth.

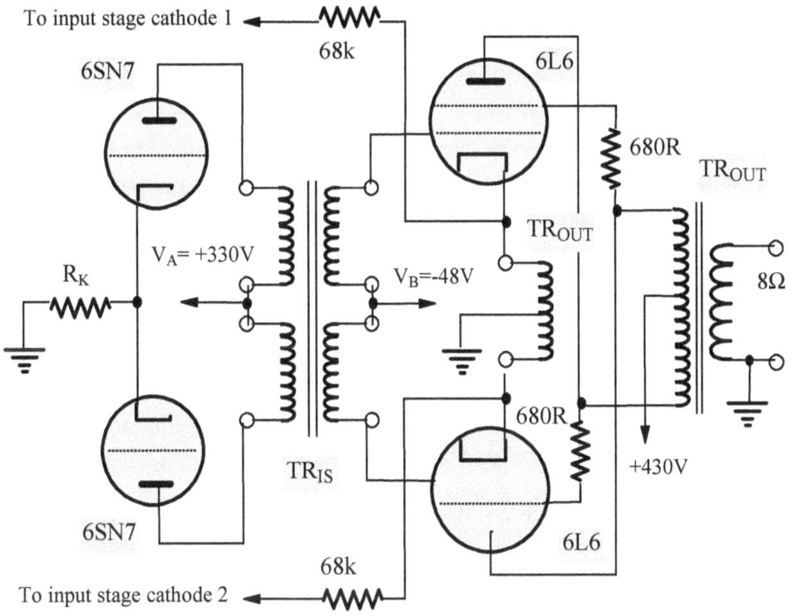

RIGHT: The driver and power stage of a McIntosh - style class B push-pull amplifier with cathode secondary winding of the output transformer and a symmetrical 1:1 interstage or driver transformer. The input stage is not shown.

Step 1: The required primary inductance L_P

$L_P = Z_P/(2*\pi*f_L) = 10{,}000/(2*\pi*10) = 10{,}000/94 = 160$ H

Step 2: Choosing the magnetic core

The two DC primary currents will cancel each other out, so there is no DC magnetization of the core and no air gap. Since $L_P = N^2\mu_R\,\mu_0 A/\ell_{MP}$, we have one equation and two unknowns, N and A (ℓ_{MP} is related to A).

Let's choose the optimal, square EI stack, using scrapless laminations, so a=S and $A=a^2$ For scrapless laminations $\ell_{MP} = 12U = 6a$ (U=a/2), so we have $L_P = N^2\mu_R\mu_0 a^2/(6a) = N^2\mu_R\mu_0 a/6$

Let's try EI84 laminations, whose U=14mm and S=a=2U=28 mm.

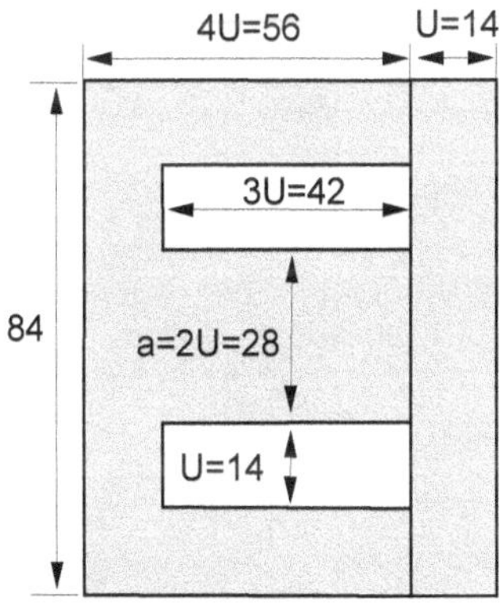

ABOVE: EI84 LAMINATIONS

Step 3: Calculating primary & secondary turns

Interstage transformers usually operate at low flux densities of 0.1T-0.3T (1,000-3,000 Gauss), so let's choose a lower B of 0.1T, corresponding to a lower μ_R of around 15,000. This lower flux density requires double the number of turns compared to higher B of 0.3T, where μ_R rises to around 30,000!

$N_1^2 = 6L_P/(\mu_R\mu_0 a) = 6*160/(15{,}000*4*\pi*10^{-7}*0.028) = 1{,}818{,}914$ so $N_1=1{,}349$ turns. Since our transformer has a 1:1 voltage ratio, $N_2=N_1$!

Step 4: Sectionalizing

The window length is WL=42 mm, useable window length or Coil Length is CL=38mm.

Let's choose #31 Heavy Formwar wire (just like McIntosh), with a diameter (including insulation) of 0.0104" or 0.26 mm and copper diameter of 0.0089" or 0.226 mm.

Theoretical maximum TPL (turns per layer): TPL_{MAX}=CL/d= 38mm/0.26mm = 146 T, and 1,349 T / 146 TPL = 9.2 layers.

Since both the primary and the secondary have to be halved, we cannot have an odd number of layers. Let's choose 12 layers and the faster bulk winding method. 9.2/12 = 0.77 or 77%, so our horizontal filing factor is only 77%, which should be achievable even by a novice or impatient winder.

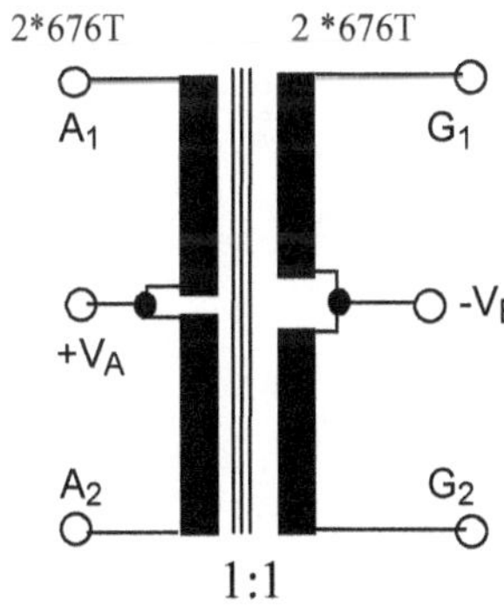

ABOVE: The "in principle" winding configuration of this interstage transformer

Skillful winders can easily achieve 90% HFF. This, of course, assumes winding by hand, not using computerized, programmable automatic winding machines.

The primary and secondary are identical, so each will have 4 sections with 1,349/4 = 338 turns in each section (1,352 turns), spread over 3 layers, or 112-113 turns in each layer.

Since the length of the winding gets progressively longer for each new section (as it gets further and further from the core, or its central leg, to be more precise), so does its DC resistance. To ensure DC and AC balance between A_1-A_2 sides, we must make sure the length of the windings is approximately equal between the two sides (A_1-V_A and A_2-V_A).

The first and the fourth section will be connected, as will the second and third sections for the other side, just as with the primary windings of push-pull output transformers.

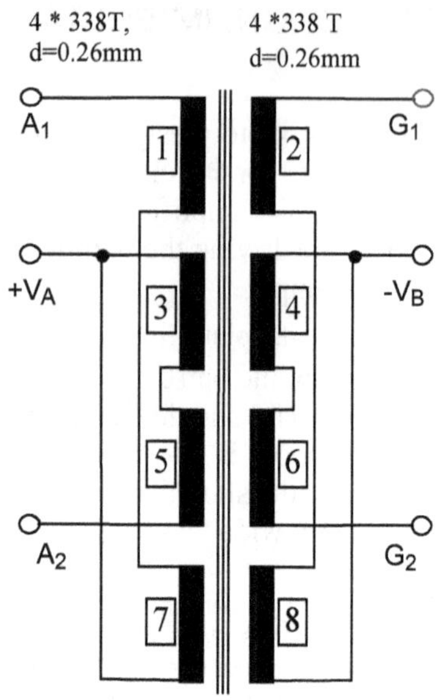

ABOVE: The complete winding diagram

Step 5: Check for vertical window fit

24 layers * 0.26mm = 6.24mm, our window height WH=14mm, minus 2 mm for the bobbin gives us a maximum of 12mm for the windings. 6.24/12 = 50.2%, so there is plenty of height left for insulation and even allowing a 15% margin for bulging, there will be plenty of clearance on top of the coil.

This coil will fit into the window even if the layered winding is used, with insulation between all layers (not just between sections as in bulk winding)!

OUTPUT AND INTERSTAGE TRANSFORMERS FOR TUBE GUITAR AMPS

- OUTPUT TRANSFORMERS FOR TUBE GUITAR AMPS
- DIY DESIGN: 5kΩ PP OUTPUT TRANSFORMER FOR GUITAR AMPS
- ANALYSIS: ORDINARY AND SIMUL-CLASS® MESA OUTPUT TRANSFORMERS
- LINE-MATCHING AUDIO TRANSFORMERS IN PUSH-PULL OUTPUT SERVICE
- INTERSTAGE TRANSFORMERS FOR TUBE GUITAR AMPS

11

OUTPUT TRANSFORMERS FOR TUBE GUITAR AMPS

Tube amplifier power stages need output transformers to efficiently couple low impedance loudspeakers, typically in the 4 to 16 ohm range, to high impedance vacuum tubes in the output stage. Power triodes have a much lower internal impedance than pentodes or beam power tubes, but even they need output transformers. For example, the venerable 300B, named the queen of triodes by audiophiles, has an internal impedance of around 800Ω .

In a push-pull output stage, most pentodes require a plate-to-plate impedance (total impedance of the primary winding) in the 3kΩ-10kΩ region. Assuming an 8Ω secondary that means output transformers' ratios need to be in the 375-1,250 region (3,000/8 =375 and 10,000/8 = 1,250).

Since the impedance ratio IR is a square of the voltage VR and turns ratio TR (VR=TR), the turns ratios between their primary and secondary windings are in the 19-35 range ($\sqrt{375}$ =19.4 and $\sqrt{1,250}$ =35.4). Notice that impedance and turns ratios don't have any unit or "dimension"; they are dimensionless ratios.

Salvaged vintage audio parts or overpriced modern replacements?

Considering their quality and modest performance demands (narrow frequency range and low power rating), the modern replacement transformers for vintage guitar amps are incredibly expensive. For example, Watts Tube Audio (and many other small and not-so-small online parts retailers) sell a replacement output transformer for Epiphone Valve Junior (made by Heyboer) for US$75 and a replacement power transformer (also by Heyboer) for US$ 115.95. One can buy a whole Epiphone Valve Junior combo amp (used) for US$116 and get *all* of its parts, including the cabinet, speaker, and chassis, not just the transformers.

If you modify and/or build tube amps regularly, this again illustrates the importance of buying cheap salvaged parts whenever an opportunity arises. You never know when you will need them. As an example, when you see a pulled vintage output transformer for a single tube such as EL84, 6V6, or ECL86 for sale at US$12.95, assuming it is in perfect working condition, buy it.

Perhaps paradoxically, many vintage transformers used better magnetic materials (laminations) than cheaply-made currently produced ones. Also, modern transformers use plastic film for insulation and plastic bobbins to carry the windings, while the vintage ones use impregnated paper insulation and cardboard or Paxolin bobbins.

Sure, plastic film (usually called Mylar®, although strictly speaking, that is just DuPont's brand name) is a superior isolating material, but paper insulated transformers sound better, warmer, softer, and more "vintage." Just as in tube amps in general, newer does not always mean better sounding when it comes to transformers in particular!

CASE STUDY: Salvaged Telefunken SE output transformers

There are still thousands of vintage tube radios, mono amps, and stereo consoles all over Europe. A few entrepreneurial recyclers (mostly from Germany) salvage transformers from such gear and sell them on eBay. This particular pair, however, was one of three identical pairs for sale by an Australian seller, only AU$30 (US$22) for two, so a very good value for a budget-conscious guitar amp constructor.

The pedantic German tube gear makers of the 1950s and 60s had a habit of printing all sorts of useful information on their transformers, something USA and UK transformer makers did not do. Here they managed to print pretty much the whole application note on the tiny transformer!

Apart from the transformer parameters such as 7kΩ primary and 3.6Ω load, 11.9H primary inductance, and f_{QU} of 79Hz, it says that the transformer was made for an EL41 output tube with 36mA idle DC current and 250V anode voltage. You now have all the info you need to decide if these oldies are suitable for your project.

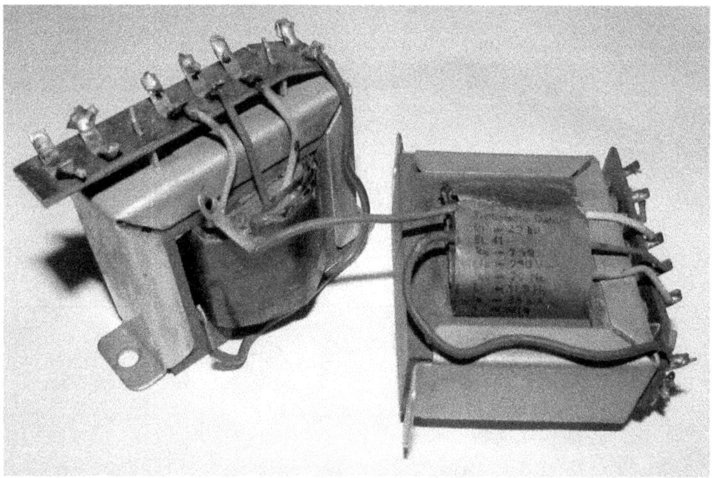

TELEFUNKEN SE OUTPUT TRANSFORMERS

- EI54 laminations
- a=18 mm, stack thickness S=20mm
- Center leg cross section A=aS = 3.6cm^2
- Power rating: P=A^2= 3.6^2= 13W
- Output tube: EL41 (I_A=36mA)
- Impedance: 7kΩ/3.6Ω
- Primary DCR: R_P=690Ω
- Primary inductance @1kHz L_P= 11.2H
- Leakage inductance @1kHz L_L=121mH
- Quality factor: QF = 11.2/0.121 = 93

The primary DCR of these baby transformers is usually quite high (690Ω in this case), meaning they were wound with many primary windings of a small diameter wire to fit so many turns into their smallish winding windows.

The power rating of the lamination stack is around 13 Watts, and the leakage inductance is quite high, 120mH, resulting in a low quality factor of less than 100. That would be shockingly bad for a hi-fi transformer but is of lesser importance for a guitar amp application.

The transformer will start attenuating very higher frequencies above, say 12 or 15 kHz, compared to higher fidelity transformers that will pass frequencies up to 25 or even 45 kHz. Those extra high frequencies usually need to be tamed down anyway; otherwise, your guitar amp may sound shrill and unpleasant.

CASE STUDY: NSC022848-T output transformers

NSC022848-T transformers are made in Taiwan and sold by the New Sensor Corporation in the USA. They've been listed as replacement output transformers for Tremolux, Pro Reverb, Vibrolux®, Bandmaster, and Vibroverb models.

Rated at 35 Watts, the primary impedance is 4kΩ PP, and the output impedance is 4Ω, so a 1,000 impedance ratio or 31.6 voltage or turns ratio.

Let's see what the power rating of their core is (at 50Hz, as per our rule-of-thumb formula). The laminations are EI75, meaning their center leg is 23mm wide. The stack thickness is 26mm wide, so the gross cross-sectional area of the center leg is A= aS = 2.3*2.6 = 5.98cm^2 . Assuming a stacking factor of 0.96, the net or effective area is A_{EF}=0.96*A= 5.74cm^2, so the power rating of the core is P=A_{EF}^2=33VA.

Notice a huge difference in DC resistance between the halves of the primary, 87 versus 67Ω! This means the DC voltage drop on one half (87Ω) will be higher, and the anode voltage of that tube will be lower. There will also be a dynamic unbalance at higher currents and with AC signals present.

CASE STUDY: Basler output transformer

Basler transformers are made in Mexico and sold on eBay. This push-pull transformer has a 4,500Ω primary impedance and 4, 8, and 16Ω secondary taps. Rated at 40 Watts, it would suit octal power tubes such as 6L6, 7027A, and EL34.

The primary inductance is very low, under 7H at 120Hz, but since its leakage inductance is extremely low, under 7mH, the resulting quality factor is very high for a guitar amp transformer, above 1,200!

CASE STUDY: VOX AC15 output transformer

Since VOX AC15 combo amps, both vintage, and reissue, are expensive, instead of buying the whole amp, we only bought this genuine reissue output transformer, firstly to measure its parameters for this case study and then to use it in one of our own push-pull guitar amps.

The transformer is very large for its 15 Watt output; its core is rated for 100 Watts! Compare that with tiny output transformers in other modern amps, for instance, the minuscule 13 Watt-rated transformer in Fender Vaporizer.

NSC022848-T MEASUREMENTS

- EI75 laminations,S=26mm
- Primary inductance : L_P= 10.25H @ 120Hz, 10.16H @ 1kHz
- Leakage inductance: L_L= 26.2 mH
- QF = 10.16/0.026 = 388
- Primary DC resistance: blue-red R_P=86.6Ω, brown-red R_P=67.1Ω
- Secondary DC resistance: R_S=0.2Ω

BASLER OUTPUT TRANSFORMER

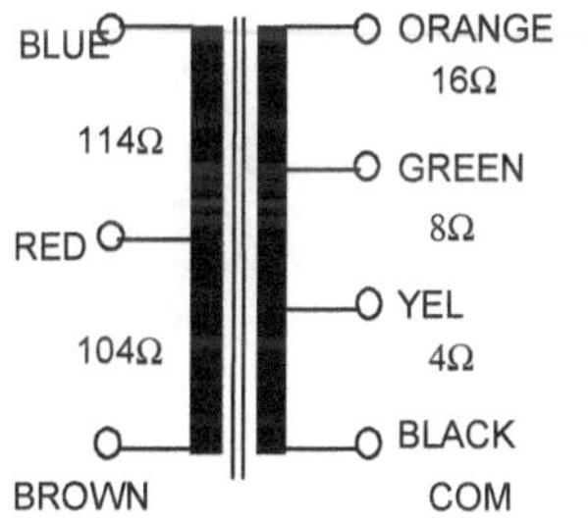

- EI76 laminations, a=25.3 mm, S=25mm
- A=aS = 6.3cm^2
- Power rating: P=A^2= 39W
- Primary inductance L_P= 6.8H@120Hz, 8.7H@1kHz
- L_L=6.9mH
- QF = 8.7/0.007 = 1,243

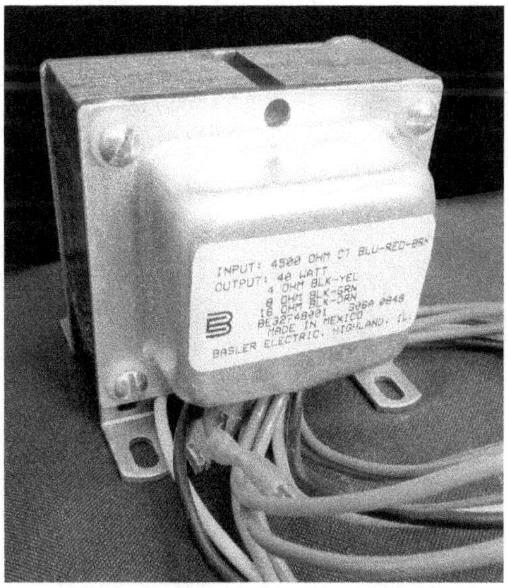

We haven't tested the output transformer for AC30, but, judging by its size, it could easily cope in a 30-50W amp. However, its primary impedance of around 8.5 kW would be too high for paralleled EL84 output stage as in AC30 or lower impedance tubes such as EL34 or 6L6.

The two primary halves are well balanced, with only 2Ω difference in DCR (DC resistance). The primary inductance is quite high, around 24 Henry, meaning this transformer's frequency range at full power will extend into very low frequencies for a full and powerful bass.

Its 100 mH leakage inductance is on the high side (anything above 25-30mH is too high), meaning its upper-frequency limit will be curtailed. At 1.2nF, the primary-to-secondary capacitance is also high, due to the relatively large lamination size used (EI76) and a thick stack (4 cm).

VOX AC15 OUTPUT TRANSFORMER:

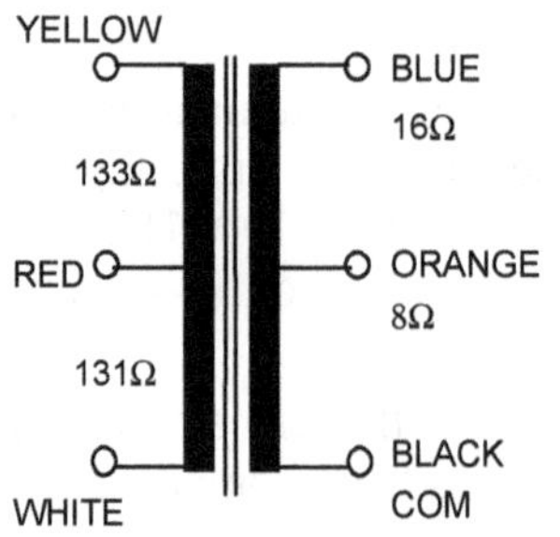

- EI76, a=25.3 mm, stack thickness S=40mm
- $A=a*S = 10.1cm^2$, $P=A^2= 10.1^2= 102$ W
- Impedance ratio (blue-black): 533
- Impedance ratio (orange-black): 1,048
- Primary impedance with 8Ω speaker: 8.4kΩ
- 1/2 Primary DCR (red-yellow): R_P=133Ω
- 1/2 Primary DCR (red-white): R_P=131Ω
- Primary inductance @1kHz L_P= 24.1H
- Leakage inductance @1kHz L_L=101mH
- Quality factor: QF = 24.1/0.101 = 239
- Primary-to-secondary capacitance: C_{PS}=1.2nF
- Primary-to-case capacitance: C_{PC}=189pF

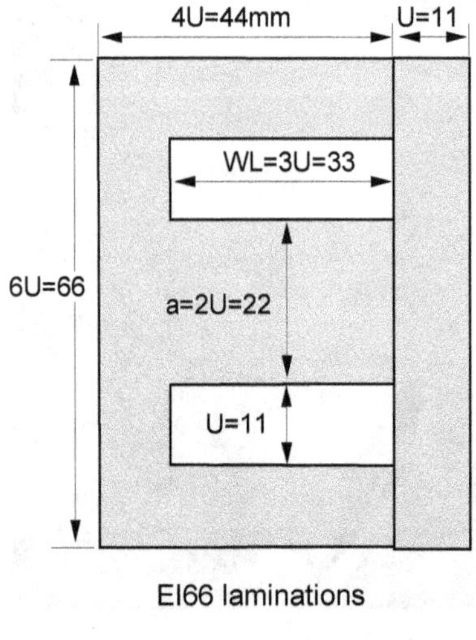

EI66 laminations

DIY DESIGN: 5kΩ PP OUTPUT TRANSFORMER FOR GUITAR AMPS

Let's design a 5kΩ "plate-to-plate" primary impedance push-pull output transformer for a pentode output stage with 6L6 tubes, to suit an 8Ω speaker (load), with 100mA maximum primary current in each half.

Choosing the laminations) and the winding arrangement

Obtaining transformer laminations, bobbins, winding wire, and insulation is surprisingly difficult in many parts of the world, even in large cities. Not many firms outside China make transformers these days!

However, there is a solution. Get some good quality small mains transformers, take them apart and rewind them as output transformers. Most use ordinary silicon steel, but some have GOSS laminations, and those will have a wider frequency range and higher primary inductance.

These 30VA (30V/1A) Taiwan-made power transformers can be converted into guitar output transformers. You get the frame, the laminations, the plastic bobbin, and often you can even reuse its winding wire!

The bobbin is divided vertically in half, but this divider can be cut and filed off. I say "vertically," although it seems horizontal in the photo because we look at the bobbin with the winding window in the horizontal plane, just as in the EI66 profile drawing on the left.

Since the frequency range of a guitar output transformer needs to be only 80 - 8,000 Hz, the cheapest output transformers do not have sectionalized windings at all, as is mandatory in hi-fi transformers. The primary winding is wound first, followed by the secondary. A better option for push-pull transformers would be to split the primary into two halves and wind the secondary between them.

ABOVE LEFT: These and similar small (30VA-rated) power transformers with low voltage secondaries are cheap and widely available. Some use better quality magnetic laminations and sound great when rewound as push-pull output transformers.

Impedance and turns ratios and wire diameters

We know the desired primary (5kΩ) and secondary impedances (8Ω); the impedance ratio is thus IR=5,000/8 = 625, and the voltage or turns ratio is TR = 25! We'll design this transformer "backward," first figuring out the secondary winding and then designing the primary.

We want the amplifier to produce a maximum of 40 watts into an 8Ω load so that the maximum secondary current will be $I_2=P_2/Z_2$ = 40/8 = 5A. The wire tables show that a 1mm diameter wire can take 2 A continuously, so two 1mm diameter wires in parallel will be OK for 4A continuously or 5 A momentarily.

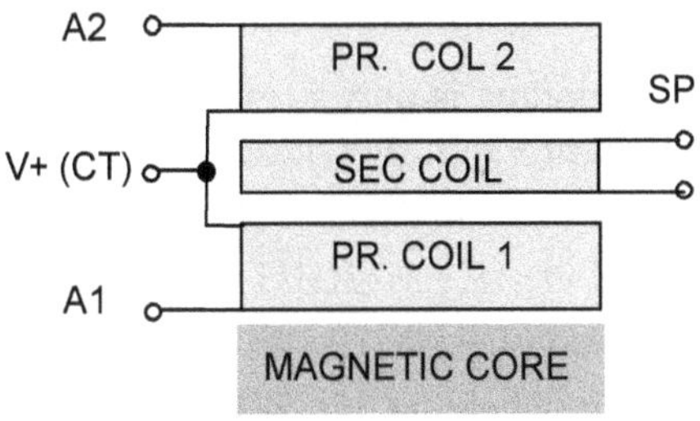

ABOVE: Each design project starts with an envisaged or initial (in principle) winding arrangement

The amp will never work continuously at its maximum power level anyway. It is easier to wind 1 mm wire than 1.5mm wire which would be needed for 5A. The larger the wire, the stiffer it is and the harder to tension it manually.

Secondary TPL, primary turns, primary current and wire diameter

We have to figure out how many turns we can fit in one layer and how many layers we need. The window length is 33mm, less 2 mm for the thickness of bobbin walls (plastic or paper frame that holds the winding) on each side, leaving us with the maximum coil length of CL = 33-4 = 29mm. We will work with 28mm.

How many turns of 1mm wire can we fit into one layer? $TPL_{MAX}= CL/d_S$ = 28/1 = 28 that is the maximum secondary turns-per-layer that we can fit. Let's work with N_2 = 24 turns, which will nicely fill the whole width of the bobbin (a horizontal fill factor of 24/28 = 86 %).

We will have 2 layers connected in parallel to double the current capacity. Impedance does not change when layers with the same number of turns are connected in parallel.

Since the TR=25, for 24 turns on the secondary we need $N_1 = TR*N_2$= 25*24 = 600 turns of the primary.

Now, if I_2=5A, the maximum primary current I_1 will be $I_1=I_2/TR$ = 5/25 = 0.2A.

From the wire tables, we see that d=0.3mm wire is good for 181mA continuous service, which seems fine for up to 200 mA occasionally. Again, a guitar amp never works at its maximum level for long.

How many turns of 0.3mm wire can we fit into one layer? $TPL_{MAX}= CL/d_S$= 28/0.3 = 93! Since we need 600 turns in total, 300 turns for each half of the primary winding, and 300/93 = 3.23, meaning we will choose 4 layers for each primary half (two sections).

TPL= 600 turns / 8 layers = 75 TPL

The chosen 75 turns versus 93 turns (theoretical maximum) ratio gives us a horizontal fill factor of 75/93 = 81%, which is great for novice winders. Anything over 90% is considered a tight fit and is not for the fainthearted.

The winding diagram

Instead of the simple 2/1 sectionalizing, you can wind one section (4 layers) of the primary, then one half (one layer) of the secondary, followed by the second half of the primary (4 layers), and finally the second half (second layer) of the secondary on top.

It is always good to have a winding with a large diameter wire on top to give the coil mechanical strength and rigidity.

Only four insulation layers are used between the windings; there is no insulation between layers. This technique is called "bulk winding" and is much faster than laying down an insulation layer after every layer!

Also, with less insulation, more turns or wire layers can be used with bulk winding, which will result in higher primary inductance and better & cleaner bass response.

Notice the winding direction; the start is always the same, from the left of the bobbin to the right and then back and forth.

The two adjacent primary taps are connected (CT or Center Tap), where the high DC voltage $+V_A$ will be brought in.

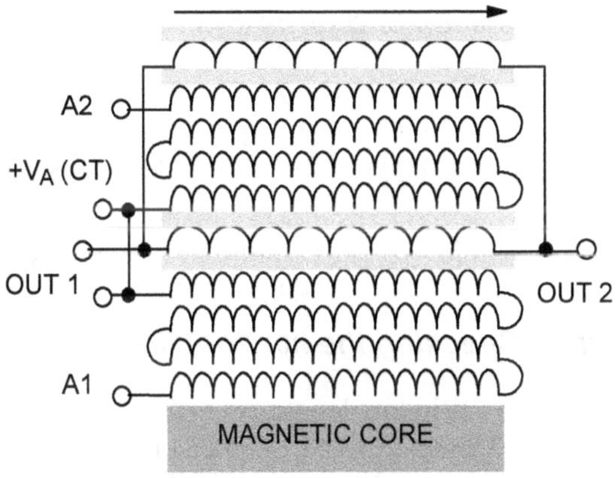

EACH SEC. SECTION: 1x24T, d_2=1.0 mm
EACH PRI. SECTION: 4x75T, d_1=0.3 mm

ABOVE: The winding diagram is the end result of the design process. It specifies all details required for the winding of the transformer's coil.

Checking for vertical fit into the window

Our window is only 11mm tall, and we lose around 2 mm on the bobbin thickness, leaving us with a maximum coil height of 9 mm. We need to check if all those layers can fit in vertically.

PRIMARY HEIGHT: PH= 8 layers * 0.3mm = 2.4mm

SECONDARY HEIGHT: SH = 2 layers * 1 mm = 2mm

Insulation height IH = 4* 0.2 mm Mylar insulation = 0.8mm

Total coil height CH = PH+SH+IH = 5.2mm. If we allow for 15% bulging factor (the winding usually bulges out in the middle), the thickest part of the winding will be 1.15*5.3 = 6.1mm tall.

The vertical fill ratio is 6.1/9 = 68%, which is great, with plenty of clearance left.

ANALYSIS: ORDINARY AND SIMUL-CLASS® MESA OUTPUT TRANSFORMERS

To get a feel of how commercial amp manufacturers (or their specialist suppliers, since most guitar amp makers don't have enough knowledge of transformer craft!) design and wind their output transformers, let's take a look at the output transformer from Mesa 290 amplifier, marked 562004R-1 EIA606-540.

Without having the actual MESA transformer in our hands, we can only go by some information from the Internet fora. The transformer was dissected, and the results were published online by one pedantic explorer so we will base our analysis on that unverified source.

Some output transformers in Mesa's Simul-Class® amplifiers have separate primary taps, so triodes and pentodes operate with different anode impedances, while others don't, meaning their pentode pair works with the same load (primary impedance) as the triode pair. That was the case here - both triode and pentode plates are connected to the same points, A1 and A2.

2:3 sectionalizing was used, the two 4 ohm secondary sections were paralleled, while the 8 Ω section was tacked-on in series.

The primary DCR was 86Ω, while its inductance was only 9H, with 16mH leakage inductance with 4Ω tap shorted to ground. Thus, the quality factor was 9/0.016= 562, which is pretty good, certainly above average for a guitar amp transformer.

108 pairs of EI96 laminations were used (3.75", 9.53 cm), with stack thickness S = 1.6" (4.06 cm). We know from tables that the width of their center leg is 3.2cm.The center leg cross section is $A=a*S = 3.2x4 = 12.8\ cm^2$, which means that transformer's magnetic core power rating is $P= A^2 = 12.8^2 = 164$ VA.

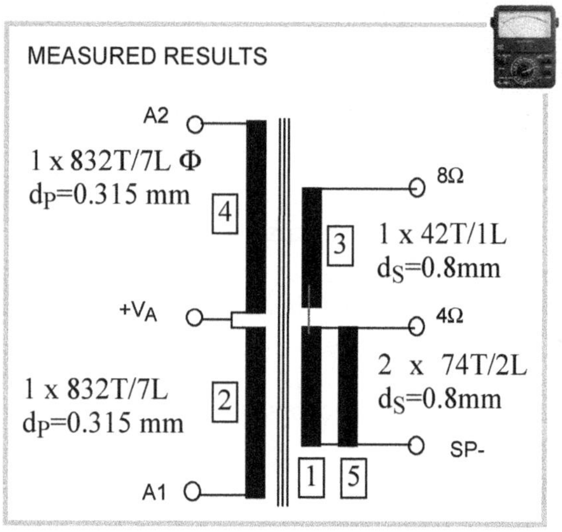

ABOVE: The measured parameters of 562004R-1 EIA606-540 transformer from Mesa 290 amplifier. The framed numbers signify winding order.

Step 1: Calculating turns and impedance ratios

$N_1 = 2* 832T = 1{,}664T$, $N_2 = 74$ T (for 4 Ω), so TR=1,664/74 = 22.49 and $IR=TR^2 = 505$, so a 4 Ω speaker would be reflected back onto the primary side as Z_P (4 Ω) = 505 x 4 = 2,022 Ω plate-to-plate. For 8 Ω output: $N_2 = 74T + 42T = 116T$, TR=1,664/116 = 14.34, $IR=TR^2 = 205.8$ and Z_P (8 Ω) = 205.8 x 8 = 1,646 Ω plate-to-plate.

We didn't get 2 kΩ, so it seems that in this case, the 8Ω tap is not exactly 8Ω but simply an approximation caused by the transformer maker's inability to fit the whole section #3 (42 turns) into one layer.

Step 2: Checking the horizontal and vertical fit of the windings

PRIMARY: Window length WL=48, Coil Length CL=44mm, so $TPL_{MAX}=CL/d_P = 44/0.355 = 124$ Turns per layer. There are 7 layers * 124 T = 868 Turns. That is the estimated maximum that could fit. Since there are 832 turns, each layer has 832/7 = 119 turns or 119/124 = 96% horizontal fill.

Anything over 90% would be too high for amateurs who wind coils by hand, but professional transformer manufacturers could achieve them if they use programmable winding machines.

I have used d=0.355 mm instead of 0.3 mm as claimed, since 0.3mm wire is only good up to about 200 mA, and 0.355 mm wire can take 250 mA continuously.

The 4 Ω secondary has 74 turns in 2 layers, 37 turns per layer, so the maximum diameter we can fit is $d_S = 44/37 = 1.18$ mm, so let's keep the 0.8 mm wire as claimed.

Let's check the vertical fit: PH= 14*0.355 = 5 mm, SH= 5*0.8 = 4 mm, CH = PH+SH= 9 mm

We have a 16mm window, less 2mm for the bobbin's thickness, so we are left with 16-2-9 = 5mm for the insulation between sections (no insulation between layers, bulk winding would be used) and the bulging factor, which is plenty.

Designing a triode/pentode output transformer based on Mesa Simul-Class® benchmark

Another enthusiast measured a few parameters of the primary of the actual Simul-Class® transformer in his Mesa amp and got the following results, which seem to confirm the previous data:

$IR = Z_P/Z_S = 3{,}300/4 = 825$ $TR = \sqrt{(IR)} = 28.7$

$N_1 = 28.7 * 74T = 2{,}125$ Turns

AP denotes the anode of the pentode, and AT is the anode of the triode.

How would one design and wind a similar homemade transformer based on that information?

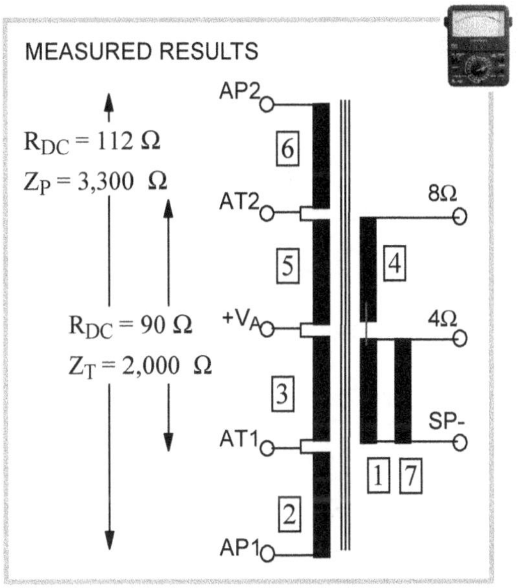

First, we need the value of plate currents. For 6L6 beam power tubes, the currents are between 116 mA (zero-signal or idle state) and 210 mA (full power) per tube. Likewise, for EL34 (triode connected) in Class A, the current is 71 - 74 mA per tube.

So, there is up to 290 mA flowing in the shared part of the primary and up to 210mA in the end sections.

This trafo has 2,125 primary turns, compared to 1,664 in the previous design. It may not be possible to fit so many turns of the 0.355mm wire, so let's start with d_P =0.315 mm primary wire.

The window length is 48mm, but allowing 4 mm for the bobbin wall's thickness and clearance, we will work with 44mm:

$TPL_{MAX} = CL/d_P = 44/0.315 = 139$ T, so we need 2,125 T / 139 T = 15.29 layers.

16 layers would demand very precise winding, if we could fit (vertically) 18 layers, that would make the winding job much easier and would result in TPL = 2,125/18 = 118 or the horizontal fill factor of 118/139 = 85%.

So, each half of the primary will have 9 layers or 1,062 turns.

Sections #3 and #5 will have 7 layers each. After 868 turns, we need our triode taps, so 832/118 = 7.05 layers, close enough to 7 full layers! So, sections #2 and #6 will have 2 layers each.

Checking vertical fit:

PH= 18*0.315 = 5.7 mm, SH= 5*0.8 = 4 mm

Total coil height without insulation: CH = PH+SH= 9.7 mm

With a 16mm high window, less 2mm for the bobbin wall thickness, so we are left with 16-2-9.7 = 4.3 mm for the insulation between sections and allowing for the bulging factor, which is just manageable.

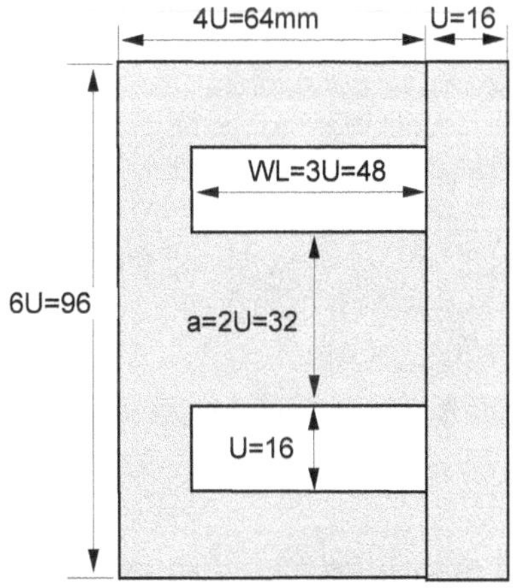

ABOVE: EI96 laminations

The winding order of the Mesa transformer is far from optimal. A better winding order is illustrated on the right.

We start with one primary section, AP1-AT1, followed by the first secondary section, the opposite inner primary section ($+V_A$-AT2), the second secondary section, the outer primary section AT2-AP2, and so on.

That way, the DC resistances of the primary halves will be almost identical since section 1 has the lowest DCR (it is the closest to the core), its MLT (mean length of a turn) is the lowest, and section 7 has the highest DCR since it is the furthest from the bobbin and the core.

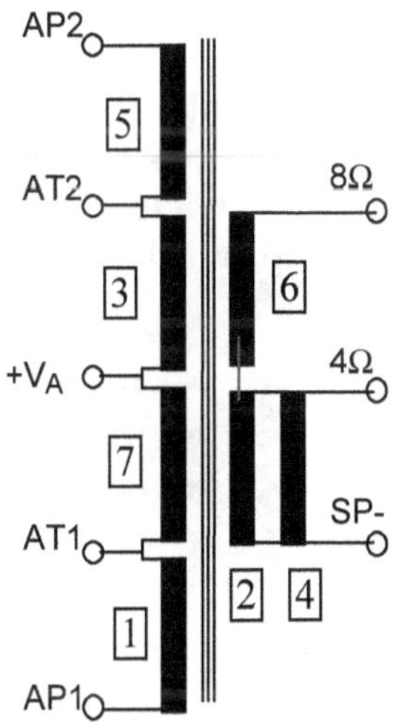

ABOVE: A better winding order for this type of push-pull transformers

LINE-MATCHING AUDIO TRANSFORMERS IN PUSH-PULL OUTPUT SERVICE

What is a 100V or 70V line system and what do line-matching transformers do?

In PA (public address) systems where long cable runs are used, connecting low impedance speakers (4-16Ω) would mean high circulating currents and thus high power losses. Instead, a constant voltage system is used, either 70V or 100V. This high voltage-low current distribution line feeds multiple speakers, each with its own line-matching or step-down transformer. It is a similar concept as in tube amplifiers, where output transformer converts high voltage low current primary signals into low voltage high current secondary signals through the loudspeaker.

Since different rooms require different power levels, the step-down transformers usually have multiple primary taps, normally marked in power levels. In the example below, speaker SP1 is connected to the 25W tap, SP2 to the 100W tap, and SP3 to the 50W tap.

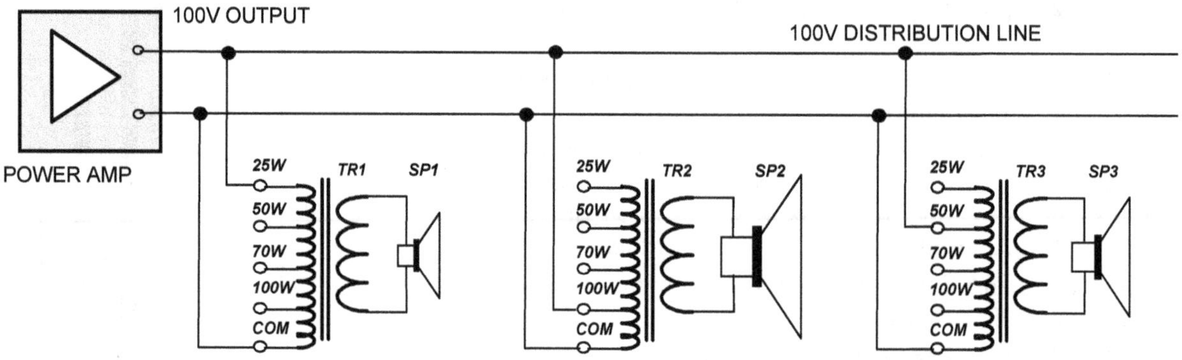

A 100V distribution system with speakers of various power ratings, all using identical step-down transformers with multiple primary taps

Making sense of line transformers' specifications

These Atlas Soundolier line step-down transformers have primary power levels marked, as all such transformers do. Unusually though, these also have corresponding resistances (impedances) marked.

How can we tell if it's a 70V or a 100V line transformer? Power is $P=V^2/R$ so $V^2=P*R$ Using any tap as an example, for instance the 10W tap: $V^2=10$ Watts*500Ω =5,000 so V= √5,000 = 70.7V, meaning it is a 70V transformer.

Secondly, notice that the power level for the 45Ω tap is not specified for some reason. However, now we know that $P=V^2/R = 70.7^2/45 =$ 111W!

Line matching transformers as push-pull output transformers

Since output transformers for tube guitar amps can get expensive, someone had the idea of using cheaper alternatives, and since line transformers usually use GOSS (grain-oriented silicon steel) laminations, which are of higher quality than those made of the ordinary (non-oriented) silicon steel, they come to mind first.

Australian retailer and distributor Altronics sells a range of 100V transformers, most likely made in Taiwan, using GOSS laminations. Let's take as an example a higher power transformer, model M-1130 and its specs: 5-40W, 100V Line, EI core transformer, frequency response: 30Hz - 20kHz ±3dB, secondary taps: 2, 4, 6 & 8 Ω, primary (power) taps: 5W, 10W, 20W, 40W.

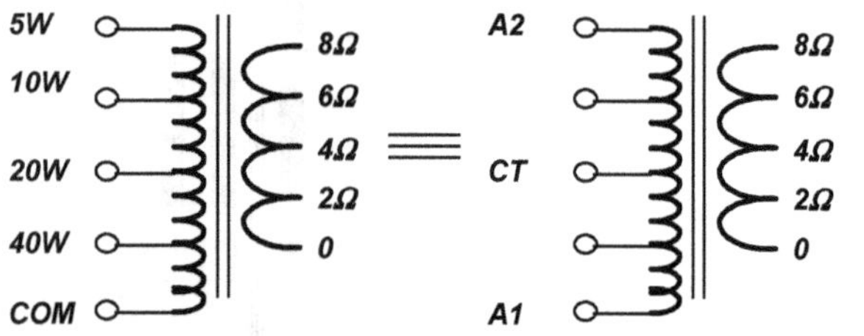

ABOVE: M-1130 terminals when used as a push-pull output transformer

Since $P=V^2/Z$, the impedance is $Z=V^2/P$, for 5W tap Z=2kΩ, for 10W tap Z=1kΩ, for 20W tap Z=500Ω, for 40W tap Z=250Ω.

For push-pull use, we need symmetrical transformers. The impedance between the center tap and each end must be 1/4 of the total anode-to-anode (end-to-end) impedance.

In this case the A-A impedance is between 0 (COM) and the end tap, which is the 5W tap.

This impedance is 2kΩ, so now we look for a tap that will have a quarter of the impedance or 500Ω. Yes, we have it - it is the 20W tap, which will be our CT.

There are also toroidal step-down line transformers, but they are of higher power ratings. As you may have noticed from the discussion above, the higher the power rating, the lower the impedance. If a line transformer has a large core, it is of a high power rating and low impedance and is thus unsuitable.

If it is of the right impedance, it is inevitably on a smaller core and thus of lower power rating, so only limited to low power push-pull amplifiers, less than 20 Watts, for tubes such as 6V6, EL84, and perhaps 6L6.

CASE STUDY: Two low power line transformers tested

M1115 is a 100V line transformer with an 8W secondary, and primary taps marked 0-1.25-2.5-5-10-15W. M1120 is the next model up, apparently rated 5 Watts higher, at 20 Watts.

Although it uses a different size magnetic core and stack thickness, its frequency response is also specified as 30Hz - 20kHz (±3dB), which seems unlikely and makes you question the truthfulness of these specs. M1120 has 4, 8, and 16Ω taps.

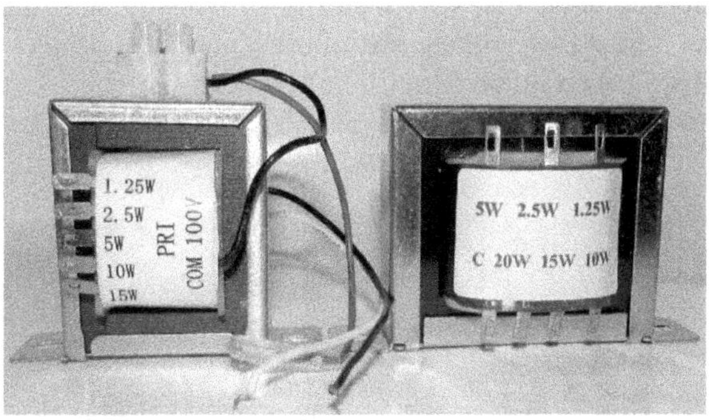

Both transformers reflect an 8Ω load as a 9kΩ anode-to-anode load onto the primary side, or, more specifically, 8.7k load for M1115 and 8.9k load for M1120.In both cases, the 5 Watt lug is the CT (Center Tap), while "0" and "1.25W" lugs are anode connections.

M1115 LINE TRANSFORMER

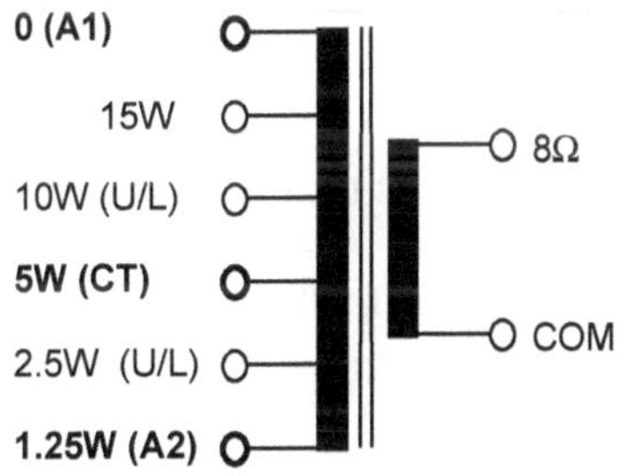

- EI48, a=16 mm, S=25mm
- $A=aS = 4cm^2$, $P=A^2= 16W$
- Z_P (A-A): 8.7kΩ
- R_{DC} 1.25-5W: 137Ω
- R_{DC} 0-5W: 72Ω
- L_P 1.25-5W: 2H2, L_P 0-5W: 2H2
- L_P=7.7H, L_L=27mH, QF = 285
- Price in 2018: AU$14.50

M1120 LINE TRANSFORMER

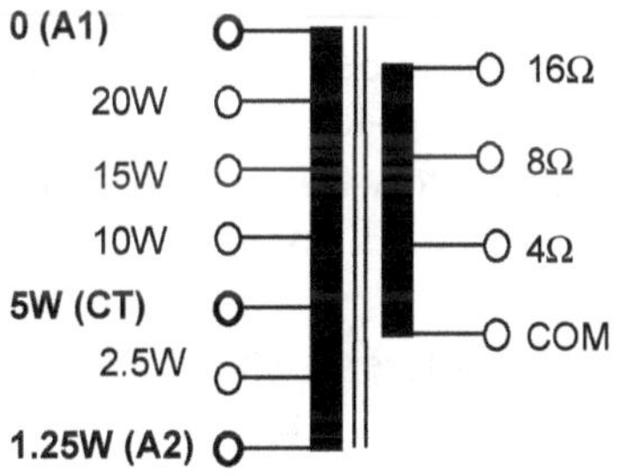

- EI57, a=19mm, S=20mm
- $A=aS = 3.8cm^2$, $P=A^2= 14.4W$
- Z_P (A-A): 8.9kΩ
- R_{DC} 1.25-5W: 99Ω
- R_{DC} 0-5W: 58Ω
- L_P 1.25-5W: 1H4, L_P 0-5W: 1H4
- L_P=4.9H, L_L=32mH, QF = 153
- Price in 2018: AU$18.95

Notice that while AC impedances of the two halves are equal, their DC resistances are very different. For instance, in the M1115 case, one side has a DCR of 72Ω, the other almost double, 137Ω!

Assuming the same anode current flowing through both, say 40mA for 6BQ5 (in idle state), the voltage drops on the two primary halves will be 2.9V and 5.5V, which is not an issue, especially not in guitar amps.

The power rating of the smaller transformer is 16 Watts, in accordance with its 15 Watt specified power output, but the 20W-rated bigger transformer has a core that can only support 14.4 Watts.

To make things worse, the bigger M1120 has a much lower primary inductance, 4.6 Henry, compared to 7.7H for M1115. Since the leakage inductances are in the same ballpark (27 and 32 mH), M1120's quality factor is much lower, 153, while that of M115 is 285. So, unless you need various secondary impedances, M1115 is a better value. As luck would have it, the remaining primary taps are also suitable for ultra-linear connections. For instance, on M1115, the 2.5W tap would be the U/L tap for one, and the 10W tap would be the U/L tap for the other tube.

A closer visual inspection indicates a different lamination color and texture for the two transformers. Lugs and bobbins are also different, so it seems that two different suppliers made them. M1120 has "VRK" stamped on the metal frame; M1115 doesn't. VRK Spectrum Co., Ltd. is a Korean company making transformers in Thailand and capacitors in their Korean and Thai plants.

CASE STUDY: ATLAS SOUND T-18 line transformer

The enclosed data sheet dates these vintage transformers back to 1972. EI66 laminations with a 24mm stack indicate a power rating of 28VA. The primary inductance is low, 2.1H @ 120Hz and only 1.45H at 1 kHz, which is sufficient for a guitar amp output transformer application.

Since 500Ω is a quarter of 2,000Ω, the 500Ω would be the center tap (CT) in the push-pull stage, and the two anodes would be connected to the 0 (COM) tap and the 2,000Ω tap. Thus, these transformers can be used as PP output transformers.

With an 8Ω load connected to the 8Ω tap, the primary impedance is around 2.3kΩ. That can be doubled to 4.6kΩ by connecting the 8Ω load to the 4Ω tap or halved to 1.15kΩ by connecting the 8Ω load to the 16Ω tap.

Such a low plate-to-plate impedance is normally not used but is needed when tubes are working with low anode voltages (24-60V_{DC}), as in our low voltage push-pull guitar amp project (see Volume 2 of my "Tube Guitar Amplifiers" book).

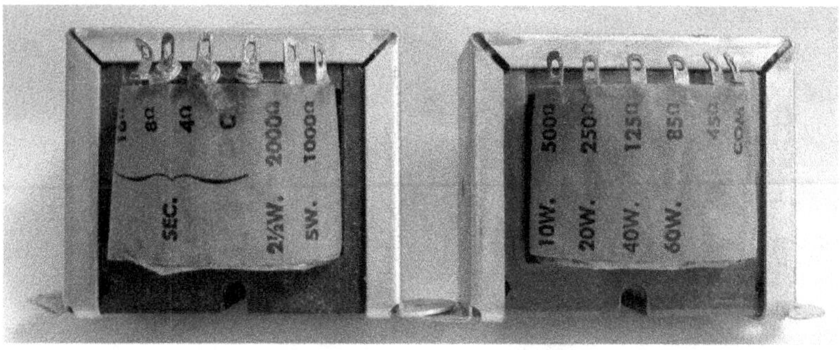

ATLAS SOUND T-18
LINE TRANSFORMER

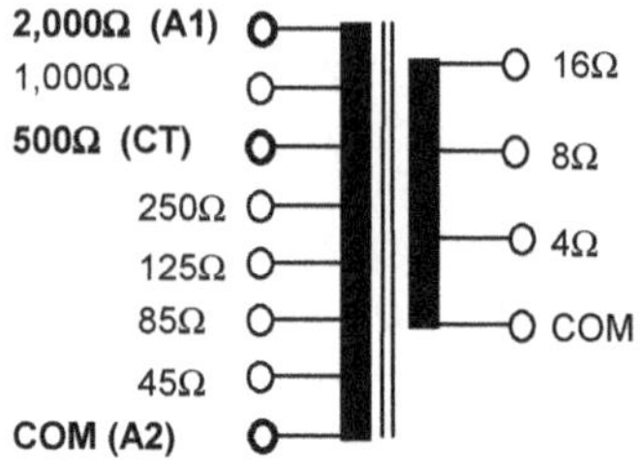

- EI66, a=2.2 cm, S=2.4cm
- A=aS = 5.3cm^2, P=A^2= 28VA
- R_{DC} (COM-CT): 17.5Ω
- R_{DC} (CT-2,000Ω): 30.8Ω
- L_P=1.45H, L_L=28mH, QF = 52
- Z_P (A1-A2): 2,330Ω

ELA-T50 LINE TRANSFORMER

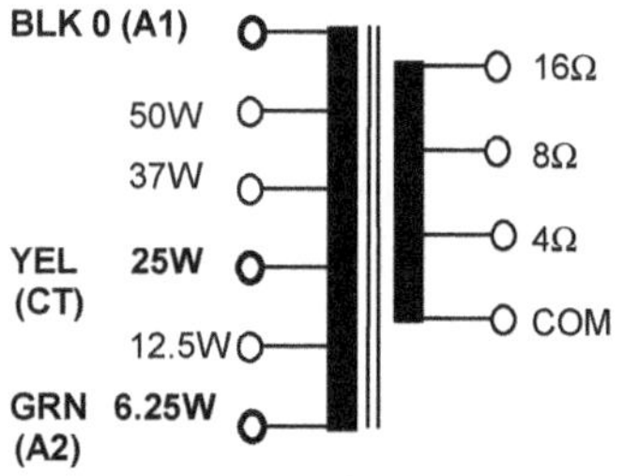

- EI66, a=2.2 cm, S=3.0cm
- A=a*S = 6.6cm^2, P=A^2= 43.5VA
- L_P=1.7H, L_L=6.7mH, QF = 253
- Z_P (A-A) = 1.6kΩ

CASE STUDY: ELA-T50 line transformer

This 50 Watt 100V line transformer sold on eBay seems to be made by or for Omnitronic in Germany. The primary plate-to-plate impedance is only 1k6, but when an 8Ω load is connected to a 4Ω tap, it increases to 4k2. The primary inductance is low, under 2H, so the bass performance will suffer.

The leakage inductance is also low, so the Q-factor is decent. The price is low, so it's worth trying it out with higher-powered output tubes such as 7591, 6L6, or paralleled EL84s for 30 Watts out.

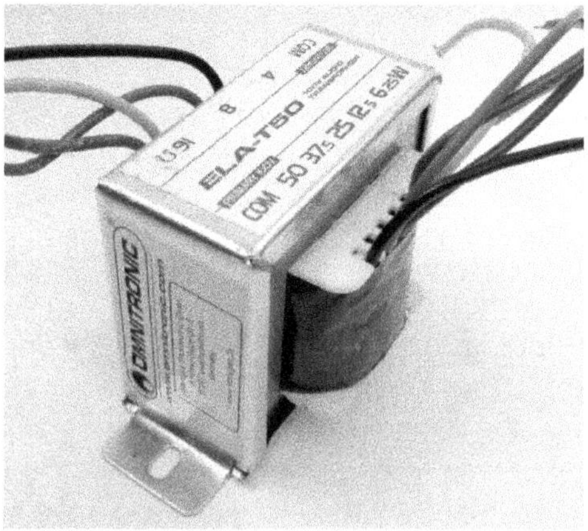

Dual-primary power transformers as output transformers

RN20 is an R-core power transformer made in China. Rated at 30VA, it has two identical primary windings for 115V_{AC} mains voltage or 230V_{AC} in series. Each of the two identical secondary windings is rated at 9V @ 1.66A.

Our idea was to use the two identical primaries as push-pull output primaries and parallel the two identical secondaries to drive the speaker load. That would give us a voltage ratio of VR= 230V/9V = 25.6 and the impedance ratio IR=VR^2 = 25.6^2 = 655. An 8Ω speaker impedance would reflect onto the primary side as 655*8 = 5.24kΩ, perfect for two 6L6 or EL34 tubes, even four 6V6 or EL84 in a parallel push-pull output stage.

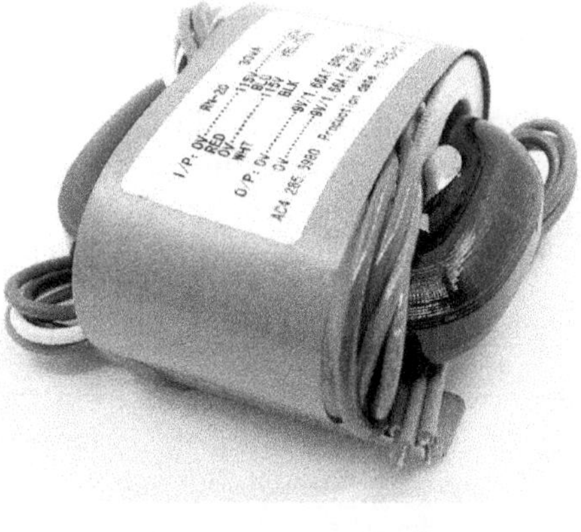

The 16Ω speaker would be a 655*16 = 10.5kΩ load on the power tubes, perfect for 6V6 or EL84 push-pull stages. Two secondaries in parallel would be able to supply 2x1.66A*9V = 29.9 Watts of power, as per the 30VA specs.

If you study the construction of R-core transformers, they always have two identical windings on two longer sides, so these can be connected in series or parallel.

The transformer weighs only 500g and measures 79 × 68 × 40 mm. Its low profile (40mm height) means it can fit even under the slimmest of chassis.

Dual primary EI transformers can also be used for this purpose, but R-core transformers are superior due to their lower magnetic leakage and the fact that windings are perfectly balanced.

Notice that the two primaries have an identical DC resistance, which would only be possible if bifilar winding was used on an EI transformer, something no mains EI transformer features.

The R-50A transformer (pictured on the right) is rated at 65VA and features two 12V/2.7A secondaries, so VR= 230/12 = 19.2 and IR=368. An 8Ω load reflects as 3kΩ onto the primary and a 16Ω speaker as a 6kΩ anode-to-anode load.

The measured parameters of these transformers, in particular, their high primary inductance and above-average quality factors, surpass most brand name output transformers for guitar amps (always of inferior EI variety), selling for $120-200, so a great value here.

As for their use in hi-fi amplifiers, the only issue is their high leakage inductance, which would result in a modestly high upper -3dB frequency limit, but try them out by all means - you may be pleasantly surprised!

DUAL PRIMARY 2 x 9V SEC. TRANSFORMER AS PP OUTPUT TRANSFORMER:

- Primary resistance WHT-BLK: 49.1Ω
- Primary resistance RED-BLU: 49.1Ω
- Primary inductance RED-BLK: L_P= 29H
- Leakage inductance L_L=36mH
- Quality factor: QF = L_P/L_L= 806

INTERSTAGE TRANSFORMERS IN GUITAR AMPLIFIERS

Input and interstage transformers were seldom used in vintage guitar amps and are unheard of in modern commercial designs. Since their narrower frequency range is not an issue in guitar amps (which need to reproduce a relatively narrow bandwidth of around 80Hz-8kHz), this is mostly due to their high cost compared to a single triodes stage with capacitive coupling.

However, they are highly regarded in hi-fi tube amps, where transformer coupling between the stages is claimed to sound better than the capacitive coupling. Most currently produced high fidelity IS transformers are very expensive, a pair typically costing $1,000-$3,000.

CASE STUDY: Thodardson 20A19 interstage transformer

Although vintage I/S designs by Triad and other yesteryears manufacturers, such as the Thordarson and STC models illustrated below, are unsuitable for hi-fi use, they can be successfully incorporated into guitar amp designs. These and similar units are regularly available on eBay, and since tube hi-fi amp builders don't want them and guitar amp builders usually copy commercial designs with other types of phase inverters, they are cheap, usually selling for under $20.

To achieve maximum possible frequency bandwidth, hi-fi interstage transformers are usually of 1:1 design, so they don't provide any voltage gain. The Thordarson transformer has a turn ratio of 1:3, so it works as a voltage amplifying stage, providing voltage gain of 3 times or 20log3 = 9.5dB.

Remember, although the voltage & turns ratio is 1:3, the impedance ratio is a square of the voltage ratio, or 1:9, so a 10kΩ primary impedance reflects onto the secondary side as 90kΩ!

ABOVE: The printing on the box and the look of the Thordarson 20A19 interstage transformer

BELOW: The winding diagram with impedances marked

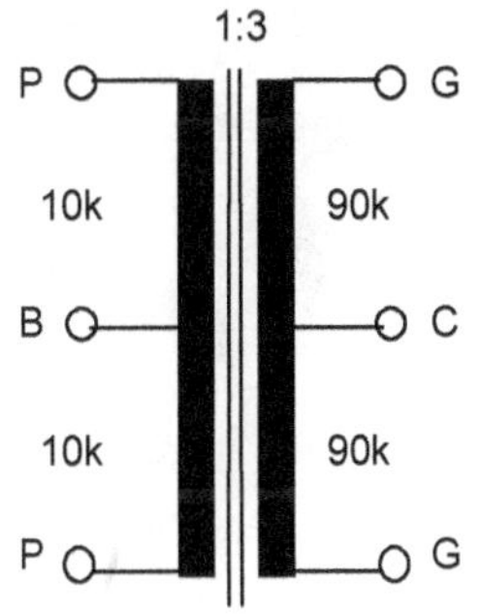

CASE STUDY: A-52-C by Chicago Standard Transformer Company

A-52-C, made in 1965 by Chicago Standard Transformer Company (also known just as STC), is a single plate to push-pull grids interstage transformer. The maximum allowed primary DC current is 10mA and the "overall turns ratio" (primary-to-secondary) is 1:2, which means (since there are two secondaries) that the voltage ratio between the primary and each secondary is 1:1. So, when used as a phase splitter, A-52-C provides no voltage gain.

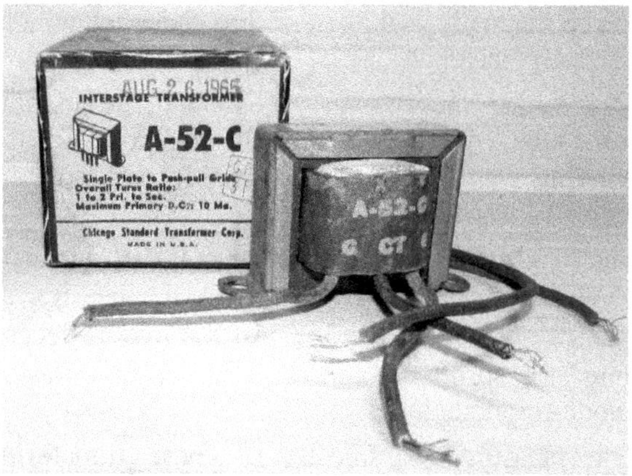

A-52-C INTERSTAGE TRANSFORMER

- EI41, a=13 mm, S=13mm, A=a*S= 1.7cm^2, P=A^2= 2.9W
- Primary DCR=485Ω, Secondary DCR=874Ω + 1,026Ω

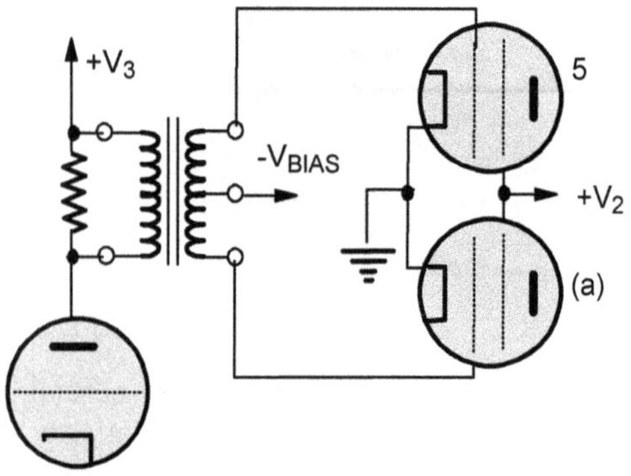

Interstage transformers as phase splitters lend themselves equally well to fixed-biased (a) and to cathode biased (b) output stages!

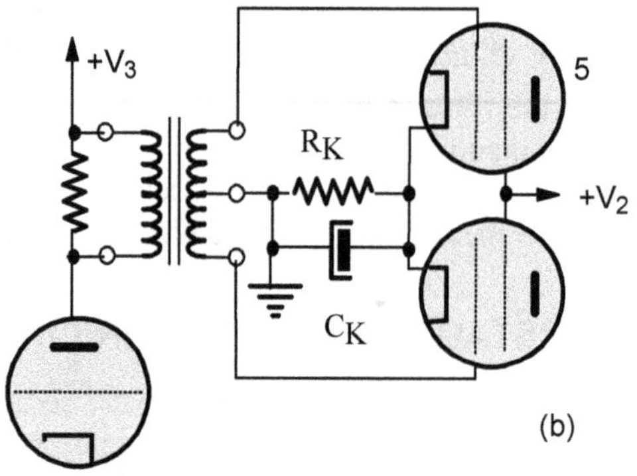

CASE STUDY: Merit A2914 interstage transformer

Just like Thodardson 20A19, this tiny vintage USA-made transformer has a turns ratio of 1:3. The instruction sheet, reproduced here, says that up to 10mA of DC current is allowed through its primary, so we can place it directly in the anode circuit of a preamp tube. Notice a huge difference in DCR values of the two secondary halves, 874Ω versus 1,026Ω. This means that one secondary section was wound first, followed by the primary winding and the second secondary section on top. The mean length of turn of the top section is much higher, resulting in its much higher DC resistance.

MERIT A2914 INTERSTAGE TRANSFORMER

- EI41, a =13 mm, S=13 mm, center leg cross section A= aS = 1.7 cm^2, power rating: P= A^2 = 2.9W
- Primary DCR: 485Ω, Secondary DCR: 874Ω + 1,026Ω

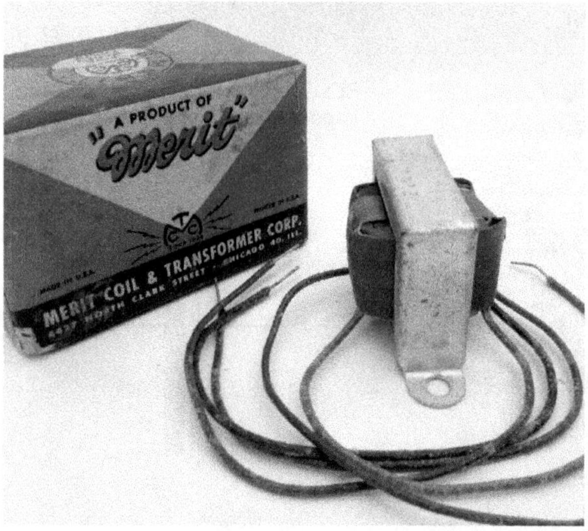

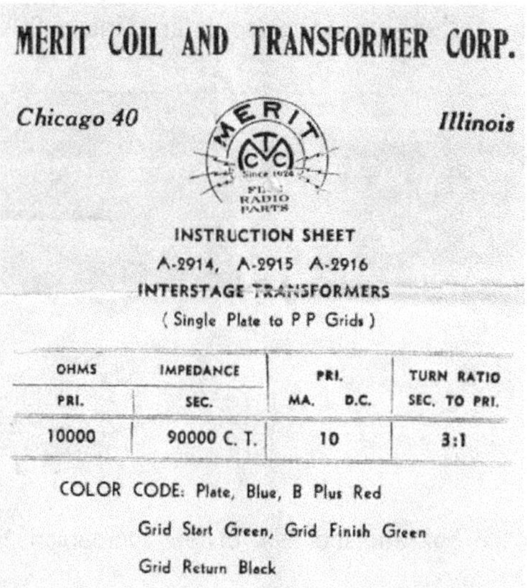

MERIT COIL AND TRANSFORMER CORP.

Chicago 40 Illinois

INSTRUCTION SHEET

A-2914, A-2915 A-2916

INTERSTAGE TRANSFORMERS

(Single Plate to P P Grids)

OHMS	IMPEDANCE	PRI.		TURN RATIO
PRI.	SEC.	MA.	D.C.	SEC. TO PRI.
10000	90000 C. T.	10		3:1

COLOR CODE: Plate, Blue, B Plus Red

Grid Start Green, Grid Finish Green

Grid Return Black

TRANSFORMER TESTS & MEASUREMENTS

- TESTING AUDIO TRANSFORMERS
- COMPARING THE PARAMETERS AND FIDELITY OF AUDIO TRANSFORMERS
- EVALUATING COMMERCIAL OUTPUT TRANSFORMERS BEFORE PURCHASE
- TESTING MAINS (POWER) TRANSFORMERS

12

Expensive test instruments aren't needed to perform the basic transformer tests and measurements. A good quality multimeter is a must-have, preferably a True-RMS type. It is also important that the multimeter be sensitive enough to measure AC and DC millivolts (mV), not just Volts. The second required instrument is a digital LCR meter. All others are only required for advanced tests.

TESTING AUDIO TRANSFORMERS

Measuring the primary and leakage inductance

In the first approximation, the easiest and fastest way to measure both the primary inductance and the leakage inductance is with a digital LCR meter.

Since inductance changes if a DC current flows through the primary (as in single-ended output transformers), a more precise way to measure these parameters would be by using an LCR bridge, especially ones with the capability to adjust the DC current through the measured coil. For estimation purposes the results without any DC flowing are close enough.

The primary inductance should be measured at the lower frequency, such as 120 Hz since its importance is only to the bass frequencies in the audio range. This measurement is performed without any load on the secondary (open secondary).

The leakage inductance should be measured using as high a test frequency as possible, which is only 1kHz in most LCR meters, but some models have a 10 kHz measurement option. Secondary terminals must be shorted for this measurement.

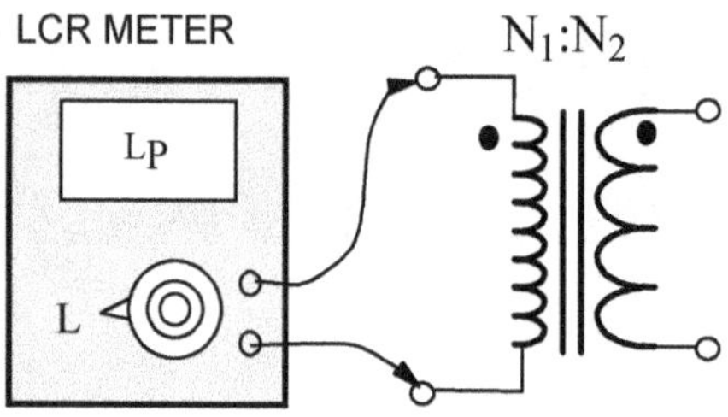

ABOVE: The primary inductance is measured with open secondary winding(s).

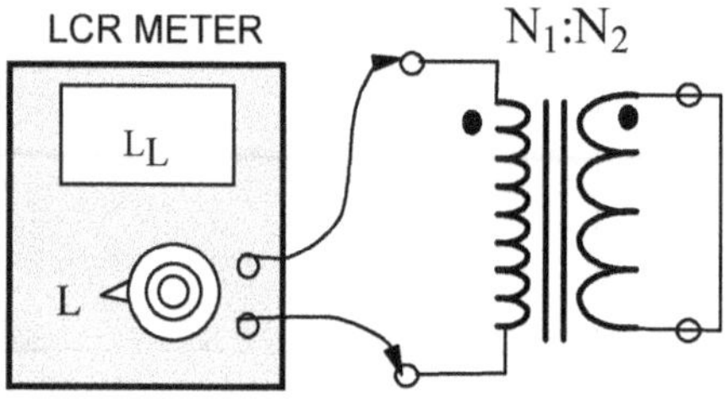

ABOVE: The leakage inductance is measured with shorted secondary winding(s).

Measuring interwinding capacitances

Transformer capacitances are important for hi-fi output and interstage transformers because these need to have an extended upper-frequency limit, way above 20 kHz.

Parasitic capacitances shunt high frequencies to the ground, reducing transformers' upper-frequency limits. Since guitar output amps don't have to go that high (8 kHz is considered a minimum), these capacitances are less important.

Capacitances should also be measured at the highest possible test frequency of your LCR meter, usually 1kHz or 10kHz.

Use crocodile clips, and perform these tests without you touching the test leads or the transformer; the capacitance between your body and the LCR meter or the transformer could affect the results.

The capacitance of the primary is measured between its ends, as is the capacitance of the secondary. Of more interest is the capacitance between the primary and secondary, and since there are at least four terminals, there are four possible measurements. The results are very close in most cases, so you don't have to perform all four tests.

The capacitance between the primary/secondary winding and the magnetic core and the metal frame or case, considered "ground" since they are usually earthed, is measured similarly. The two ends of a winding may have different capacitance to the ground, so you may need to test both.

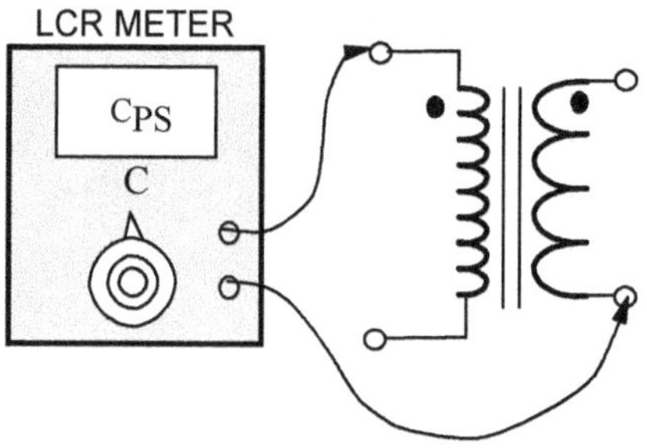

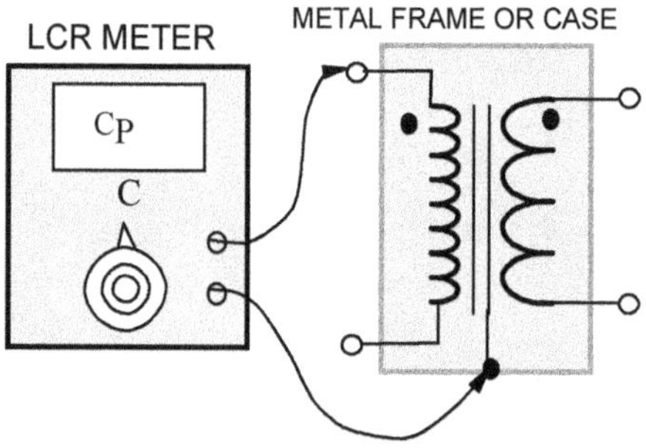

ABOVE: The primary-to-secondary and primary-to-ground (metal case or core) capacitance measurements

How to determine transformer's primary impedance

When set on "R" or resistance range, digital LCR meters measure AC impedance, not DC resistance. So, if you terminate a transformer with its nominal load R_L (usually 4, 8, or 16Ω) using a dummy (resistive) load, the "R" reading will be its reflected impedance onto the primary side.

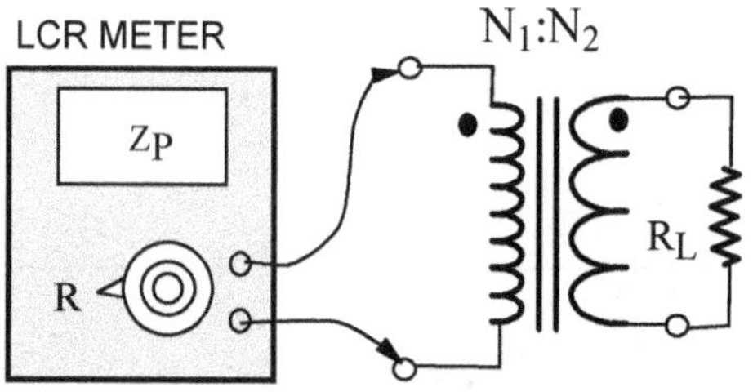

ABOVE: The primary impedance is measured with secondary loaded by its nominal load

Perform this test on both 120 and 1kHz frequencies of a typical LCR meter to see how much the two impedances differ due to transformers' complex behavior and frequency- dependent nature of its parameters.

The measurement of primary and secondary DC resistances is trivial, using an analog or digital multimeter set on "Ohms".

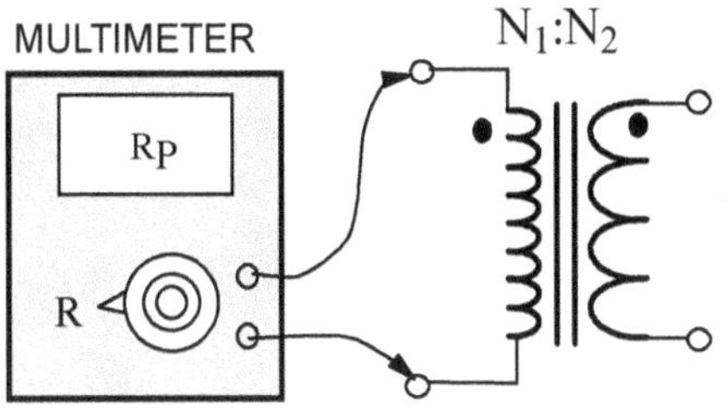

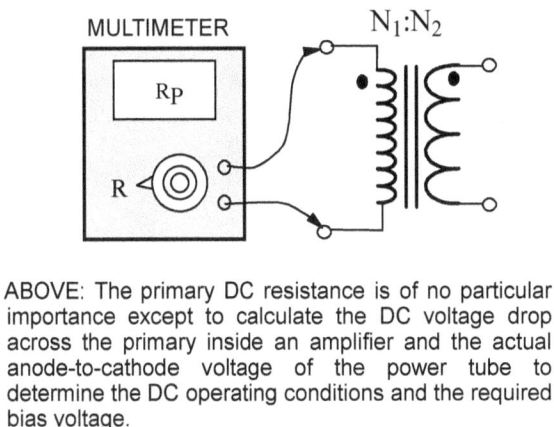

ABOVE: The primary DC resistance is of no particular importance except to calculate the DC voltage drop across the primary inside an amplifier and the actual anode-to-cathode voltage of the power tube to determine the DC operating conditions and the required bias voltage.

Digital LCR meters

A digital LCR (inductance-capacitance-resistance) meter is a must-have instrument for transformer and amplifier builders. The Taiwan-made Escort LCR meter and the identical Tenma model 72-960 are the mid-1990s technology. This auto-ranging instrument measures parameters at two frequencies, 120Hz, and 1kHz.

The 120Hz frequency was chosen since it's double the 60Hz mains frequency in the USA, and that is the frequency of the fully rectified AC ripple in a power supply. This LCR meter displays both values simultaneously (C & D or L & Q).

It also measures "resistance," but that is somewhat misleading. The "resistance" is measured with an AC signal, so it's the impedance modulus at the two frequencies.

The B&K Precision LCR meter can also measure parameters at 10kHz, which is a handy feature when measuring transformer leakage inductance.

Both have slots into which component leads can be inserted. Alternatively, you can use test leads terminated with crocodile clips, included with both instruments. The Escort unit chews through 9V batteries very quickly. Luckily, there is a provision for an external DC supply connection that can save you a small fortune on replacement batteries!

COMPARING THE PARAMETERS AND FIDELITY OF AUDIO TRANSFORMERS

One quick way to assess the quality of a transformer's core is to measure its primary inductance at 120 Hz and 1 kHz, as all digital LCR meters do (some also have a 10kHz test frequency). The smaller the difference between the measured values at those two frequencies, the better suited the laminations are for audio use.

The oscillographic methods outlined below are more precise and do what the LCR method cannot do, and that is to evaluate the linearity and losses of the transformer's laminations, which can be deduced from the shape of its hysteresis curve and the magnitude and waveform of the magnetizing current. The test setup is very similar. While identical on the primary side, the main difference is the way an oscilloscope is hooked up to the secondary.

Observing transformer's magnetizing current on an oscilloscope

Once the primary winding is energized, a sinusoidal primary voltage (either the mains voltage in power transformers or a sine test signal for audio transformers) establishes a sinusoidal magnetic flux through the magnetic core. The flux lags behind the voltage by 90 degrees, so the flux is at its peak when the primary voltage drops to zero (point A).

A waveform of the primary (magnetizing) current can be drawn if point-by-point construction for a specific hysteresis loop is carried out. Due to the nonlinear dependence between the flux and the current, its waveform will be significantly distorted. If displaying the hysteresis curve (the next experiment) is too complex or time-consuming, this simpler test setup that displays the magnetizing current can also be used as a proxy, and meaningful comparisons between different transformers can still be made.

The wider and more tilted the hysteresis curve, the more distorted the magnetizing current and the higher the insertion losses and harmonic distortion of the audio transformer.

A variable autotransformer (popularly called Variac®) is used to adjust the amplitude of the mains signal used for this test. Its output is fed into an isolation (1:1) transformer for safety reasons. A small value (1 to 10 Ω) resistor is connected in series with the primary winding. The voltage drop across it is proportional to the exciting current I_1, and this voltage is observed on the oscilloscope. The transformer under test is left unloaded (open secondary).

The mains voltage is used as a signal source in case of mains (power) and tube interstage and output transformers, which operate with similar primary voltage amplitudes (100-400V).

A function generator needs to be used for low-level audio transformers such as MC step-up, and input types since very low amplitudes (under 1Volt) are needed. In that case, neither the Variac nor the isolation transformer is required.

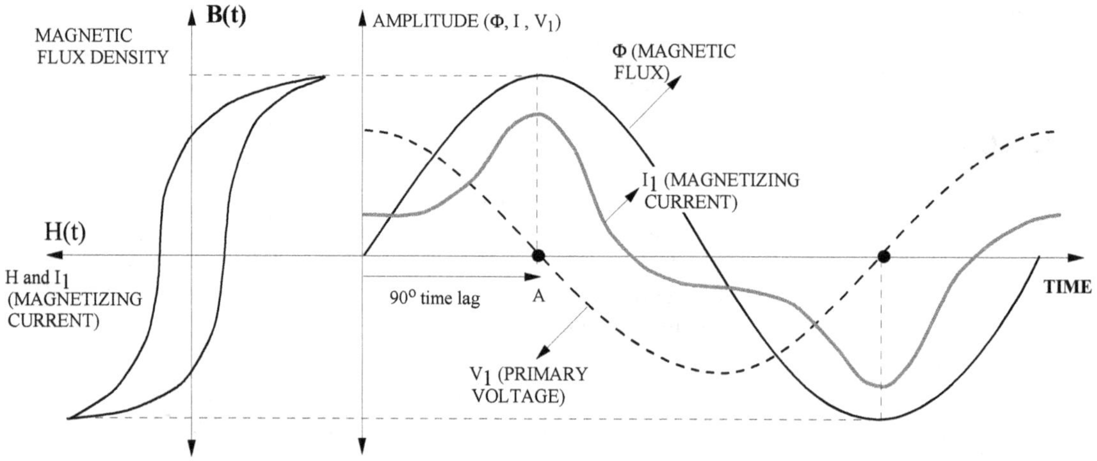

ABOVE: While the primary voltage and the magnetic flux it produces are undistorted sine waves, the primary magnetizing current isn't. Due to the hysteresis curve, it has the waveform illustrated.

RIGHT: The test setup for displaying the waveform of the magnetizing current on an oscilloscope

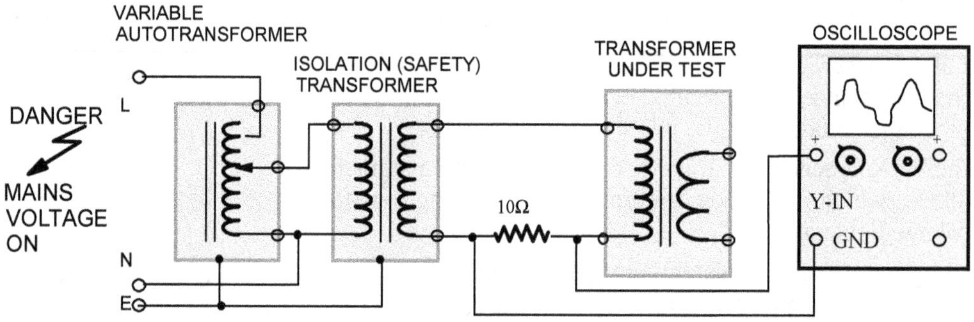

Estimating transformer's laminations quality by observing its hysteresis loop on an oscilloscope

The hysteresis curve of a particular transformer's core (lamination stack) has H, magnetic field strength or "magnetomotive force" on the X (horizontal) axis, and B, magnetic flux density as a dependent variable on the vertical (Y) axis. Since H is a direct function of the primary current of the transformer without any load, if we connect a small value (1 ohm to 10 ohms) resistor in series with the primary winding, the voltage drop across such resistor will be proportional to the exciting current I_1 and thus to H (just as in the previous test). This voltage should be applied to the horizontal input of the oscilloscope, which will work in X-Y mode (its time base generator will be disconnected).

Since the voltage induced in the secondary winding is proportional to the rate of change of magnetic flux Φ. This is expressed mathematically as a derivative ($V_2 = N_2 d\Phi/dt$), where N_2, the number of secondary turns, is a constant. The faster the rate of magnetic flux's change and the higher the number of secondary turns, the higher the induced secondary voltage!

B is proportional to magnetic flux Φ. To get B, we must perform the opposite mathematical operation onV2. The opposite of differentiation is integration, so we need to integrate the secondary voltage V_2. A simple RC network can perform that task under the proviso that its time constant τ=RC is much larger than the period of the test signal.

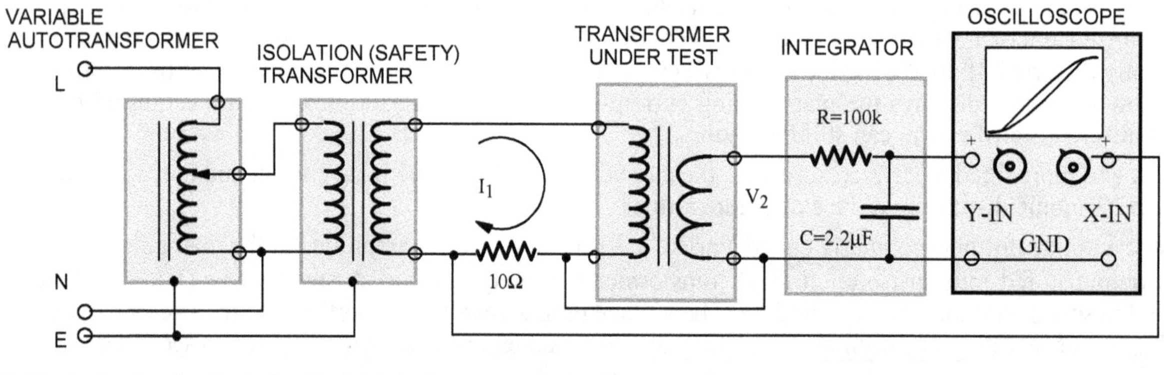

ABOVE: The test setup for displaying the hysteresis curve on an oscilloscope

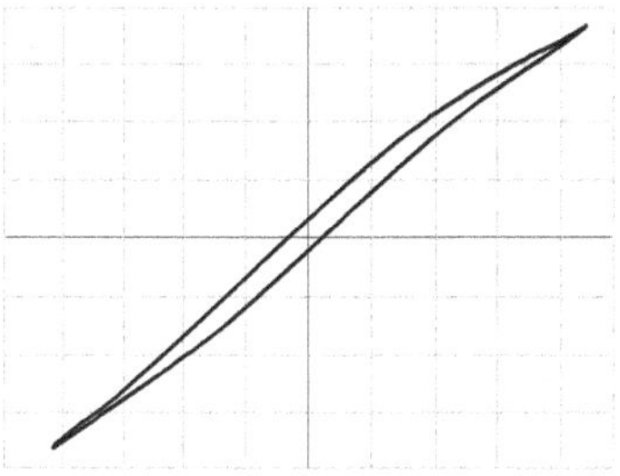

Hysteresis loop of a good quality output transformer with GOSS laminations

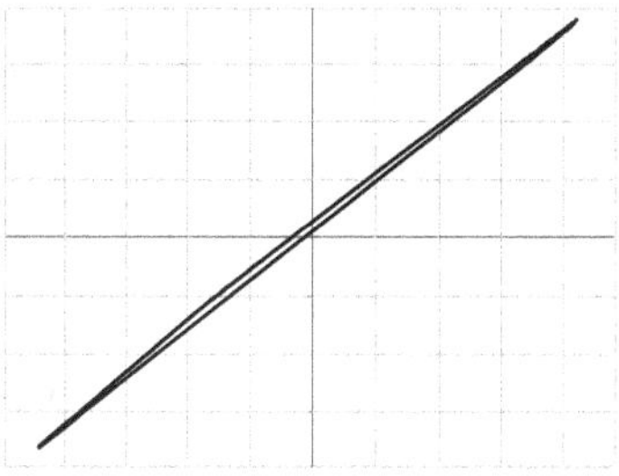

Hysteresis loop of a top quality output transformer with extremely small losses and very low distortion

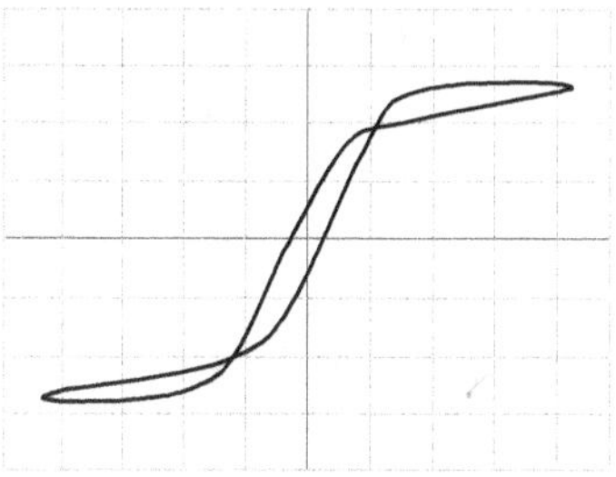

Distortion due to improperly selected RC integrator (ωRC product is too small).

In other words, we must ensure that $\tau = RC >> 1/\omega$ or that $\omega RC >> 1$! Angular frequency ω (omega) is $\omega = 2\pi f$! If this condition is not satisfied, the display will not be a true representation of the hysteresis curve; the curve will be "folded" or warped, as in the illustration above.

If the mains voltage is used for this test, a frequency of 50 Hz (or 60 Hz in some countries) needs to be used in the formulas. Let's see what we get with our values $R=100k\Omega$ and $C=2.2\mu F$: $\omega RC = 2\pi fRC = 2*\pi*50*10^5*2.2*10^{-6} = 10*\pi*2.2 = 69$, which is much larger than 1 (69>>1), so the hysteresis curve will be displayed properly.

Going back to the illustrative examples above, even the example on the far left isn't bad at all; the curve is fairly narrow (low losses) and fairly straight (low distortion). The example in the middle is even better; only the highest fidelity audio transformers using EI laminations or C-cores of superior quality will have such a linear hysteresis curve.

A few pointers regarding the test circuit. The capacitor used in the integrator network should be a low loss film type, polyester, or polypropylene. The test should start with the Variac in zero output position. Its dial should be slowly raised until the nominal primary voltage is reached. The vertical sensitivity control of the oscilloscope and the time base will need to be adjusted until a display such as those illustrated is obtained.

Due to the polarities of the signals involved, the display will be a mirror image unless the invert button on the scope is pressed (activated), which flips the signal brought to the scope's X-input.

If the scope's sensitivity cannot be lowered enough to display a large enough curve (budget scopes of low sensitivity or very low magnetizing current amplitudes), the 10-ohm series resistance will need to be increased.

Not all vintage audio transformers deserve their stellar reputation!

In April 1952, The Naval Research Laboratory of the US Navy prepared a report titled "Laboratory Tests of Some of the Popular Audio Amplifiers," in which they compared the design and performance of a dozen or so commercial tube amplifiers. Some amplifiers were constructed by their staff in-house using different output transformers available for purchase at the time, including Acrosound TO-300 and Peerless S-265-Q models.

They tested output transformers inside the same Williamson-type amplifiers; two such oscillograms are shown on the next page, together with the A-f characteristics.

While its upper -3dB limit was around a respectable 70kHz, at low power levels, Acrosound TO-300 exhibited an incredibly high resonant ultrasonic peak, which was detectable even at higher output levels. This manifested itself in a high overshoot at the rising edge of a square wave and low damping.

Peerless S-265-Q, on the other hand, showed no such ultrasonic instability, reproduced square wave signals better, and had a much higher upper -3dB limit!

The moral of the story is? While vintage output transformers generally fetch very high prices today, not all were created equal. As mentioned previously, the quality factor is the best single indicator of a transformer's quality.

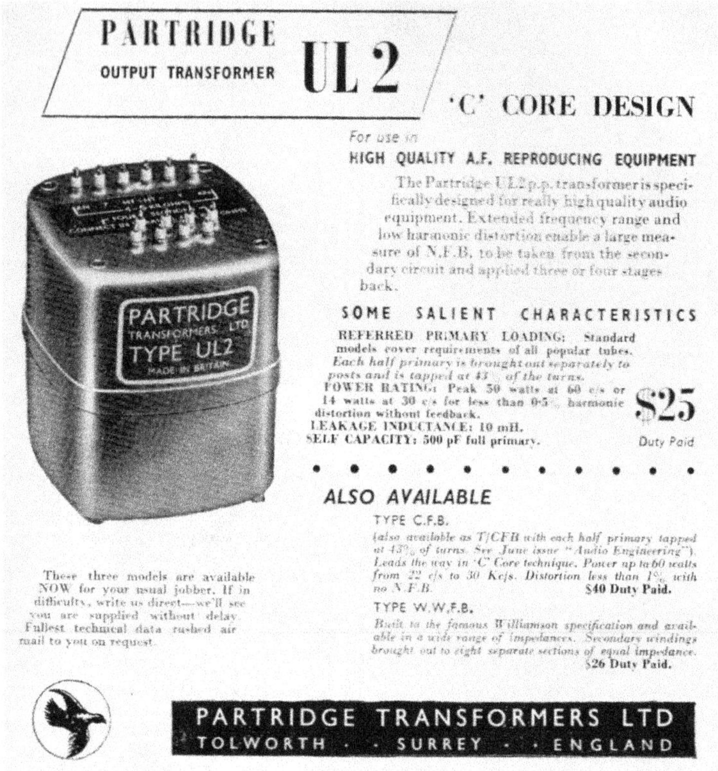

For the Partridge UL2 model, the primary inductance is 140H, and since the leakage inductance is only 10mH, its quality factor is $Q_F=L_P/L_L=14{,}000$!

Such a good result is most likely a consequence of their use of C-cores instead of inferior EI-laminations and advanced sectionalizing and winding techniques.

Notice the very keen competitive pricing of the Partridge UL2 model, which, despite being imported from the UK, was selling in the USA at around the same price ($25) as the smaller and inferior Acrosound TO-300 transformer.

LEFT: Partridge advertisement, December 1954, © Audio magazine

BELOW LEFT: The A-f characteristics of a Williamson amplifier with Peerless S-265-Q output transformer and the fidelity of a 20 kHz signal at its output.

BELOW RIGHT: The A-f characteristics of an ultra-linear amplifier with Acrosound TO-300 output transformer and the fidelity of a 10 kHz signal at its output.

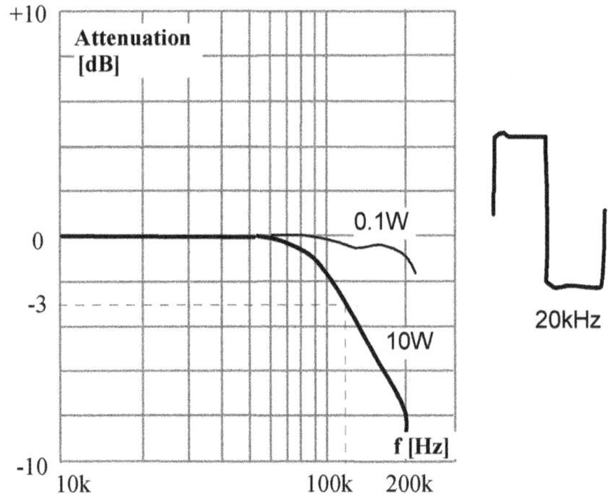

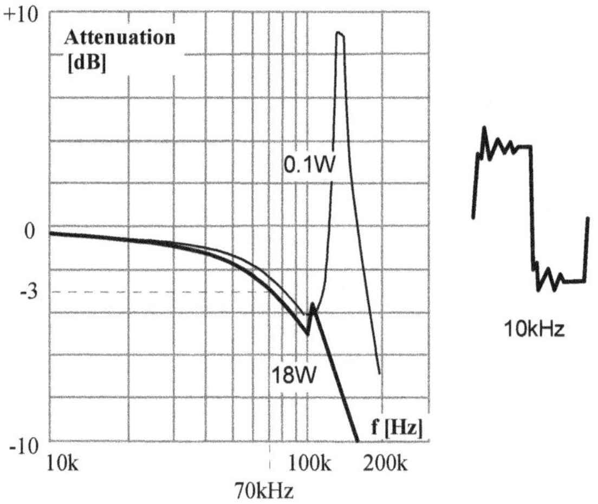

Quick power transformer lamination quality test using an LCR meter

Finding transformer laminations in many parts of the world is almost impossible. However, EI mains transformers with 6-24V secondaries are widely available. Most use low-grade laminations (cheap 3% non-oriented silicon steel), but some use GOSS materials, and you can use those to wind audio transformers. How can you tell them apart?

First, learn to recognize 0.5mm lamination thickness from 0.35mm. The 0.5mm laminations are not grain-oriented, so forget those. Some 0.35mm laminations are GOSS, but not all. Use this quick test to find out which is which. Using a digital LCR meter, measure the primary inductance L_P of the mains transformer at 120 Hz, then measure it at 1kHz. The closer the two values are to each other, the better suitable are the laminations for audio use!

Say you get 9.2H at 120H and only 2.1H at 1kHz. These laminations would not perform well in an audio transformer. In contrast, another transformer may have an L_P of 6.7H at 120Hz and 5.6H at 1kHz. These laminations could be suitable for use in output or interstage transformers.

A simple way to plot the impedance curve of an audio transformer

The transformer under test is terminated by its nominal impedance, in this case, 8Ω. DC voltage source VDC supplies the required DC current to the primary winding (for single-ended transformers only, of course), 100mA, while the AC signal source provides the test signal of a suitable amplitude. The frequency of such source must be continuously variable from 10Hz to 80+ kHz!

The same current flows through the series resistor R and the primary impedance Z_L. The voltage drops across them, measured by an AC voltmeter, are proportionate to their impedances: $V_L/V_R=Z_L/R$ so $Z_L=V_L/RV_R$.

Instead of one switchable voltmeter, a more elegant and faster method is to use two voltmeters, one across the primary winding, the other across resistor R. The voltmeter(s) must be capable of operation up to at least 80 kHz, so cheap multimeters are out of the question.

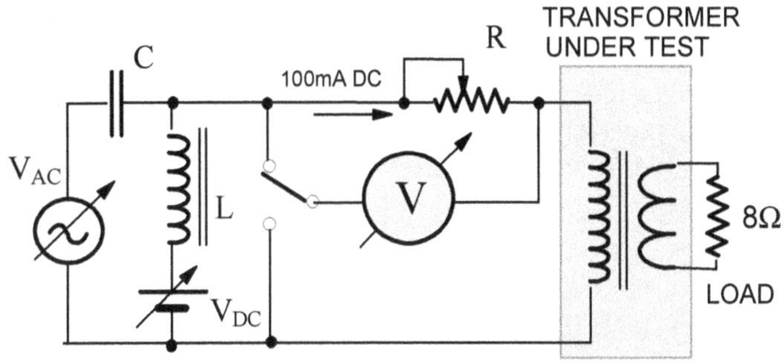

RIGHT: Test setup for manual (point-by-point) plotting of the impedance versus frequency curve of a SE output transformer. For push-pull transformers, the DC source and the choke are not needed.

EVALUATING COMMERCIAL OUTPUT TRANSFORMERS BEFORE PURCHASE

How to scrutinize transformer data and manufacturers' claims

We selected three transformers from the same manufacturer, all of the same physical size, which was an important consideration for this exercise. Their center-leg cross-section is the same, meaning their power handling capabilities are the same. Only physical dimensions were specified in the datasheet (except stack thickness), so how did we determine the lamination size?

First, find the transformer's largest dimension (laminations only, without end bells), 86 mm in this case. Then, find the laminations' width, in this case, 72 mm. Finally, to determine if the laminations are of the scrapless (wasteless) size (the ratio of width to length of 5/6 = 0.8), divide 72/86 = 0.837. Since the manufacturer is from the USA, it is possible that an "imperial" size EI85.73 (L112) laminations were used, 85.73mm height *5/6 = 71.44 mm width, which is close to the possibly rounded specified figure of 72 mm! The stack thickness is 5cm.

MODEL	UBT-1	UBT-3	UBT-2
Primary impedance Z_P [Ω]	1k6	3k0	4k8
Maximum DC plate current I_{PMAX} [mA]	160	110	110
Maximum output power P_{MAX} [W]	13-18	13-18	13-18
Primary resistance R_1 [Ω]	165	286	432
Primary inductance L_1 [H]	8	17	29

Notice that the choice of primary impedances seems related: 1,600Ω, 2*1,600Ω = 3,200Ω, and 3*1,600Ω = 4,800Ω! Sure, specified is 3,000Ω but 3,200Ω seems suspiciously close.

The center-leg (CL) cross-sectional area for EI85.73 (L112) laminations with a 5cm stack is A=2*1.43*5=14.3 cm^2, so A_{EF} is around 13.7 cm^2. The empirical formula $A_{EF}=k\sqrt{(P/f_L)}$ gives us k=17 for P=13W and f_L=20Hz, and k=14.4 for P=18W and f_L=20Hz. Since k should be 20-30, the available power transferred to the load at 20Hz would be below the specified 13-18 W. The center leg cross-section is simply too small!

One thing is immediately obvious: The higher the primary impedance, the higher the primary resistance and the primary inductance. This is easily explained; to get a higher impedance ratio, a higher number of primary windings is needed (N_1). The primary inductance goes up since it is proportional to N_1^2. More primary windings also mean that the length of the primary winding increases, so its DC resistance must also go up.

Comparing output transformers

We took two output transformers for single-ended 300B tube amplifiers, one made by One Electron in the USA, the other wound by us, and measured their primary and leakage inductances.

Our transformer had a much higher primary inductance, which means its bass performance will be superior. However, One Electron has a lower leakage inductance (4.1 mH versus 23 mH). The explanation lies mostly in the size of the laminations used, EI114 versus EI85.73 (L112) used by One Electron. Larger laminations mean more primary turns, higher primary inductance, and larger leakage inductance.

SIDE-BY-SIDE: OUTPUT TRANSFORMERS

	Ours	One Electron UBT-1
Primary inductance L_1 @ 120Hz and 1,000Hz	15.5 H, 18.5H	5.8 H, 4.5 H
Leakage inductance L_L @ 1,000Hz	23 mH	4.1 mH
L_L as % of L_1 @ 1,000Hz	0.124	0.091

L_L divided by L_P is the relative ratio used as an indicator of the overall frequency range of a transformer, 0.124% for ours and 0.091% for One Electron, meaning that the frequency range of an OE transformer will be 124/91 = 36% wider. Remember that One Electron is a specialist transformer maker, and we are not. They have access to better quality laminations, superior wire, and professional winding machines, while we wind everything by hand on the simplest winder possible. It would be a sad indictment of their product if our transformer measured better than theirs.

We should have used One Electron's UBT-3 model for this comparison, which is much closer in terms of primary impedance, but we purchased a pair of UBT-1 s, and the aim of this exercise was educational, to see how various parameters of output transformers impact their measured values, not to determine which design is superior, or which transformer sounds better in an amplifier.

Information gathering from manufacturers' data sheets

Even if you don't intend to wind your own transformers, as a designer and builder of tube amps, you still need an intuitive feeling for their parameters and behavior since you will have to evaluate the specs of commercial units or you will need to specify the required parameters if you order custom-made transformers from transformer makers.

In this educational exercise, we'll use four typical lower-cost output transformers. There are more expensive models from manufacturers such as James, Bertolucci, Sophia Electric, Hashimoto, Audio Note, Sowter, Lundahl, ElectraPrint, MagneQuest, etc. These four models have approximately the same primary impedance 3.0-3.6 kΩ. They range in price from $80 to $160 each, a price ratio of 2:1!

MODEL	One Electron UBT-3	Hammond 1630SEA	Transcendar TT-311-OT	Edcor CXSE25-8-3K
Retail price each (Jan 2015)	US$99.00	US$159.00	US$80.00	US$88.00
Laminations (M6 GOSS):	EI85.73 (L112)	EI114	EI96	EI114
Stack thickness ST [cm]	5.0 (2")	6.35	5.0 (2")	5.00 (2")
Primary impedance Z_P [Ω]	3k0	3k5	3k5	3k0
Nominal secondary impedance [Ω]	4-8-16	4-8-16	8	8
Max.. plate current I_P [mA]	110	135	100	200
Rated output power P_{OUT} [W]	13-18	30	10	25
Primary resistance R_1 [Ω]	286	110	332	84
Primary inductance L_1 [H]	17	42	25	48
Frequency range	20Hz-40kHz*	20Hz-20kHz**	20Hz-100kHz***	20Hz-20kH****
Weight [lbs/kg]	5 (2.25)	11 (5.0)	4 (1.8)	9.75 (4.4)
Cost per kg	US$44/kg	US$32/kg	US$44.4/kg	US$20/kg
Center-leg cross-sectional area [cm^2]	14.3	24.1	16	19.0

* +/- 1 db at unspecified power level

** No specific range, generically specified as "at 20Hz-20kHz at full rated power (+/-1dB max.., ref. 1 kHz)"

*** +/- 1 dB at 1 watt

**** no voltage or power level specified, "<1dBu"

Criterion #1: Cost/weight ratio

The most expensive model by Hammond weighs 5.0 kg, the cheapest (by Transcendar) 1.8kg, so when the price per unit of weight is calculated, we get $32 per kg for Hammond and $44.4 per kg for Transcendar. At $20 per kg, Edcor is the cheapest of the four.

Criterion #2: Center-leg cross-sectional area. The largest, by far, is Hammond, followed by Edcor.

Using these two criteria, Edcor is the best value for money, followed by Hammond, with Transcendar in 3rd place and One Electron last. This correlates well with the specified primary inductance figures, the only "anomaly" being that Edcor specified a slightly higher L_P than Hammond, 48 versus 42 Henries.

How to determine the turns & impedance ratios of mains and audio transformers

You have a salvaged or unmarked transformer and want to determine its turns ratio(s). Since tube amplifiers' output transformers are always of a step-down type, their primary voltages are always high, in the order of magnitude of mains voltages (100V to 240V, depending on the country).

The first step is to carefully connect its primary to the mains outlet and measure the primary and secondary AC voltages. You can assume that black is the reference terminal or "0 Ω," so all secondary measurements are between black and one of the other three colors. In step two, calculate voltage ratios and square them to get impedance ratios.

Finally, multiply the nominal secondary loads (4, 8, and 16 Ω) by their impedance ratios to get their reflected primary impedances.

We had The results and calculations are in the table. All voltages are AC (RMS) values.

The nominal primary impedance of One Electron single-ended transformer with 3 secondary taps, declared by the manufacturer, is 1.6kΩ, but our three reflected impedances, measured as 1,479 Ω, 1,504 Ω, and 1,490 Ω, indicate that this is a transformer with 1,500Ω primary and not 1,600Ω!

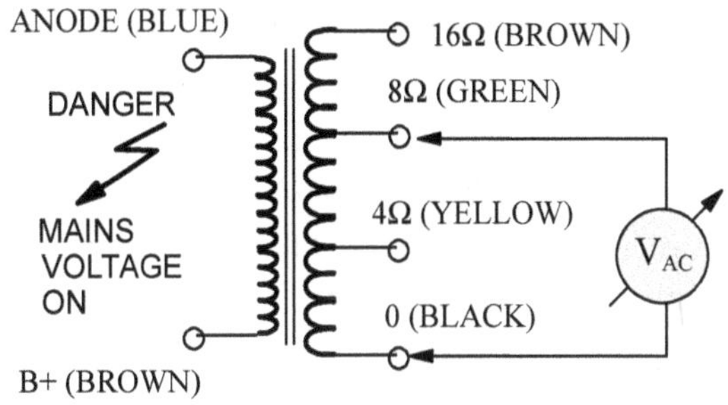

ABOVE: Using the mains voltage on output transformer's primary to measure secondary voltages

How to compare disparate frequency range specifications: the "dB graph"

Transformer and amplifier manufacturers specify the frequency limits of their products in various ways; some use -1dB figures, others -2dB, although most use -3dB bandwidths since that gives the impression of the widest frequency range when superficially compared to -1 and -2dB specs of their competitors.

An audio transformer is a bandpass filter, and the RL-type first-order model can approximate its low-frequency behavior. First-order systems have a roll-off slope of 20dB per decade or 6dB per octave. A decade is 10x lower or higher frequency (a ratio of 10:1), and an octave is a ratio of 2:1 (double or half the frequency).

The "dB graph" shows the low-frequency response of an audio transformer or a whole amplifier in decibels. The X-axis is linear, expressed as a ratio of frequency and -3dB frequency f_L.

Y-axis is attenuation in dB, relative to A_0 (mid band level), a log scale.

Very few manufacturers specify their transformers' frequency range as -1dB since it looks bad, others as -2dB. This used to be more common in the 60s, much less so now, with -3dB being the most common spec today. This graph can help you convert these disparate specs to the same level so you can compare them!

> LOW FREQUENCY ATTENUATION
>
> $A_{LF} = 20\log(V_{IN}/V_{OUT}) = 10\log[(1 + (f_L/f)^2]$
>
> For the RL high-pass filter model of a transformer, attenuation at low frequencies in dB is $A_{LF} = 20\log\sqrt{[(1 + (f_L/f)^2]} = 10\log[(1 + (f_L/f)^2]$

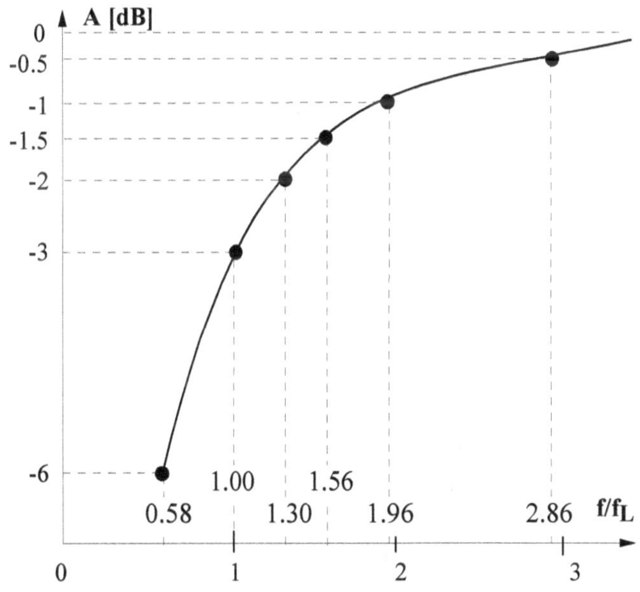

ABOVE: By using multiplication factors, this universal dB graph enables you to convert and compare attenuation figures expressed in different dB levels.

The way it used to be: Ferguson push-pull output transformers

A vintage Australian-made Ferguson "Medium Fidelity" push-pull transformer offers three choices of primary impedances (7, 8, and 10 kΩ) and has the following specs: P=15 W, 40-30,000 Hz (-2dB), L_P=50H, L_L=29 mH.

"Medium Fidelity" most likely refers to a relatively high f_L of 40 Hz and a relatively low f_U of 30kHz. 40Hz is the -2 dB frequency, which is roughly $1.3f_L$ (from the "dB graph), the -3 dB frequency would be $f_L \approx 40/1.3 = 30.8$ Hz.

Although the high-frequency behavior of audio transformers isn't as simple or clear-cut as their low-frequency attenuation, we can approximate that it also follows the first-order curve (although it is often a second-order filter).

The upper -3dB frequency can be estimated from the -2dB figure, $f_U \approx 1.3 * 30{,}000 = 39$ kHz.

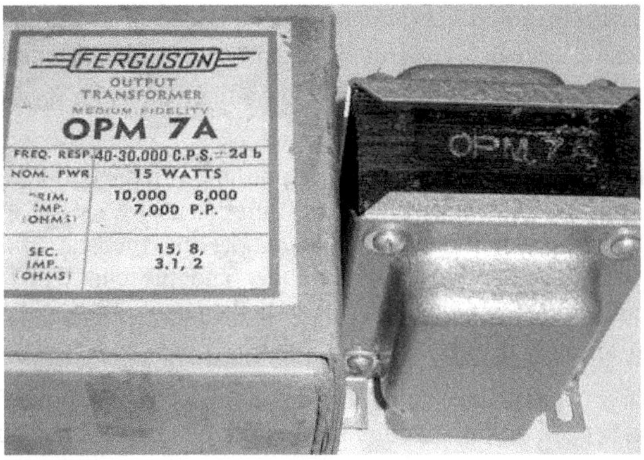

TESTING MAINS (POWER) TRANSFORMERS

Measuring transformer's no-load current and power losses

BOB'S BLACK BOX (A SIMPLE POWER METER) DIY PROJECT

Bob's Black Box (or BBB for short) is a simple watt meter, or, more precisely, a VA-meter (apparent power). The load current flows through the 10 Ω resistor. The voltage drop measured by an AC voltmeter between test points TP1 and TP2 is proportional to this current.

The load current is $I_L=V/10$, and by multiplying this current by the mains voltage, we get the power draw in VA.The diagram shows an Australian power outlet, but any international type can be used. There is a fuse in the active (line) feed for safety reasons, the test points are on the neutral side, and a double-pole switch is used for both line and neutral connections.

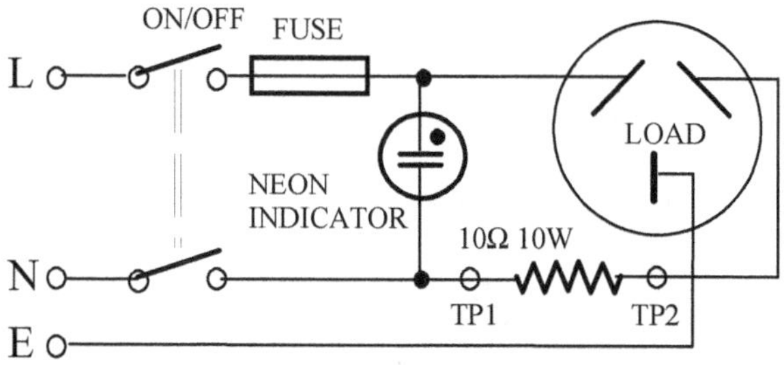

BBB is not useful just for powering up and testing amplifiers; we can also measure their no-load or magnetizing currents and voltage ratios.

EXAMPLE: What is the power consumption in VA of a mains transformer without load (open secondaries) if the measured voltage drop on Bob's Black Box is 0.527 V_{AC} and the measured mains voltage is 247 V_{AC}?

The current is 0.527 V / 10 ohms = 0.0527 Amperes or 52.7 mA. The power draw (losses) are 247V*0.0527A = 13VA.

If you are winding your own power transformer, once it's finished, connect it to BBB (without any load on the secondary). If the magnetizing current is between 40 and 100mA (depending on the lamination quality and transformer's size) and there is no buzzing, leave it powered up.

If there is any problem (insulation breakdown or overheating), the fuse in BBB will eventually blow. If the transformer is still idling happily after 8 hours, it is ready for installation into an amplifier!

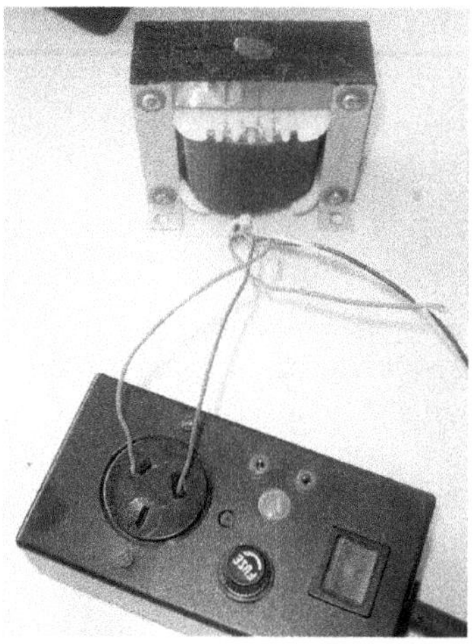

Measuring the current capacity of the high voltage winding - Method #1

Once we have a transformer in our hands, there are two ways to determine the maximum current loading. This quick method works quite well for currents up to 200 mA and does not require any transformer loading.

In Step 1, with the primary disconnected from mains (transformer de-energized), measure the DC resistance of the half of the secondary (assuming a secondary with CT). With one vintage power transformer, we got 82.2Ω and 76.4Ω. Unless they are bifilar-wound, the two halves of the winding will not have equal DC resistances.

In step 2, energize the transformer from the mains and measure the AC voltage across the same half of the secondary. We measured 386V. Then, $I_{MAX} = 25V/R = 25*386/82.2 = 117$ mA.

Measuring the current capacity of the high voltage winding - Method #2

Once you estimate the maximum secondary current (an educated guess based on the secondary wire diameter), for example, 100mA, calculate the value of the load and the power dissipated on it. The secondary voltage is 386V in this case (between CT, point "B" and one end of the HV secondary winding), so $R_L=V_0/I_{EST} = 386/0.1 = 3,860\Omega$.

Calculate the power dissipation on the load as $P=V_0^2/R_L = 386^2/3,860 = 38.6W$. Thus, we would need a 50W-rated 5kΩ rheostat R_X.

Lower the rheostat's resistance until the voltage under load drops 10% from the unloaded voltage level, in this case $V_{90\%}=0.9*386 = 348V$. Power down the circuit, disconnect the rheostat and measure its resistance R_X (between the slider and the end B). The maximum current that winding can supply is then $I_{MAX}=V_{90\%}/R_X$.

With $R_X = 3,520\Omega$, $I_{MAX}=348/3,020 = 115$ mA!

If you don't have such a rheostat, hook up 5-20W rated fixed resistors in series-parallel combinations until you get a 10% voltage drop.

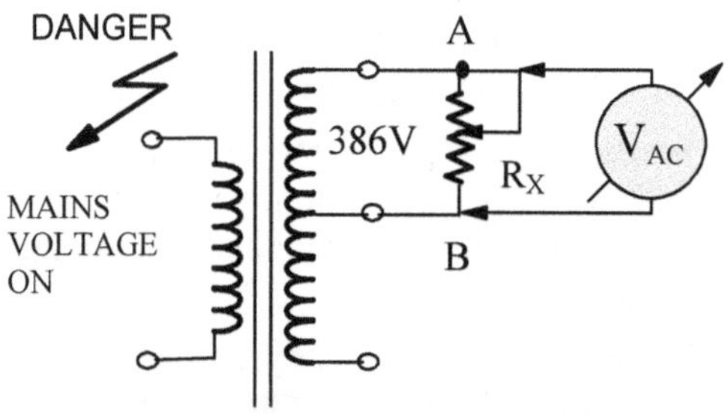

Transformer phasing check

In most applications, it is crucial to connect two or more secondaries of a transformer the right way so that their voltages are in phase. Say you have a power transformer with two primaries, and you either have to connect them in parallel for 115V operation or in series for 230V mains. How to determine which terminals are "in phase"?

You'll need an LCR meter set on "L" or inductance. There are a few possible test combinations, but the principle is the same. For instance, join wires or terminals Y & U and measure the inductance of this series connection at the ends, terminals X & V.

Say you got a low reading (in mH). That means the two windings now oppose each other; they are "out-of-phase. Terminals with the "dot" are in phase. So, if you need a series connection for 230V mains, you should join Y & V and use X &U as mains terminals, or join U & X and use Y &V as mains terminals. For a parallel or 115V connection, join X&V as one input and Y&U as the other.

If your reading was high, a few Henrys, typically 5-10H, points X and U are in phase (as are Y and V), so the two windings "support" one another and send their magnetic fluxes the same way. Thus, bridge U &Y and use X & V for mains connection in series.For a parallel 115V connection, join X&U as one input and Y&V as the other. The same testing method can be used for secondary windings and audio (output) transformers.

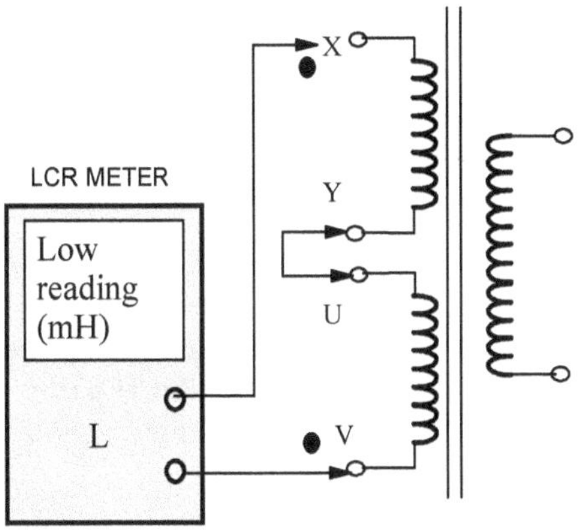

ABOVE & BELOW: The phasing check using an LCR meter to measure inductance

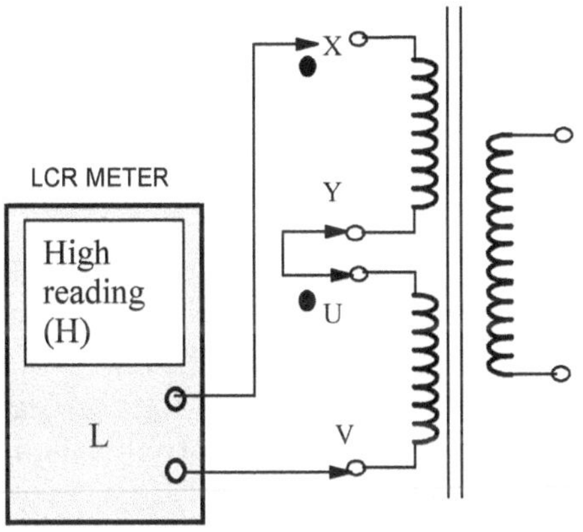

Determining TPV (Turns-Per-Volt) of an existing transformer

If a power transformer had its heater secondary wound last (topmost), which has been the most common coil arrangement over the years, it is possible to carefully open the final insulation layer (always paper for older and plastic for newer transformers) and count the number of windings. This makes it easy to determine TPV used by its designer simply by visual means.

This vintage German power transformer from a small tube amplifier didn't even have such a top insulation layer, making the heater winding visible.

Some units had see-through Mylar insulation. There are 36 turns for 6.8V (unloaded heater voltage), so the TPV figure was 36/6.8 = 5.3 Turns-Per-Volt.

Some transformers are encapsulated, or the low voltage secondary winding cannot be seen or is not accessible, so in those cases, the visual method will not work.

The following simple experimental method will enable you to determine TPV by simple voltage measurements. If there is a gap of 1mm or more between the coil and the laminations, feed through a piece of magnet wire, enough for 3-5 turns (four are illustrated). The wire can be any diameter.

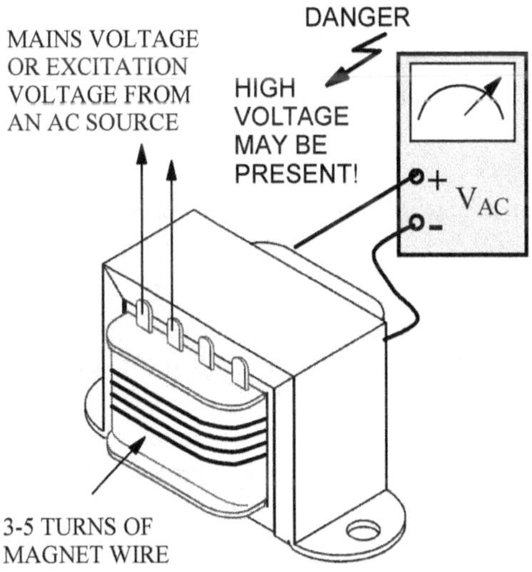

Connect a multimeter on the "AC Volts" function to the ends of this temporary winding. If it's an output transformer of a step-down kind (such as a tube output transformer) or a mains transformer, connect its primary to the mains voltage and note the reading of the voltmeter.

Turns-per-volt is easily calculated as $TPV=N/V_{AC}$, where N is the number of turns you have used for this test (in this example, N=4), and V_{AC} is the measured voltage. Say you measured $1.25V_{AC}$, then TPV= 4/1.25 = 3.2.

Alternatively, use a function generator (on sine voltage setting, 50-60 Hz frequency) on one of the transformer's windings.

DANGER - TRANSFORMER TURNS-PER-VOLT TESTS INVOLVE LETHAL VOLTAGES!

Perform this test under extreme caution since high voltages will be present on some windings. Be careful where you connect the "excitation" voltage; if you connect, say, $8V_{AC}$ from a function generator onto a 6.3V heater winding and the transformer has a 240V primary, you will get 240*8/6.3 = 305 V_{AC} on the primary!

INCREMENTAL INDUCTANCE TESTER

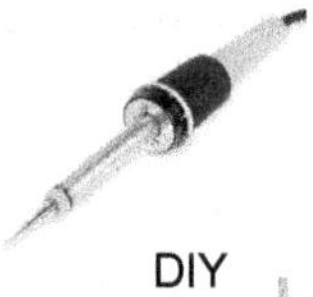

DIY PROJECT

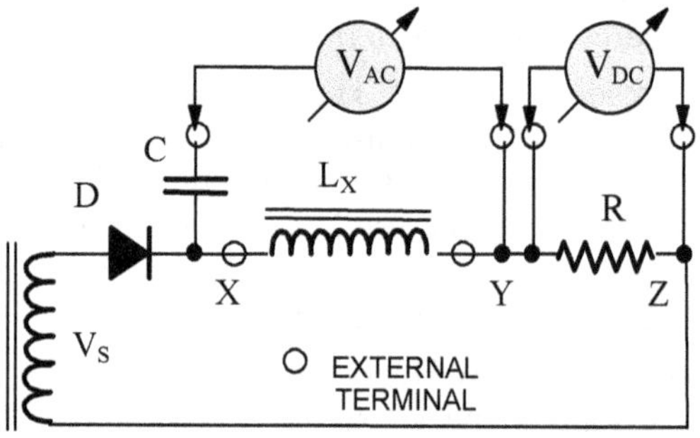

The inductance of a power supply filtering choke or any choke or audio transformer designed to operate with a DC current flowing through its (primary) winding is called an incremental inductance. Due to the presence of DC current, the permeability of the choke lamination stack is reduced, and so is its inductance. Thus, the actual inductance such a choke will have in a rectifier filter will be lower than the inductance measured by a digital LCR meter, which is inductance without any DC current.

Instead of using a complex bridge setup, this method determines incremental inductance by comparing the AC voltage drop across the choke and the DC voltage drop across a resistor of a known resistance R.

Another advantage of this circuit is that it is almost identical to the actual power supply a choke would be used in an amplifier. In this case, half-wave rectification is used, but if you wish, replace a single diode here with a diode bridge and use full-wave rectification. The test principle remains the same.

The same current flows through both components, and the voltage drops are proportionate to their impedances: $V_L/V_R=X_L/R$ and since $X_L=\omega L$ we have $L=V_L/V_R * R/\omega$.

Normally R would be a power rheostat (10-20W) used to adjust $V_L=V_R$ and whose scale would be calibrated to read L directly. However, such rheostats are hard to get and expensive, so we used a trick, making $R=\omega$ or $R/\omega=1$, thus simplifying the ratio to $L=V_L/V_R$!

An AC voltmeter connected directly across the choke will measure a voltage drop proportionate to the choke's total impedance (including its ohmic resistance), not just its inductive reactance, introducing an error of a few %. A series capacitor is used to eliminate this error, which also separates the AC voltmeter from the DC circuit.

The AC ripple frequency for half-wave rectifiers is the same as the mains frequency, so ω=2pf = 314 and R=314Ω (for 60Hz countries, use R=377 Ω)!

To determine the required secondary voltage, we must consider that power supply filtering chokes have L of up to 20H and DC resistance of between 10 and 200Ω, while their DC current ratings are usually below 200-250 mA.

This means that the transformer's nominal secondary voltage V_S needs to be around 200 V_{AC} and have 200-250mA current capability. C should be 2.2μF/250V_{AC}, and the resistor R should have a power rating of 10 Watts.

For example, a choke with L=5H and a DC resistance of 85W, which was probably good for up to 150mA, was tested at three secondary voltages, 43V, 65V, and 190V. The higher the secondary voltage, the higher the DC current as well. The inductance dropped from 4.7H at around 20mA to 3.7H at 110mA!

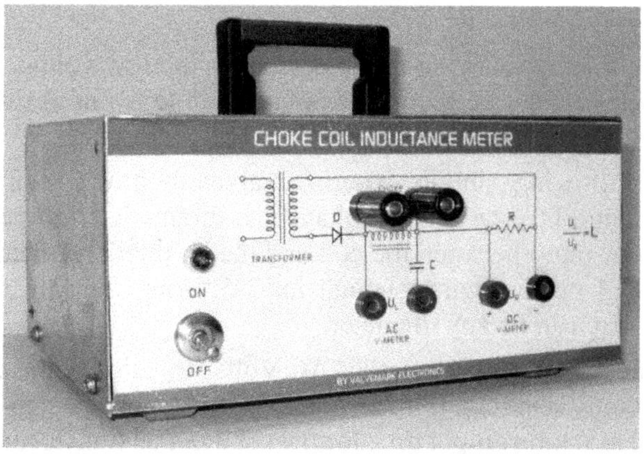

ABOVE: Our prototype incremental inductance tester

BELOW: Test results of a 5H - 85Ω choke

V_S [V_{AC}]	V_L [V_{AC}]	V_R [V_{DC}]	L [H]	I_{DC} [mA]
43	32	6.8	4.70	21.7
65	42	9.6	4.38	30.6
190	130	35	3.71	111.5

THE END MATTER

- SINGLE-ENDED OUTPUT TRANSFORMER DESIGN SEQUENCE
- PUSH-PULL OUTPUT TRANSFORMER DESIGN SEQUENCE
- REFERENCE DATA
- FURTHER READING
- INDEX

13

SINGLE-ENDED OUTPUT TRANSFORMER DESIGN SEQUENCE

A) PRIMARY & SECONDARY NUMBER OF TURNS (FOR PENTODES & BEAM POWER TUBES)

1. DECIDE ON MAXIMUM POWER & LOWER -3 dB FREQUENCY: **P, f_L**
2. FIND V_A & I_A FOR THE CHOSEN OUTPUT TUBE & CALCULATE THE NOMINAL PRIMARY IMPEDANCE, $Z_P=V_A/I_A$
3. CALCULATE THE EFFECTIVE CROSS-SECTION REQUIRED $A_{EF}=25*\sqrt{(P/f_L)}$
4. CHOOSE THE LAMINATION SIZE AND STACK THICKNESS: **a, S**
5. CALCULATE THE PARALLEL IMPEDANCE $R_P=R_I \| Z_P$ AND THE REQUIRED PRIMARY INDUCTANCE L_P FOR THE CHOSEN f_L: $L_P=R_P/(2*\pi*f_L)$, WHERE R_I = INTERNAL RESISTANCE OF THE OUTPUT TUBE
6. CALCULATE THE NUMBER OF PRIMARY TURNS: $N_1= 1{,}000*\sqrt{(10*L_P/Z_P)}$
7. CALCULATE THE NUMBER OF SECONDARY TURNS: $N_2= N_1/\sqrt{IR}$, WHERE $IR=Z_P/Z_L$

A) PRIMARY & SECONDARY NUMBER OF TURNS (FOR TRIODES AND ULTRALINEAR CONNECTION)

1. DECIDE ON MAXIMUM POWER, LOWER -3 dB FREQUENCY & PRIMARY IMPEDANCE: **P, f_L, Z_P**
2. CALCULATE THE EFFECTIVE CROSS-SECTION REQUIRED $A_{EF}=25*\sqrt{(P/f_L)}$
3. CHOOSE THE LAMINATION SIZE, STACK THICKNESS & MAXIMUM AC FLUX DENSITY: **a, S, B_{MAX}**
4. CALCULATE PRIMARY AC VOLTAGE FOR MAXIMUM POWER: $V_1=\sqrt{(P/Z_P)}$
5. CALCULATE THE NUMBER OF PRIMARY TURNS: $N_1= V_1*10^{-4}/(4.44*f_L*B_{MAX}*A_{EF})$
6. CALCULATE THE NUMBER OF SECONDARY TURNS: $N_2= N_1/\sqrt{IR}$, WHERE $IR=Z_P/Z_L$

B) PRIMARY & SECONDARY CURRENTS

1. CHOOSE OR FIND FROM OUTPUT TUBE DATA SHEETS PRIMARY (ANODE) DC CURRENT: I_A
2. CALCULATE THE AC COMPONENT OF THE PRIMARY CURRENT FOR FULL POWER: $i_1= \sqrt{(P/Z_P)}$
3. CALCULATE TOTAL PRIMARY CURRENT $I_1= I_A+i_1$
4. CALCULATE SECONDARY CURRENT AT MAXIMUM POWER $I_2= \sqrt{(P/R_L)}$

C) WIRE SIZING

1. FROM WINDOW LENGTH & HEIGHT (WL, WH) CALCULATE COIL LENGTH & MAXIMUM COIL HEIGHT: **CL = WL- 2B, CH_{MAX} = WH-B**, WHERE B=BOBBIN THICKNESS
2. CALCULATE THE MAX. DIAMETER OF THE SEC. WIRE SO THE WHOLE SECONDARY FITS IN ONE LAYER: $d_{2MAX}=CL/N_2$
3. CHOOSE THE CURRENT DENSITY (J) AND CALCULATE THE MAXIMUM SECONDARY CURRENT WITH d_{2MAX}: $I_{2MAX}=J*(d_{2MAX}/1.13)^2$
4. ASSUMING ALL SECONDARY SECTIONS WILL BE PARALLELED, CHOOSE THE NUMBER OF SECONDARY LAYERS (SL) SO THAT I_2 IS SMALLER THAN $SL*I_{2MAX}$
5. CALCULATE THE DIAMETER OF THE PRIMARY WIRE: $d_1=1.13*\sqrt{(I_1/J)}$

D) SECTIONALIZING & LAYERING

1. CALCULATE MAXIMUM NUMBER OF PRIMARY TURNS THAT CAN FIT IN ONE LAYER: $TPL_{1MAX} =CL/d_1$
2. CHOOSE TPL_1 SO IT'S SMALLER THAN $0.9*TPL_{1MAX}$ AND IT GOES NICELY (EVEN NUMBER OF TIMES) INTO N_1
3. CALCULATE THE NUMBER OF PRIMARY LAYERS (PL): $PL=N_1/TPL_1$
4. FINALIZE THE NUMBER OF SECTIONS & THEIR INTERLEAVING ARRANGEMENT

E) CHECKING VERTICAL FIT

1. FROM LAMINATION SIZE DETERMINE THE MAXIMAL COIL HEIGHT CH_{MAX}
2. CALCULATE TOTAL PRIMARY WINDING HEIGHT **PH** AND SECONDARY WINDING HEIGHT **SH**
3. CHOOSE THE INSULATION TYPE(S) AND THICKNESS, THEN CALCULATE THE TOTAL INSULATION THICKNESS (HEIGHT): **IH**
4. CALCULATE THE TOTAL NETT COIL HEIGHT: **CH= PH + SH + IH**
5. CALCULATE COIL HEIGHT WITH BOWING (BULGING) ALLOWANCE OF 15%: $CH_B= 1.15*CH$
6. IS CH_B SMALLER THAN $0.9*CH_{MAX}$?

YES - THE BASIC DESIGN IS FINISHED

NO - CHOOSE ONE OR MORE OF THE FOLLOWING OPTIONS:

1. Reduce TPV and the total number of primary and secondary turns by using higher level of flux density and/or by increasing the cross-sectional area by increasing S, the stack thickness
2. Reduce wire sizes (use smaller diameter wire).
3. Choose a larger lamination, to increase the window size.

PUSH-PULL OUTPUT TRANSFORMER DESIGN SEQUENCE

A) PRIMARY & SECONDARY NUMBER OF TURNS

1. DECIDE ON MAXIMUM POWER, PRIMARY IMPEDANCE (ANODE-TO-ANODE) & LOWER -3 dB FREQUENCY: **P, Z_P, f_L**
2. CALCULATE THE EFFECTIVE CROSS-SECTION REQUIRED **$A_{EF}=20*\sqrt{(P/f_L)}$**
3. CHOOSE THE LAMINATION SIZE AND STACK THICKNESS: **a, S**
4. CALCULATE ANODE-TO-ANODE AC VOLTAGE FOR MAXIMUM POWER: **$V_{AA}=\sqrt{(P/Z_P)}$**
5. CHOOSE MAXIMUM FLUX DENSITY **B_{MAX}**
6. CALCULATE THE NUMBER OF PRIMARY TURNS: **$N_1= V_{AA}*10^{-4}/(4.44*f_L*B_{MAX}*A_{EF})$**
7. CALCULATE THE NUMBER OF SECONDARY TURNS: **$N_2= N_1*\sqrt{(R_L/Z_P)}$**

B) PRIMARY & SECONDARY CURRENTS

1. CHOOSE OR FIND FROM OUTPUT TUBE DATA SHEETS PRIMARY (ANODE) DC CURRENT: I_A
2. CALCULATE THE AC COMPONENT OF THE PRIMARY CURRENT FOR FULL POWER: **$i_1=\sqrt{(P/Z_P)}$**
3. CALCULATE TOTAL PRIMARY CURRENT **$I_1= I_A+i_1$**
4. CALCULATE SECONDARY CURRENT AT MAXIMUM POWER **$I_2=\sqrt{(P/R_L)}$**

C) WIRE SIZING

1. FROM WINDOW LENGTH & HEIGHT (WL, WH) CALCULATE COIL LENGTH & MAXIMUM COIL HEIGHT: **CL = WL- 2B, CH_{MAX} = WH-B**, WHERE B=BOBBIN THICKNESS
2. CALCULATE THE MAX.. DIAMETER OF THE SEC. WIRE SO THE WHOLE SECONDARY FITS IN ONE LAYER: **$d_{2MAX}=CL/N_2$**
3. CHOOSE THE CURRENT DENSITY (J) AND CALCULATE THE MAXIMUM SECONDARY CURRENT WITH d_{2MAX}: **$I_{2MAX}=J*(d_{2MAX}/1.13)^2$**
4. ASSUMING ALL SECONDARY SECTIONS WILL BE PARALLELED, CHOOSE THE NUMBER OF SECONDARY LAYERS (SL) SO THAT I_2 IS SMALLER THAN $SL*I_{2MAX}$
5. CALCULATE THE DIAMETER OF THE PRIMARY WIRE: **$d_1=1.13*\sqrt{(I_1/J)}$**

D) SECTIONALIZING & LAYERING

1. CALCULATE MAXIMUM NUMBER OF PRIMARY TURNS THAT CAN FIT IN ONE LAYER: **$TPL_{1MAX}=CL/d_1$**
2. CHOOSE TPL_1 SO IT'S SMALLER THAN $0.9*TPL_{1MAX}$ AND IT GOES NICELY (EVEN NUMBER OF TIMES) INTO N_1
3. CALCULATE THE NUMBER OF PRIMARY LAYERS (PL): **$PL=N_1/TPL_1$**
4. CHOOSE SIMPLE (UNBALANCED) OR SUPERIOR (BALANCED) SECTIONALIZING
5. FINALIZE THE NUMBER OF SECTIONS & THEIR INTERLEAVING ARRANGEMENT

E) CHECKING VERTICAL FIT

1. FROM LAMINATION SIZE DETERMINE THE MAXIMAL COIL HEIGHT CH_{MAX}
2. CALCULATE TOTAL PRIMARY WINDING HEIGHT **PH** AND SECONDARY WINDING HEIGHT **SH**
3. CHOOSE THE INSULATION TYPE(S) AND THICKNESS, THEN CALCULATE THE TOTAL INSULATION THICKNESS (HEIGHT): **IH**
4. CALCULATE THE TOTAL NETT COIL HEIGHT: **CH= PH + SH + IH**
5. CALCULATE COIL HEIGHT WITH BOWING (BULGING) ALLOWANCE OF 15%: **CH_B= 1.15*CH**
6. IS CH_B SMALLER THAN $0.9*CH_{MAX}$?

YES - THE BASIC DESIGN IS FINISHED

NO - CHOOSE ONE OR MORE OF THE FOLLOWING OPTIONS:

1. Reduce TPV and the total number of primary and secondary turns by using higher level of flux density and/or by increasing the cross-sectional area by increasing S, the stack thickness
2. Reduce wire sizes (use smaller diameter wire).
3. Choose a larger lamination, to increase the window size.

REFERENCE DATA

Transformer lamination sizes - EI type

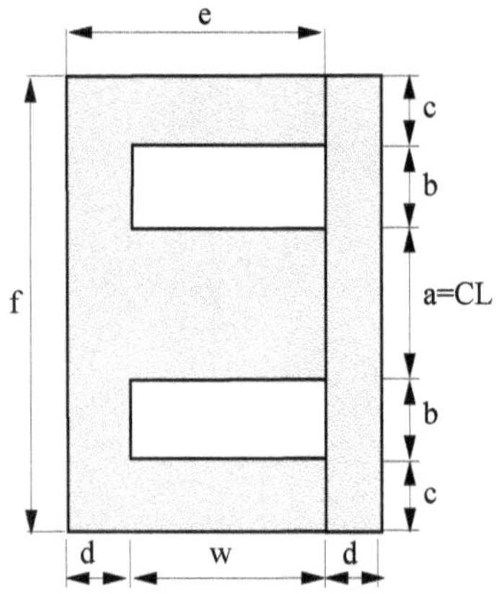

Dimensions and magnetic path length for all EI sizes. a = center leg width (CL), b = window height, w = window width, S = stack thickness (not shown), W = Window area W= b*w, A = gross core area A = a*S

For scrapless EI laminations (gray shaded in tables) b = c = g and a = 2b = 2c = 2g

ABOVE: EI lamination sizes. Wastless sizes are highlighted in gray.

TYPE	f	e	a	c	b	w	d
EI 19	19	12.5	5	2.5	4.5	10	2.5
EI 24	24	15	6	3	6	12	3
EI 25.4	25.4	16.1	6.4	3.1	6.4	13.1	3
EI 28	28	21	8	4	6	17	4
EI 35	35	24.5	9.6	5	7.7	19.5	5
EI 41	41	27	13	6	8	21	6
EI 48	48	32	16	8	8	24	8
EI 54	54	36	18	9	9	27	9
EI 57	57	38	19	9.5	9.5	28.5	9.5
EI 60	60	40	20	10	10	30	10
EI 66	66	44	22	11	11	33	11
EI 73	73	49	23	11.5	13.5	34.5	11.5
EI 74	74	51	23	11.5	14	34	14
EI 75	75	51.5	23	12	14	37.5	14
EI 76.2	76.2	50.8	25.4	12.7	12.7	38.1	12.7
EI 84	84	56	28	14	14	42	14
EI 85.8	85.8	57.2	28.6	14.3	14.3	42.9	14.3
EI 86	86	59	26	13	17	46	13
EI 88	88	60	29.4	13.3	16	44	16
EI 90	90	60	30	15	15	45	15
EI 96	96	64	32	16	16	48	16
EI 100	100	67	32	16	18	51	16
EI 105	105	70	35	17.5	17.5	52.5	17.5
EI 111	111	76	35	17.5	20.5	55.5	20.5
EI 114	114	76	38	19	19	57	19
EI 120	120	80	40	20	20	60	20
EI 133.2	133.2	88.8	44.4	22.2	22.2	66.6	22.2
EI 152.4	152.4	101.6	50.8	25.4	25.4	76.2	25.4
EI 181.2	181.2	120.8	60.4	30.2	30.2	90.6	30.2
EI 192	192	128	64	32	32	96	32

Copper winding wire table (imperial sizes)

AWG	diameter d (mils)	diameter d (mm)	single coated d (mm)	heavy coated d (mm)	Resistance Ω/1,000 ft.	Resistance mΩ/m	Weight (lb/1,000 ft)	Weight (kg/1,000m)
10	101.9	2.59	2.64	2.68	1.0	3.28	31.4	46.728
11	90.7	2.30	2.35	2.39	1.261	4.14	24.9	37.055
12	80.8	2.05	2.10	2.14	1.588	5.21	19.8	29.465
13	72.0	1.83	1.87	1.91	2.001	6.56	15.7	23.364
14	64.1	1.63	1.67	1.71	2.524	8.28	12.4	18.453
15	57.1	1.45	1.49	1.53	3.181	10.43	9.87	14.688
16	50.8	1.29	1.33	1.37	4.020	13.18	7.81	11.662
17	45.3	1.15	1.19	1.22	5.054	16.56	6.21	9.2415
18	40.3	1.02	1.06	1.09	6.386	20.93	4.92	7.3128
19	35.9	0.91	0.95	0.98	8.046	26.37	3.90	5.8038
20	32.0	0.81	0.85	0.88	10.13	33.20	3.10	4.6133
21	28.5	0.72	0.76	0.79	12.77	41.85	2.46	3.6609
22	25.3	0.64	0.68	0.70	16.20	53.10	1.94	2.8870
23	22.6	0.57	0.60	0.63	20.30	66.53	1.55	2.3066
24	20.1	0.51	0.54	0.57	25.67	84.13	1.22	1.8156
25	17.9	0.45	0.48	0.51	32.27	105.77	0.97	1.4435
26	15.9	0.40	0.43	0.45	41.02	134.45	0.765	1.1384
27	14.2	0.36	0.39	0.41	51.44	168.60	0.61	0.9078
28	12.6	0.32	0.34	0.37	65.31	214.-06	0.481	0.7158
29	11.3	0.29	0.31	0.33	81.21	266.17	0.387	0.5759
30	10.0	0.25	0.27	0.29	103.7	339.88	0.303	0.4509
31	8.9	0.23	0.25	0.26	130.9	420.00	0.24	0.3572
32	8.0	0.20	0.22	0.24	162.0	530.97	0.194	0.2887
33	7.1	0.18	0.20	0.22	205.7	674.19	0.153	0.2277
34	6.3	0.16	0.18	0.20	261.3	856.43	0.12	0.1786
35	5.6	0.14	0.16	0.18	330.7	1,083.9	0.0949	0.1412
36	5.0	0.127	0.147	0.16	414.9	1,359.9	0.0757	0.1127
37	4.5	0.114	0.132	0.145	512.1	1,678.4	0.0613	0.0912

FURTHER READING

IN ENGLISH

Audio Transformer Design Manual, Robert G. Wolpert, 1989, 108 pages
Capacitors, Magnetic Circuits, and Transformers, Leander Matsch, 1964, 350 pages
Components Handbook, Vol. 17, M.I.T. Radiation Laboratory Series, J. F. Blackburn (ed.), 1964, 626 pages
Electronic and Radio Engineering, Frederick E. Terman, 1955, 1078 pages
Electronic Designers' Handbook, Landee, Davis and Albrecht, 1957
Electronic Transformers and Circuits, Reuben Lee, 1955, 349 pages
Magnetic Circuits and Transformers, M.I.T. Electrical Engineering Staff, 1943, 718 pages
Practical Transformer Design Handbook, Eric Lowdon, 1989, 389 pages
Radio Engineer's Handbook, F. Terman, 1943, 1,021 pages
Radiotron Designer's Handbook, F. Langford Smith, 4th edition, 1952, 1,482 pages
Transformers For Electronic Circuits, Nathan R. Grossner, 1983, 468 pages

IN FRENCH

Calcul des petits transformateurd, R. Beyaert, 1958, 398 pages
La construction des petits transformateurs, Marthe Douriau, 1963, 218 pages

IN GERMAN

Der Übertrager der nachrichtentechnik, Gunther H. Domsch, 1953, 192 pages
Telefunken Laborbuch, Band I, 1970, 404 pages
Theorie der Spulen und Übertrager, Richard. Feldtkeller, 1963, 186 pages
Theorie und Praxis des Röhrenverstärkers, Peter Dieleman, 2007, 253 pages

IN ITALIAN

I piccoli trasformatori : calcolo e costruzione ad uso degli elettricisti, Mario Pierazzuoli, 1949, 146 pages
I trasformatori tipo radio e simili, Dr. Ing. Enrico Baldoni, 1955, 182 pages
La construzione e il calcolo dei piccoli trasformatori, Ernesto Carbone, 1966, 80 pages
Teoria e calcolo dei piccoli trasformatori, Giacomo Giuliani, 1948, 80 pages
Valvole e trasformatori per Hi-Fi, Gieffe, 2004, 280 pages

IN CROATIAN

Transformatori i prigušnice, Prof. Dr. Ing. Tihomil Jelakovic, 1960, 212 pages

IN SERBIAN

Hi-Fi Lampaška Pojacala, Nikola Vukušic, 2011, 726 pages

INDEX

A

B

C

D

E

F

G

H

I

J

L

M

N

O

P

INDEX, cont.

www.ingramcontent.com/pod-product-compliance
Lightning Source LLC
Chambersburg PA
CBHW082306210326
41598CB00028B/4458
* 9 7 8 0 9 8 0 6 2 2 3 8 6 *

INDEX, cont.

R

S

T

T, cont.

U

V

W

www.ingramcontent.com/pod-product-compliance
Lightning Source LLC
Chambersburg PA
CBHW082306210326
41598CB00028B/4458